B Cell Protocols

METHODS IN MOLECULAR BIOLOGY™

John M. Walker, SERIES EDITOR

293. **Laser Capture Microdissection:** *Methods and Protocols,* edited by *Graeme I. Murray and Stephanie Curran, 2005*

292. **DNA Viruses:** *Methods and Protocols,* edited by *Paul M. Lieberman, 2005*

291. **Molecular Toxicology Protocols,** edited by *Phouthone Keohavong and Stephen G. Grant, 2005*

290. **Basic Cell Culture,** *Third Edition,* edited by *Cheryl D. Helgason and Cindy Miller, 2005*

289. **Epidermal Cells,** *Methods and Applications,* edited by *Kursad Turksen, 2004*

288. **Oligonucleotide Synthesis,** *Methods and Applications,* edited by *Piet Herdewijn, 2004*

287. **Epigenetics Protocols,** edited by *Trygve O. Tollefsbol, 2004*

286. **Transgenic Plants:** *Methods and Protocols,* edited by *Leandro Peña, 2004*

285. **Cell Cycle Control and Dysregulation Protocols:** *Cyclins, Cyclin-Dependent Kinases, and Other Factors,* edited by *Antonio Giordano and Gaetano Romano, 2004*

284. **Signal Transduction Protocols,** *Second Edition,* edited by *Robert C. Dickson and Michael D. Mendenhall, 2004*

283. **Bioconjugation Protocols,** edited by *Christof M. Niemeyer, 2004*

282. **Apoptosis Methods and Protocols,** edited by *Hugh J. M. Brady, 2004*

281. **Checkpoint Controls and Cancer, Volume 2:** *Activation and Regulation Protocols,* edited by *Axel H. Schönthal, 2004*

280. **Checkpoint Controls and Cancer, Volume 1:** *Reviews and Model Systems,* edited by *Axel H. Schönthal, 2004*

279. **Nitric Oxide Protocols,** *Second Edition,* edited by *Aviv Hassid, 2004*

278. **Protein NMR Techniques,** *Second Edition,* edited by *A. Kristina Downing, 2004*

277. **Trinucleotide Repeat Protocols,** edited by *Yoshinori Kohwi, 2004*

276. **Capillary Electrophoresis of Proteins and Peptides,** edited by *Mark A. Strege and Avinash L. Lagu, 2004*

275. **Chemoinformatics,** edited by *Jürgen Bajorath, 2004*

274. **Photosynthesis Research Protocols,** edited by *Robert Carpentier, 2004*

273. **Platelets and Megakaryocytes, Volume 2:** *Perspectives and Techniques,* edited by *Jonathan M. Gibbins and Martyn P. Mahaut-Smith, 2004*

272. **Platelets and Megakaryocytes, Volume 1:** *Functional Assays,* edited by *Jonathan M. Gibbins and Martyn P. Mahaut-Smith, 2004*

271. **B Cell Protocols,** edited by *Hua Gu and Klaus Rajewsky, 2004*

270. **Parasite Genomics Protocols,** edited by *Sara E. Melville, 2004*

269. **Vaccina Virus and Poxvirology:** *Methods and Protocols,* edited by *Stuart N. Isaacs, 2004*

268. **Public Health Microbiology:** *Methods and Protocols,* edited by *John F. T. Spencer and Alicia L. Ragout de Spencer, 2004*

267. **Recombinant Gene Expression:** *Reviews and Protocols, Second Edition,* edited by *Paulina Balbas and Argelia Johnson, 2004*

266. **Genomics, Proteomics, and Clinical Bacteriology:** *Methods and Reviews,* edited by *Neil Woodford and Alan Johnson, 2004*

265. **RNA Interference, Editing, and Modification:** *Methods and Protocols,* edited by *Jonatha M. Gott, 2004*

264. **Protein Arrays:** *Methods and Protocols,* edited by *Eric Fung, 2004*

263. **Flow Cytometry,** *Second Edition,* edited by *Teresa S. Hawley and Robert G. Hawley, 2004*

262. **Genetic Recombination Protocols,** edited by *Alan S. Waldman, 2004*

261. **Protein–Protein Interactions:** *Methods and Applications,* edited by *Haian Fu, 2004*

260. **Mobile Genetic Elements:** *Protocols and Genomic Applications,* edited by *Wolfgang J. Miller and Pierre Capy, 2004*

259. **Receptor Signal Transduction Protocols,** *Second Edition,* edited by *Gary B. Willars and R. A. John Challiss, 2004*

258. **Gene Expression Profiling:** *Methods and Protocols,* edited by *Richard A. Shimkets, 2004*

257. **mRNA Processing and Metabolism:** *Methods and Protocols,* edited by *Daniel R. Schoenberg, 2004*

256. **Bacterial Artifical Chromosomes, Volume 2:** *Functional Studies,* edited by *Shaying Zhao and Marvin Stodolsky, 2004*

255. **Bacterial Artifical Chromosomes, Volume 1:** *Library Construction, Physical Mapping, and Sequencing,* edited by *Shaying Zhao and Marvin Stodolsky, 2004*

254. **Germ Cell Protocols, Volume 2:** *Molecular Embryo Analysis, Live Imaging, Transgenesis, and Cloning,* edited by *Heide Schatten, 2004*

253. **Germ Cell Protocols, Volume 1:** *Sperm and Oocyte Analysis,* edited by *Heide Schatten, 2004*

252. **Ribozymes and siRNA Protocols,** *Second Edition,* edited by *Mouldy Sioud, 2004*

251. **HPLC of Peptides and Proteins:** *Methods and Protocols,* edited by *Marie-Isabel Aguilar, 2004*

250. **MAP Kinase Signaling Protocols,** edited by *Rony Seger, 2004*

METHODS IN MOLECULAR BIOLOGY™

B Cell Protocols

Edited by

Hua Gu, PhD

Department of Microbiology, Columbia University
College of Physicians and Surgeons, New York, NY

Klaus Rajewsky, MD

The Center for Blood Research,
The Harvard Medical School, Boston, MA

HUMANA PRESS ✳ TOTOWA, NEW JERSEY

Library of Congress Cataloging-in-Publication Data

B cell protocols / edited by Hua Gu, Klaus Rajewsky.
    p. ; cm. -- (Methods in molecular biology ; v. 271)
    Includes bibliographical references and index.
    ISBN 1-58829-347-5 (alk. paper)
    1. B cells--Laboratory manuals.
    [DNLM: 1. B-Lymphocytes--Laboratory Manuals. WH 25 B364 2004] I. Gu, Hua, PhD. II. Rajewsky, K., 1936- III. Series: Methods in molecular biology (Clifton, N.J.) ; v. 271.
    QR185.8.B15B2127 2004
    616.07'98--dc22
                                                                2004002869

Preface

B-lymphocyte development and function remains an exciting area of research for those interested in the physiology and pathology of the immune system in higher animals. While recent advances in genetics and cellular and molecular biology have provided a large spectrum of powerful new experimental tools in this field, it is both time consuming and often very difficult for a student or just any bench-side worker to identify a reliable experimental protocol in the ocean of the literature.

The aim of *B Cell Protocols* is to provide a collection of diverse protocols ranging from the latest inventions and applications to some classic, but still frequently used methods in B-cell biology. The authors of the various chapters are all highly qualified scientists who are either the inventors or expert users of these methods. Their extensive experience in mastering a particular method provides not only the step-by-step details of a reproducible protocol, but also useful troubleshooting tips that readers will appreciate in their daily work. We hope that this book will be helpful for both beginning and experienced researchers in the field in designing or modifying an experimental approach, and exploring a biological question from multiple angles.

Hua Gu, PhD
Klaus Rajewsky, MD

Contents

Preface v

Contributors ix

1 Characterization of B Lymphopoiesis in Mouse Bone Marrow
and Spleen
Richard R. Hardy and Susan A. Shinton 1

2 Characterization of B-Cell Maturation
in the Peripheral Immune System
Rita Carsetti 25

3 Identification of B-Cell Subsets:
An Exposition of 11-Color (Hi-D) FACS Methods
**James W. Tung, David R. Parks, Wayne A. Moore,
Leonard A. Herzenberg, and Leonore A. Herzenberg** 37

4 Analysis of B-Cell Life-Span and Homeostasis
Irmgard Förster 59

5 Differentiation of B Lymphocytes From Hematopoietic Stem Cells
Paulo Vieira and Ana Cumano 67

6 Analysis of Lymphocyte Development and Function
Using the RAG-Deficient Blastocyst Complementation System
Faith M. Young, Carl A. Pinkert, and Andrea Bottaro 77

7 Conditional Gene Mutagenesis in B-Lineage Cells
Stefano Casola 91

8 Analysis of the Germinal Center Reaction
and In Vivo Long-Lived Plasma Cells
Kai-Michael Toellner, Mahmood Khan, and Daniel Man-Yuen Sze 111

9 T-Cell-Dependent Immune Responses, Germinal Center
Development, and the Analysis of V-Gene Sequences
Claudia Berek 127

10 Introduction of Genes Into Primary Murine Splenic B Cells
Using Retrovirus Vectors
Kuo-I Lin and Kathryn Calame 139

11 Immunoglobulin Class Switching: *In Vitro Induction and Analysis*
Sven Kracker and Andreas Radbruch 149

12 In Vitro and In Vivo Assays of B-Lymphocyte Migration
Chantal Moratz and John H. Kehrl 161

13   Analysis of Antigen-Specific B-Cell Memory Directly Ex Vivo
     Louise J. McHeyzer-Williams and Michael G. McHeyzer-Williams ... 173

14   B-Cell Signal Transduction: *Tyrosine Phosphorylation,*
     Kinase Activity, and Calcium Mobilization
     Tilman Brummer, Winfried Elis, Michael Reth, and Michael Huber... 189

15   Isolation of Lipid Rafts From B Lymphocytes
     **Anu Cherukuri, Shiang-Jong Tzeng, Arun Gidwani, Hae Won Sohn,
     Pavel Tolar, Michelle D. Snyder, and Susan K. Pierce 213**

16   Molecular Single-Cell PCR Analysis of Rearranged Immunoglobulin
     Genes As a Tool to Determine the Clonal Composition
     of Normal and Malignant Human B Cells
     Ralf Küppers 225

17   Analysis of B-Cell Immune Tolerance Induction
     Using Transgenic Mice
     Helen Ferry and Richard J. Cornall 239

18   Analysis of B-Cell Signaling Using DT40 B-Cell Line
     Tomoharu Yasuda and Tadashi Yamamoto 261

19   B-Cell Development and Pre-B-1 Cell Plasticity In Vitro
     Antonius G. Rolink 271

Index 283

Contributors

CLAUDIA BEREK • *Deutsches Rheuma-Forschungszentrum (DRFZ), Berlin, Germany*

ANDREA BOTTARO • *Departments of Medicine and Microbiology and Immunology, University of Rochester School of Medicine and Dentistry; James P. Wilmot Cancer Center at the University of Rochester Medical Center, Rochester, NY*

TILMAN BRUMMER • *Department of Molecular Immunology, Biology III, University of Freiburg and Max-Planck-Institute for Immunobiology, Freiburg, Germany*

KATHRYN CALAME • *Department of Microbiology, Columbia University, New York, NY*

RITA CARSETTI • *Research Center, Ospedale Bambino Gesú, Rome, Italy*

STEFANO CASOLA • *The CBR Institute for Biomedical Research, Harvard Medical School, Boston, MA*

ANU CHERUKURI • *Laboratory of Immunogenetics, National Institute of Allergy and Infectious Diseases, National Institutes of Health, Rockville, MD*

RICHARD J. CORNALL • *Nuffield Department of Clinical Medicine, Oxford University; John Radcliffe Hospital, Oxford, UK*

ANA CUMANO • *Unité du Developpment des Lymphocytes, Institut Pasteur, Paris, France*

WINFRIED ELIS • *Department of Molecular Immunology, Biology III, University of Freiburg and Max-Planck-Institute for Immunobiology, Freiburg, Germany*

HELEN FERRY • *Nuffield Department of Clinical Medicine, Oxford University; John Radcliffe Hospital, Oxford, UK*

IRMGARD FÖRSTER • *Institute for Medical Microbiology, Immunology, and Hygiene and Department of Internal Medicine II, Technical University of Munich, Munich, Germany*

ARUN GIDWANI • *Laboratory of Immunogenetics, National Institute of Allergy and Infectious Diseases, National Institutes of Health, Rockville, MD*

RICHARD R. HARDY • *Institute for Cancer Research, Fox Chase Cancer Center, Philadelphia, PA*

LEONARD A. HERZENBERG • *Department of Genetics, Stanford University Medical School, Stanford, CA*

LEONORE A. HERZENBERG • *Department of Genetics, Stanford University Medical School, Stanford, CA*

MICHAEL HUBER • *Department of Molecular Immunology, Biology III, University of Freiburg and Max-Planck-Institute for Immunobiology, Freiburg, Germany*

JOHN H. KEHRL • *B Cell Molecular Immunology Section, Laboratory of Immunoregulation, National Institute of Allergy and Infectious Diseases, National Institutes of Health, Bethesda, MD*

MAHMOOD KHAN • *MRC Centre for Immune Regulation, Medical School, University of Birmingham, Birmingham, UK*

SVEN KRACKER • *Deutsches Rheuma-Forschungszentrum (DRFZ), Berlin, Germany*

RALF KÜPPERS • *Tumor Research, Institute for Cell Biology, University of Essen Medical School, Essen Germany*

KUO-I LIN • *Department of Microbiology, Columbia University, New York, NY*

LOUISE J. MCHEYZER-WILLIAMS • *Department of Immunology, The Scripps Research Institute, La Jolla, CA*

MICHAEL G. MCHEYZER-WILLIAMS • *Department of Immunology, The Scripps Research Institute, La Jolla, CA*

WAYNE A. MOORE • *Department of Genetics, Stanford University Medical School, Stanford, CA*

CHANTAL MORATZ • *B Cell Molecular Immunology Section, Laboratory of Immunoregulation, National Institute of Allergy and Infectious Diseases, National Institutes of Health, Bethesda, MD*

DAVID R. PARKS • *Department of Genetics, Stanford University Medical School, Stanford, CA*

SUSAN K. PIERCE • *Laboratory of Immunogenetics, National Institute of Allergy and Infectious Diseases, National Institutes of Health, Rockville, MD*

CARL A. PINKERT • *Department of Pathology and Laboratory Medicine, University of Rochester School of Medicine and School of Dentistry; Center for Aging and Developmental Biology, Aab Institute of Biomedical Sciences, University of Rochester Medical Center, Rochester, NY*

ANDREAS RADBRUCH • *Deutsches Rheuma-Forschungszentrum (DRFZ), Berlin, Germany*

MICHAEL RETH • *Department of Molecular Immunology, Biology III, University of Freiburg and Max-Planck-Institute for Immunobiology, Freiburg, Germany*

ANTONIUS G. ROLINK • *Division of Molecular Immunology, Pharmazentrum, University of Basel, Basel, Switzerland*

Susan A. Shinton • *Institute for Cancer Research, Fox Chase Cancer Center, Philadelphia, PA*

Michelle D. Snyder • *Laboratory of Immunogenetics, National Institute of Allergy and Infectious Diseases, National Institutes of Health, Rockville, MD*

Hae Won Sohn • *Laboratory of Immunogenetics, National Institute of Allergy and Infectious Diseases, National Institutes of Health, Rockville, MD*

Daniel Man-Yuen Sze • *Institute of Haematology, Royal Prince Alfred Hospital, Camperdon, Australia*

Kai-Michael Toellner • *MRC Centre for Immune Regulation, Medical School, University of Birmingham, Birmingham, UK*

Pavel Tolar • *Laboratory of Immunogenetics, National Institute of Allergy and Infectious Diseases, National Institutes of Health, Rockville, MD*

James W. Tung • *Department of Genetics, Stanford University Medical School, Stanford, CA*

Shiang-Jong Tzeng • *Laboratory of Immunogenetics, National Institute of Allergy and Infectious Diseases, National Institutes of Health, Rockville, MD*

Paulo Vieira • *Unité du Development des Lymphocytes, Institut Pasteur, Paris, France*

Tadashi Yamamoto • *Department of Oncology, Institute of Medical Science, University of Tokyo, Tokyo, Japan*

Tomoharu Yasuda • *Department of Cell Regulation, Medical Research Institute, Tokyo Medical and Dental University, Tokyo, Japan*

Faith M. Young • *Departments of Medicine, Pathology and Laboratory Medicine, Microbiology and Immunology, and Pediatrics, University of Rochester School of Medicine and Dentistry; James P. Wilmot Cancer Center at the University of Rochester Medical Center, Rochester, NY*

1

Characterization of B Lymphopoiesis in Mouse Bone Marrow and Spleen

Richard R. Hardy and Susan A. Shinton

Summary

This chapter provides information on the application of flow cytometry for analysis of B-cell development, describing in detail the particular surface proteins that can serve as markers for recognizing distinct stages in this process. These cell fractions range from just prior to initial heavy chain rearrangement, the germline pro-B stage, through D-J rearranged pro-B and heavy chain expressing pre-B stages, to the maturing surface BCR positive B-cell stages. It also outlines assays for the characterization of these cells, including procedures for testing functional lineage restriction, determination of rearrangement status, analyses of gene expression at the ribonucleic acid (RNA) and protein level, and assessment of cell cycle state.

Key Words

B-cell development; lymphopoiesis; rearrangement; gene expression; flow cytometry; pro-B cell; pre-B cell.

1. Introduction

B cells develop in a highly ordered process from pluripotent stem cells. In the mouse, this occurs in the liver prior to birth and shifts to the bone marrow shortly thereafter. Over the past 15 yr a great deal has been learned through the study of mutant mice in which this process is disrupted-for example, in animals that fail to efficiently rearrange their immunoglobulin loci and so do not produce heavy- or light-chain proteins (severe combined immunodeficient [SCID] mice) (1–3). Careful characterization of such mice and their normal wild-type counterparts has led to the elucidation of a stepwise set of stages (**Fig. 1**) delineated by distinct cell-surface phenotypes using multiparameter flow cytometry (4,5). In turn, this has facilitated further studies, characterizing key processes that facilitate the growth and differentiation at key stages in this process (6–10).

From: *Methods in Molecular Biology, vol. 271: B Cell Protocols*
Edited by: H. Gu and K. Rajewsky © Humana Press Inc., Totowa, NJ

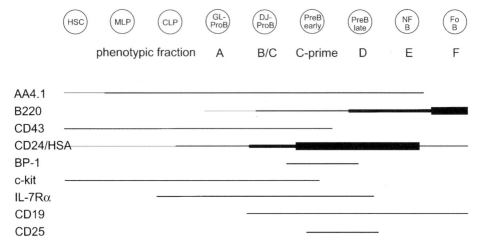

Fig. 1. Diagram of hematopoietic cell stages along the B lineage in mouse bone marrow. Phenotypic fraction letters are shown under the cell type. Surface proteins that can serve as markers for isolation by flow cytometry are also indicated. HSC, hematopoietic stem cell; MLP, multilineage progenitor; CLP, common lymphoid progenitor; GL-Pro-B, germline Pro-B; DJ-Pro-B, DJ rearranged Pro-B; Pre-B, cytoplasmic μ+; NF B, newly formed B cell; Fo B, follicular-type recirculating B cell.

Here, we indicate some approaches for resolving these stages and also describe assays that may prove useful in characterizing these cells. Flow cytometry is a developing technology, particularly in terms of the numbers of simultaneous measurements that can be made, so recommendations for basic identification using widely available conventional instrumentation are provided. More complete analysis, now possible with recently available instruments, is also described, as this level of multiparameter analysis, measuring eight or more cell-surface proteins, will likely become routine over the next few years.

Simply identifying and quantitating these cell populations may suffice for a preliminary characterization of B-cell development in a new mutant mouse, but at least some degree of functional analysis and/or measurement of gene expression will greatly add to the power of the investigation. Therefore a variety of assays and procedures for characterizing these cell populations are also presented, with a focus on those that work well with the relatively small numbers of cells that can be isolated by electronic cell sorting.

2. Materials

1. Mice: source of bone marrow (Balb/c or C57BL/6); recipients (SCID, Rag-null, Ly5-congenic B6). Timed matings of Rag-null/Common γ-chain double-deficient

mice, required for generation of fetal thymic lobes for fetal thymus organ culture (FTOC) or high-oxygen saturation (HOS) assays, where embryos at 15 d postcoitus are optimal. Appropriate room and cages, food, and bedding.

2. Staining medium: deficient RPMI medium (Irvine Scientific), 10 mM HEPES, 3% fetal bovine serum (FBS), and 0.1% NaN$_3$; phosphate-buffered saline (PBS): 0.01 M sodium phosphate, pH 7.2, 0.15 M NaCl.

3. Cell preparation components: red cell lysis solution: 0.165 M NH$_4$Cl; scissors, forceps, microscope slides, and Nitex cloth for cell preparation; hemocytometer slide for cell counting; plastic disposable pipets; micropipettors and disposable tips; centrifuge tubes (15 and 50 mL) for use in cell preparation and washing; fluorescence-activated cell sorting (FACS) tubes or microcentrifuge (Eppendorf) tubes for staining and analysis; flexible 96-well polyvinyl chloride (PVC) plates for staining; syringes and needles.

4. Fluorescent-labeled antibodies; for mouse-specific reagents, an excellent source is BD Biosciences/Pharmingen. An alternative with a large inventory of mouse-specific antibodies is eBioscience. All reagents must be protected from light and stored at 4°C.

5. Petri dishes, 96-, 24-, and 6-well culture plates; Terasaki plates (NUNC, no. 163118); 96-well V-bottom plates (Costar, no. 3894).

6. Cell-transfer items: antibiotic solution, consisting of 500 µg/mL neomycin and 12.8 µg/mL polymyxin B added to a water bottle; heat lamp for warming mice prior to injection.

7. Avertin anesthetic for intrathymic transfer (IT) assay: stock solution is 25 g tribromoethanol dissolved in 15.5 mL tertamyl alcohol and warmed to dissolve. Store viscous solution at room temperature. Working solution is 0.5 mL stock added to 39.5 mL sterile physiologic saline (0.85% NaCl), again warmed to obtain homogenous solution.

8. Methylcellulose culture: MethoCult growth-factor supplemented–methylcellulose medium (StemCell Technologies); 35- and 100-mm Petri dishes.

9. Stromal culture: S17 stromal cell line or equivalent; growth factors (all from R&D Systems): recombinant mouse interleukin (IL)-7, FMS-like tyrosin kinase (Flt)-3, stem cell factor (SCF); 24-well plates; 96-well plates; T25 cell culture flasks; RPMI-1640 medium, FBS, glutamine, gentamycin, HEPES, 2-mercaptoethanol (medium components). Trypsinization solution: 0.05% trypsin/0.53 mM ethylenediaminetetraacetic acid (EDTA) (Gibco-BRL). Stromal culture medium: RPMI-1640, 5% FBS, heat inactivated, 2 mM L-glutamine, 50 mg/L gentamycin, 10 mM HEPES, 5 × 10^{-5}M 2-ME.

10. FTOC special items: medium is RPMI-1640, 10% FBS, heat inactivated, 0.1 mM nonessential amino acids, 1.0 mM minimum essential medium (MEM) sodium pyruvate, 2 mM L-glutamine, 100 U/mL penicillin and 100 µg/mL streptomycin (Pen–Strep), 5 × 10^{-5}M 2-ME; collagen hemostatic sponge (Helistat 7.5 cm × 10 cm × 5 mm, no. 3410-ZX from Integra LifeSciences Corp.); Millipore membrane filter (5.0 µM TMTP [Millipore, no. TMTP01300]).

11. HOS culture special items: plastic sealable bags (Tedlar 5-L bags with Jaco valve in center, [Sassco, no. TB 12J]); sealer apparatus; cylinder of HOS gas (70% O_2, 25% N_2, 5% CO_2); growth factors for HOS cultures (IL-7, SCF).

12. Polymerase chain reaction (PCR) general components: PCR oligos; 10X PCR buffer: 500 mM KCl, 200 mM Tris-HCl, pH 8.3, 25 mM MgCl$_2$, and 1 mg/mL bovine serum albumin (BSA); deoxynucleotide-triphosphates (dNTPs); *Taq* polymerase (Promega).

13. Deoxyribonucleic acid (DNA)–PCR assay: Tris-buffered saline (50 mM Tris-HCl, 150 mM NaCl, pH 8.0); buffer A (0.5% sodium lauroyl sarkosinate,10 mM EDTA, 50 mM Tris-HCl, pH 8.0), containing 0.5 mg/mL proteinase K; low-gelling temperature agarose; Tris-EDTA (TE) buffer: 10 mM Tris, pH 8.0, 1 mM EDTA; ribonuclease (RNase); *Eco*RI restriction enzyme.

14. Reverse transcription PCR (RT–PCR) assay: 1.5X solution D: 5 g guanidium thiocynate, 0.5 mL 10% sarkosyl, 0.25 mL 1M sodium citrate, 70 μL 2-mercaptoethanol (ME), 3 mL DepC H$_2$O. Heat at 42°C for 5–10 min to dissolve. Store at room temperature. Acid phenol (Sigma, no. P-4682). Chloroform-isoamyl alcohol; glycogen; isopropanol; MeHgOH; 2-ME; random hexamers (Pharmacia, Piscataway, NJ); RNase inhibitor (Promega, Madison, WI); Maloney murine leukemia virus (M-MLV) reverse transcriptase.

15. PCR-blotting: Hybond-N membrane (Amersham, Arlington Heights, IL); hybridization and washing solutions (prehybridization: 50% formamide (Fluka, Sigma-Aldrich, no. 47671), 5X sodium saline citrate (SSC), 5X Denhardt's blocking solution, 50 mM sodium phosphate, pH 7.0, 250 μg/mL salmon sperm DNA, 500 μg/mL transfer RNA (tRNA), 0.1% sodium dodecyl sulfate (SDS); hybridization: 50% formamide (Fluka, Sigma-Aldrich, no. 47671), 5X SSC, 1X Denhardt's, 20 mM sodium phosphate, pH 7.0, 100 μg/mL salmon sperm DNA, 250 μg/mL tRNA, 0.1% SDS, 10% dextran sulfate (Sigma-Aldrich, no. D8906, sodium salt); wash: twice with 2X SSC, 0.5% SDS; high stringency—0.2X SSC, 0.5% SDS). Radiolabeling components and radionuclide.

16. Cytoplasmic staining: Caltag Fix and Perm Kit (GAS-003), including Reagent A and Reagent B.

17. Western blotting: Robert lysis buffer (2X stock): 40 mM Tris-HCl, pH 8, 275 mM NaCl, 20% glycerol, 2% Nonidet P-40 (NP40); 6X loading buffer: 12% SDS, 60% glycerol, 600 mM dithiothreitol (DTT), 360 mM Tris-HCl, pH 6.8, 0.006% bromophenol blue; protease inhibitors: 2 mM phenylmethylsulfonyl fluoride (PMSF) (200 mM stock), 2 mM aprotinin (100 mM stock); tyrosine phosphatase inhibitor: 2 mM sodium vanadate (100 mM stock).

18. Cell cycle: propidium iodide (PI) solution (for 10 mL): 0.5 mL 20X PI (20X stock is 1 mg/mL PI in PBS for a final concentration of 50 μg/mL); 1000 U (1 mL) RNase A (this is a 1/10 dilution of a 10mg/mL stock in TE buffer, heat activated); 8.5 mL buffer (1 g glucose in 1 L PBS).

19. BrdU detection denaturing solution: 0.15 M NaCl, 4.2 mM MgCl$_2$, 10 μM HCl (pH carefully adjusted to 5.0) and containing 100 U/mL deoxyribonuclease (DNase).

20. General equipment: refrigerated centrifuge, possibly with plate carriers for staining using plates; refrigerators; ice maker; vortex mixer; water baths; humidifier, CO_2-gassed incubator.
21. Microscope for determining cell counts; inverted microscope for cell culture; dissecting microscope (stereo).
22. Equipment for PCR analysis: gel boxes and power supply for agar electrophoresis; ultraviolet (UV) transilluminator/photodocumentation setup; programmable thermal cycler; pressure-blotter apparatus; hybridization oven and bottles; UV crosslinker; SpeedVac; phosphorimager for scanning and quantitating PCR blots.
23. For Western blot analysis: sodium dodecyl sulfate-polyacrylamide gel electrophoresis (SDS-PAGE) setup and power supply; electroblot apparatus.
24. For phenotypic analysis: flow cytometer; minimum: FACScan/FACSCalibur; optimal: LSR II or equivalent.
25. For purification and functional analyses: flow cytometer with single-cell deposition unit; minimum: analog FACS-Vantage or MoFlo; optimal: digital FACS-DiVa or FACS-Aria.
26. Cesium irradiator for cell-transfer experiments.

3. Methods

3.1. Cell Preparation and Isolation/Analysis

1. Mice are sacrificed by cervical dislocations or CO_2 inhalation. All buffers are ice-cold and operations are performed on ice.
2. a. For preparation of bone marrow, the long bones are removed, wiped to eliminate adhering tissue, and the bone marrow flushed with deficient RPMI staning medium using a 1-mL syringe fitted with a 25-gage needle.
   b. For spleen cell preparation, after placing spleen in a Petri dish containing 10 mL staining medium, spleen tissue is gently ground between frosted glass microscope slides, then cells are pelleted by centrifugation.
3. Cell suspensions are filtered through Nitex cloth to eliminate clumps prior to staining.
4. Red blood cells are eliminated by treatment with 1.5 mL red cell lysis solution on ice for 2 min, adding 1.5 mL staining medium, and then cells are pelleted through a cushion of 0.5 mL newborn or FBS.
5. After resuspending in 1.5 mL staining medium, cells are again filtered through Nitex, counted using a hemocytometer slide, and resuspended in an appropriate volume of staining medium to give a concentration of 25×10^6/mL (so that 20 µL contains 5×10^5 cells).

3.2. Cell-Surface Staining: Subset Definition

Here we present the original fractionation of mouse bone marrow using four-color flow cytometry (**Fig. 2A**) as a minimal approach for analyzing development. It should be noted that the earliest fraction expressing CD45R/B220, termed fraction A (Fr. A) is contaminated with non-B lineage cells if only iden-

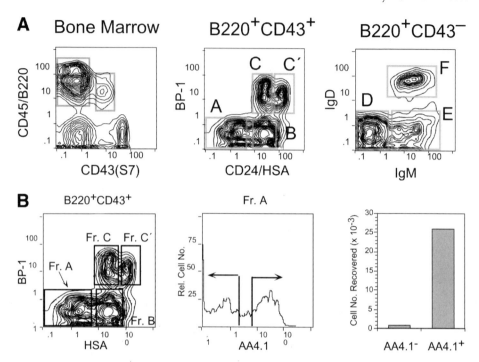

Fig. 2. (**A**) Basic four-color flow cytometry delineation of stages in mouse B-cell development in bone marrow. (**B**) Indication that Fr. A cells are not homogenous, as only the AA4.1⁺ fraction can grow in stromal cell culture.

tified by staining with B220 and CD43, but lacking CD24/HSA. This contamination is clearly shown by fractionation with AA4.1 (**Fig. 2B**) and performing a functional analysis (growth on stromal cells, *see* **Subheading 3.4.3.**). A very thorough analysis of the early stage cell fractions is obtained by 10-color analysis (nine surface stains plus PI dead-cell exclusion), as shown in **Fig. 3**. Peripheral maturation through transitional stages to mature stages in spleen can be done with four-color analysis (**Fig. 4**). See **Notes 1–6** for important recommendations relating to analysis and isolation of cell fractions by flow cytometry.

3.2.1. Using Four- or Five-Color (Minimum) Flow Cytometry

Earliest stages of B lymphopoiesis: B220(6B2)/CD19/AA4.1/Ly6c; for delineating pre-pro-B cells (germline pro-B) select B220⁺CD19⁻AA4⁺/Ly6c⁻ cells (*11–13*).

Intermediate stages: B220/CD43(S7)/CD24(HSA; 30F1)/BP-1; B220⁺CD43⁺ cells can be separated into HSA intermediate BP-1⁻ and BP-1⁺; these are distinguishable from HSA high BP-1⁺ cells (*5,14*).

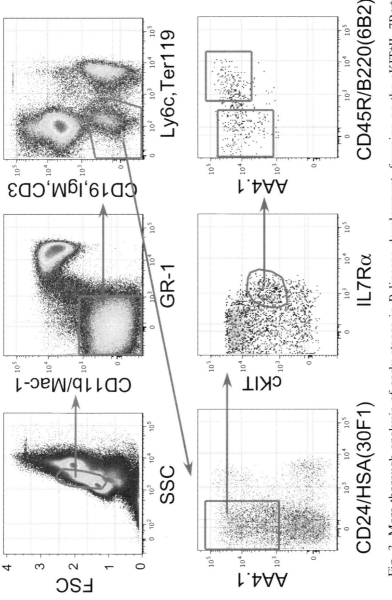

Fig. 3. More thorough analysis of early stages in B lineage development, focusing on the cKIT$^+$IL-7Rα^+ cells, some of which express low levels of CD45R/B220 (Fr A).

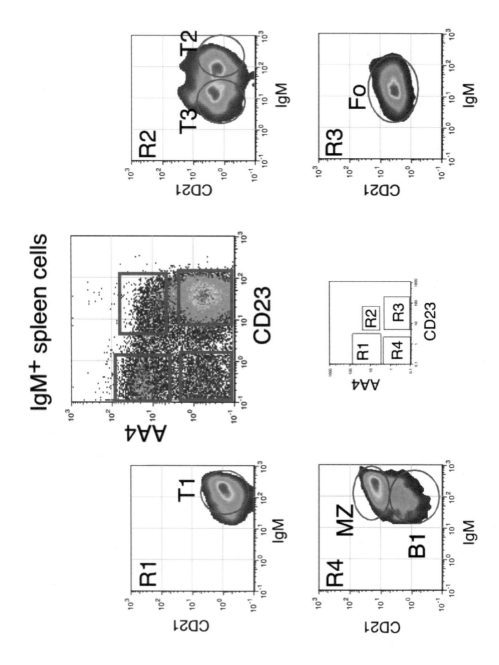

8

Later stages: B220/CD43(S7)/IgM/IgD or B220/CD43(S7)/IgM/AA4.1; $B220^+CD43(S7)^-$ cells can be separated into those lacking IgM and those that are IgM^+. The IgM^+ cells can be further separated into immature cells lacking IgD (or expressing AA4.1) and the mature recirculating cells expressing IgD (or lacking AA4.1).

Peripheral stages: IgM/CD21/CD23/AA4.1; IgM^+ cells can be discriminated into earlier $AA4^+$ stages and more mature $AA4^-$ stages *(15)*. The $AA4^+$ fraction can be further subdivided into earlier $CD23^-$ and later $CD23^+$ subsets. The $AA4^-$ fraction can be subdivided into $CD23^+CD21^+$ follicular mature B cells, $CD23^-CD21^{low}$ B-1 type B cells, and $CD23^-CD21^{+++}$ marginal zone type B cells.

3.2.2. Using 8- to 12-Color (Optimal) Flow Cytometry

Earliest stages of B lymphopoiesis: Mac1(CD11b)/GR1/Ter119/Ly6c/CD19/CD3/IgM/CD24HSA(30F1)/AA4.1/cKIT/IL-7α/B220(6B2); this combination works well for delineating CLP *(16)* and Fr. A stages; eliminate cells expressing Mac1(CD11b), GR1, Ter119, Ly6c, CD19, CD3, IgM; select cells expressing AA4.1 with low levels of HSA(30F1); display cKIT/IL-7Rα and select cells with intermediate cKIT expression that are also IL-7Rα^+. Display B220 histogram: cells lacking B220 are CLP; cells expressing B220 are Fr. A

Intermediate/later stages: Mac1(CD11b)/GR1/Ter119/Ly6c/CD3/CD24HSA (30F1)/AA4.1/CD19/CD43/BP-1/IgM; eliminate cells expressing Mac1(CD11b), GR1, Ter119, Ly6c, CD3; select cells expressing AA4.1 and CD19; IgM–$CD43^+$ cells expressing intermediate levels of HSA can be subdivided into BP-1^- and BP-1^+ subsets; cells expressing the highest levels of HSA are predominantly BP-1^+; IgM^-CD43^- cells are small pre-B cells; IgM^+CD43^- cells can be subdivided into AA4.1^+ newly formed immature B cells and AA4.1^- mature recirculating B cells. Rare IgM^+CD43^+ B cells belong to the B-1 B-cell subset *(17)*.

Peripheral stages: B220/CD19/IgM/IgD/CD21/CD23/AA4.1/HSA(CD24)/CD43(S7)/CD5; $CD19^+IgM^+$ expression delineates the B-cell population. AA4 expression divides the immature ($AA4^+$) vs mature ($AA4^-$) stages *(15)*. As described above, these major fractions can be further subdivided using CD23 and CD21. The additional surface markers are very helpful in discriminating heterogeneity within the main subsets and also help to better define these populations.

Fig. 4. *(see opposite page)* Peripheral B-cell maturation can be resolved by four-color analysis with IgM, AA4.1, CD21, and CD23. IgM^+ cells can be separated into immature/transitional (T1–T3) stages that express AA4.1 and mature stages that lack it. Transitional stages can be subdivided, based on expression of CD23 and IgM levels. Mature stages can be subdivided based on expression of CD23 and CD21. Fo, follicular B cells; MZ, marginal zone B cells; B1, B-1 B cells.

3.3. Staining Procedure

1. Selected reagents are diluted in staining medium to the appropriate concentration (determined by titering) so that 25 µL contains all the stains necessary for that sample.
2. Stain combinations are plated in a 96-well PVC flexible round-bottom plate.
3. Add 20 µL of cells to each well, incubate plate 20 min on ice, then wash three times with staining medium; for the first wash add 100 µL; for each successive wash, use 150 µL.
4. The cells are pelleted in the plate using a refrigerated centrifuge equipped with plate carriers. Cells will pellet in 3 min at 280*g* (1200 rpm).
5. After the final wash step, cells are resuspended in 100 µL of staining medium and transferred to fluorescence-activated cell-sorting (FACS) tubes (Falcon 2052) containing 100 µL staining medium plus 25 µL of 10X PI (10 µg/mL).
6. For two-step staining, 2nd step reagent (typically fluorescent-labeled avidin to detect a biotin-labeled first step) in 25 µL is added to the cells resuspended in 20 µL of staining medium. After the 2nd incubation on ice, cells are washed as in **step 3** and transferred to FACS tubes. Throughout this procedure keep samples well chilled and protected from light.

3.4. Functional Assays

3.4.1. Functional Analyses: Cell Transfer

The lineage potential of a precursor population can be tested by transferring the cells into an immunodeficient or irradiated host *(18,19)*. Test for lymphoid engraftment can most easily employ SCID or Rag-deficient hosts that have been lightly irradiated (300–400 rad). Test for myeloid/lymphoid engraftment can be done using lethally irradiated Ly5-congenic mice, where Ly5-congenic bone marrow is also provided to spare recipients when precursors are not anticipated to repopulate all hematopoietic lineages *(12)*. A variation on this procedure that tests directly for T-lineage potential, even if cells cannot migrate properly to the thymus, is intrathymic injection.

3.4.1.1. CELL TRANSFER ASSAY

1. Irradiate recipients 1 d before injection. For CB17 SCID or Rag1-null (BALB background) provide 300 rad; for Ly5-congenic (C57B/L6 background) provide 900 rad. Use neomycin polymyxin B solution in the water bottle.
2. Sort cells into 0.4 mL complete medium.
3. Pellet cells by centrifugation at 280*g* (1500 rpm) for 7 min at 4°C.
4. Resuspend in PBS in a volume that allows 0.2 mL to be intravenously injected per mouse. For the Ly5-congenic transfer, each sample also includes 10^5 unfractionated host bone marrow.
5. Warm mice with heat lamp for approx 5 min before iv injection.

6. Tissues are harvested (bone marrow, spleen, thymus, peritoneal cavity washout, lymph nodes) 3 wk after transfer for differentiated precursors and 4–8 wk after transfer for more primitive progenitors that may require a longer time to differentiate. Reconstitution is monitored by flow cytometry, assaying for B and T cells in SCID/Rag1-null recipients and for appropriate Ly5 allele (vs myeloid/B/T markers) in the Ly5-congenic transfer.

3.4.1.2. INTRATHYMIC INJECTION

1. Use 4-wk-old recipients, as older animals do not recover from anesthesia as well. Inject 0.02 mL avertin/g body weight intraperitoneally; for example, 0.3 mL for a 15-g mouse. Anesthesia takes effect in a few minutes and lasts for approx 10 min.
2. Tape animal to board, head at front.
3. By hooking paper clip to upper teeth, spread neck area.
4. Swab with cotton ball soaked in 70% ethanol.
5. Make small cut in skin over top of rib cage and enlarge incision to view neck and top of rib cage.
6. Lift rib cage with forceps and make small cut with fine scissors at top of rib cage, slightly to the right.
7. With forceps, pull ribs back to reveal thymus, the white organ pulsing through the opening.
8. Inject 50 µL of cell suspension into thymus and then close skin incision with wound clip.
9. Return mouse to cage with warming lamp until recovered from anesthesia.
10. Analyze reconstituted cells as for iv transfer described in **Subheading 3.4.1.1., step 6**.

3.4.2. Methylcellulose Culture

Growth in culture medium containing methylcellulose and supplemented with various cytokines can be used to determine myeloid and erythroid lineage potential of sorted cells *(12)*. Colonies developing from individual cells can be evaluated morphologically and enumerated in terms of colony-forming units (CFU) as granulocyte (CFU–G), macrophage (CFU–M), granulocyte/macrophage (CFU–GM), erythroid (burst-forming unit [BFU]–E and CFU–E), or pluripotent mixed (CFU–GEMM). In addition, colonies can be picked with a Pasteur pipet and analyzed by flow cytometry, using antibodies capable of distinguishing erythroid and myeloid (macrophage and granulocyte) lineage cells. These include Ter119 for erythroid and Mac-1, Gr-1, and F480 for myeloid cells.

3.4.2.1. METHYLCELLULOSE CULTURE ASSAY

1. Use the MethoCult growth-factor supplemented methylcellulose medium. The medium is thawed following the manufacturer's directions, bubbles are allowed to escape the medium, and 3-mL aliquots are dispensed using a syringe fitted with

a 16-gage needle (medium is viscous) and stored frozen until just prior to use. Thaw overnight in a refrigerator or at room temperature.

2. Typically, 10^4–10^5 cells in 0.3 mL Iscove's medium (2% FBS) are added to 3 mL of MethoCult medium, then 1.1 mL is plated into each of two 35-mm culture dishes. After cell addition, vortex and let stand for 5 min for bubbles to escape. Dispense using a 3-mL syringe fitted with a 16-gage needle.

3. Gentle rotation of the dishes spreads the medium, and then they are placed in a 100-mm culture dish, together with a third 35-mm dish containing 3 mL sterile H_2O (no cover) for extra humidity.

4. Incubate 12 d in a 37°C 5% CO_2-gassed humidified incubator, then count colonies of distinct morphology using a microscope.

3.4.3. Stromal Cell Culture

A variety of stromal cell lines have been described that support the survival, proliferation, and differentiation of early B lineage precursor cells. These cultures also can be designed to support the development of myeloid lineage cells, allowing clonal assay of B/myeloid potential. Growth of the S17 line is contact inhibited, making its use in such assays particularly straightforward (*20*). Cells are passaged by trypsinization and can be plated in 24- or 96-well flat-bottom plates for assay. For B lineage assay, RPMI culture medium is supplemented with IL-7. Although myeloid growth is also possible in such cultures, this may depend on the lot of FBS, and so IL-3 can be used (in addition to IL-7) for a B/myeloid bipotential assay (*21*). Cultures in 24-well plates can be seeded with 10–40×10^3 cells and analyzed after 4 d for proliferation and differentiation using flow cytometry. Smaller cell numbers (from a few thousand down to individual cells) can be analyzed in 96-well plates. Clonal growth can usually be observed in about a week.

3.4.3.1. STROMAL CULTURE ASSAY

1. A confluent T25 flask can be passaged to 2–4 new flasks but still provide sufficient cells to seed 8 "center" wells of a 24-well plate (1 mL each) or all wells of a 96-well plate (150 µL each). Bring cells into suspension by treatment with 2 mL trypsinization solution (*see* **Subheading 2., item 9**) at 37°C for 5 min. It may be necessary to shake the flask to dislodge the cells from the bottom. Add an equal volume of growth medium and pipet vigorously several times to detach all cells and dissociate cells into single-cell suspension.

2. Pellet cells for 5 min at 280g (1200 rpm).

3. Resuspend cells in an appropriate volume of growth medium: 5 mL/T25 flask, 8 mL/24-well plate, 15 mL/96-well plate; thus, to generate 2 new T25 flasks and a 96-well plate, resuspend in 25 mL. A multichannel pipettor will greatly expedite transfer of cells to a 96-well plate.

4. Cultures are maintained in a humidified, 5–10% CO_2-gassed incubator at 37°C. Test wells are ready when cells are confluent; medium should be changed periodically (every 4–7 d) if cells are not used within 3–5 d of plating.
5. Medium is replaced just before sorting with fresh medium containing growth factors: 100 U/mL recombinant mouse IL-7 for simple B lineage growth; add SCF (10 ng/1 mL) and Flt-3 (10 ng/mL) for analysis of earlier stage precursors.
6. Incubate 4–7 d in a 37°C 5% CO_2-gassed humidified incubator.
7. Growth is scored by microscopic observation; wells with colonies can be harvested (most S17 cells will be lost in this process) by washing the well with a micropipet. Suspended cells can be counted and phenotyped using flow cytometry analysis.

3.4.4. Fetal Thymic Organ Culture

Classically, T-lineage development in vitro requires a complex microenvironment that has been difficult to mimic using simple cell line cultures, even when supplemented with various cytokines. Instead, fetal thymic lobes rendered alymphoid by treatment with drugs (such as deoxyguanosine) that eliminate proliferating cells can serve as excellent hosts for allowing development of limited numbers of purified cell fractions. More recently, this procedure has been simplified by the use of fetal thymic lobes from mutant mice lacking capacity to generate lymphocytes (Rag deficient) and also NK cells (γ chain deficient), eliminating the requirement for initial culture of lobes in deoxyguanosine *(22)*. There is also a report that a transfected adherent cell line allows differentiation of T-lineage cells on a monolayer, which would provide a much simpler assay system for study of T-lineage potential *(23)*.

3.4.4.1. PREPARATION OF FTOC PLATES

1. Place 4.5 mL FTOC media in each well of a 6-well tissue-culture plate.
2. Cut collagen hemostatic sponge (*see* **Subheading 2., item 10**) sterilely into 1-cm × 1-cm squares.
3. Place one square in each well with rough side up—helps keep membrane filter from sliding off.
4. Sterilely place one Millipore membrane filter with shiny side down onto center of collagen sponge.
5. Wrap plate with saran wrap and place in incubator until needed.

3.4.4.2. FETAL THYMIC LOBE DISSECTION

1. Remove embryos from sacrificed pregnant mice that are 15 d postcoitus into an empty 100-mm × 15-mm Petri dish.
2. Cut open placenta of each embryo and allow it to slide into a Petri dish filled with RPMI plus Pen–Strep.

3. Rinse recovered embryos two more times in RPMI plus Pen–Strep by transferring them to a clean dish of medium using blunt-ended forceps. Pick them up from the waist down so as not to damage the neck/lobes architecture. Place embryos on ice and keep covered with media.
4. Remove thymic lobes under dissection microscope using forceps. Place dissected lobes onto previously prepared FTOC plate. Lobes can be harvested 1 d before use.

3.4.4.3. CULTURE

1. Sort selected populations into 0.6 mL FTOC medium. After sorting, spin at 280g (1800 rpm) 7–10 min at 4°C. Remove supernatant using a Gilson P1000 Pipetman. Flick tube to resuspend cells; keep on ice.
2. Prepare Terasaki plate by placing 25 μL FTOC medium in each well to be used. Prepare one plate for each fraction. Wrap plate in saran wrap.
3. Under dissecting microscope and forceps, place one lobe in each Terasaki well. Check that lobes are placed properly in wells.
4. Close lid and turn plate upside down.
5. Wrap plate in saran wrap and return to incubator for 2 d.
6. After 2 d, transfer lobes to collagen sponges/filters in 6-well plates. Use Gilson P200 Pipetman set to 25 μL to transfer lobes to filters. Place 6–8 lobes in each filter.
7. Wrap plates in saran wrap and incubate 1–2 wk before analysis.

3.4.4.4. LOBE ANALYSIS

1. Add 100 μL staining medium to small Petri dish (Falcon 3001) and place lobe in medium using forceps.
2. Using gasket end of a 1-mL syringe, mash lobes to make a single cell suspension.
3. Add 500 μL staining medium, and transfer cell suspension to a 1.5-mL micro-centrifuge tube, filtering through nylon mesh.
4. Wash dish with another 500 μL staining medium and transfer to same tube.
5. Pellet cells at 350g (2000 rpm) for 7 min at 4 °C. Remove supernatant using a P1000. Flick to resuspend cells. Add stains diluted in 25 μL staining medium. Incubate on ice for 20 min. Wash once with 1 mL staining medium. Add second step in 25 μL staining medium. Incubate on ice for 15 min. Wash once with 1 mL staining medium. Resuspend in 150μL staining medium. Transfer to FACS tube and add 25 μL 10X PI for analysis.

3.4.5. High Oxygen Submersion Culture/Multilineage Progenitor (HOS/MLP) Assay

Katsura and coworkers have described a novel variation on the FTOC system that permits assay of multiple cell lineages in addition to T cells *(24–26)*. One modification in this system is to carry out the culture of fetal thymic lobes submerged in medium equilibrated with a high oxygen atmosphere (60–70%). The other difference is to supplement the medium with a variety of cytokines that promote the growth of B and NK lineage cells (IL-7 and IL-2) and of hematopoietic precursors (SCF, IL-3). In this system, individual micromanipulated cells can be assayed for potential to develop along multiple lineages.

3.4.5.1. HOS/MLP Assay Procedure

1. Prepare V-bottom plate with 200 μL FTOC medium containing IL-7 and SCF added at 10 ng/mL each. Add 200 μL sterile H_2O to surrounding wells to maintain humidity.
2. Add 1 lobe to each well and centrifuge plate at 120g for 5 min at room temperature.
3. Clone 1 cell/well using cell-sorter cloning apparatus or by micromanipulation of a small number of sorted cells.
4. Centrifuge plate at 120g for 5 min after adding cells.
5. Place plate into Tedlar bag and heat-seal edges.
6. Fill bag with 70% O_2, 25% N_2, 5% CO_2 gas. Open Jaco valve and expel gas. Refill and expel once more. Fill bag again to 2/3 full.
7. Place in 37°C incubator.
8. After 7 d replace medium. Remove 175 μL with P200 micropipettor and replace with fresh FTOC medium containing IL-7 and SCF.
9. Analyze by flow cytometry after 14 d, testing for B cells (CD19), T cells (Thy-1 and IL-2R), NK cells (NK-1 and/or DX5), and myeloid/dendritic cells (CD11b, CD11c, GR-1).

3.5. Rearrangement Status: Deletion; Generation (DNA-PCR)

The key process for B-cell development is rearrangement of immunoglobulin heavy and light chain genes, eventually resulting in expression of the B-cell antigen receptor. Thus, assessment of heavy and light chain locus context can be a useful way to characterize developing B lineage cells. PCR-based methods have been described that test for the loss of germline (unrearranged) DNA sequence *(5)* or the production of newly generated (rearranged) sequences *(7,14,27–30)*. Rearrangement is, for the most part, an ordered process, with heavy chain preceding light chain and D-J preceding V-DJ. Thus one can delineate a B lineage stage—defined on the basis of surface phenotype—that has extensive D-J, but little V-DJ rearrangement, and a distinct stage that has VDJ, but not light chain rearrangement.

3.5.1. Loss of Germline Assay Procedure

1. $1–2 \times 10^5$ cells are sorted directly into a microcentrifuge tube.
2. Cells are washed in Tris-buffered saline at 4°C and then digested with proteinase K for 2 h at 50°C in buffer A (*see* **Subheading 2., item 13**) containing 1% low-gelling temperature agarose.
3. After digestion, samples are allowed to gel on ice for 5 min, dialyzed against TE (3 changes in 36 h), and then stored at 4°C.

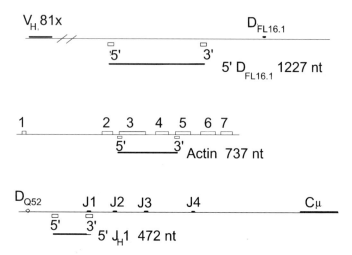

Fig. 5. A diagram of the region upstream of DFL16.1 and of the J locus is shown, together with primers used for amplification of these segments in the PCR assay for loss of germline. The α-actin gene is also shown, along with the primers used for amplification.

4. Prior to use, DNA samples are melted at 65°C, treated with 10 ng RNAase and 10 U *Eco*RI restriction enzyme (37°C for 4–16 h), then gelled on ice and redialyzed three times against ddH$_2$O. Agarose containing the digested DNA is melted at 65°C, diluted 1:5 with 65°C ddH$_2$O (final volume 150 μL), and stored at 4°C.
5. One-fifth of the sample is analyzed by PCR using three sets of oligos (D: GCCTGGGGA GTCACTCAGCAGC and GTGTGGAAAGCTGTGTATCCCC; JH: GGTCCCTGCGCCCCAGACA and CCCGGACAGAGCAGGCAGGTGG; actin: GGTGTCATGGTAGGTATGGGT and CGCACA ATCTCACGTTCAG). These three sets are designated 5′DFL16.1, actin, and 5′J$_H$1; amplified regions are shown in **Fig. 5**.
6. For light chain (κ) rearrangement, a region 5′ of J$_k$1 can be amplified (ATGTA CTTGTGGATGCAGAGGCTG and CCTCCACCGAACGTCCAC CAC) together with the actin fragment. In addition to DNA template, reactions contained PCR buffer, 50 μ*M* each of 4 dNTPs, 100 p*M* for each oligo, and 2 U of *Taq* polymerase.
7. Conditions for PCR are denaturation at 95°C for 1 min, annealing at 63°C for 30 s, and polymerization at 72°C for 1.5 min.
8. Withdraw aliquots at 18, 20, and 22 cycles for separate analysis to ensure that amplification is within the linear range, and care is taken to use relatively comparable levels (within a threefold range) of DNA.
9. Then separate 10 μL of the PCR samples by 1.5% agarose gel electrophoresis, stained with ethidium bromide, photographed, and blotted onto nylon membrane (*see* **Subheading 2.**, step 15).

10. Filters are UV cross-linked, prehybridized for 1–3 h, and then hybridized overnight (at 42°C) with probes prepared from the PCR products.
11. Membranes are washed (twice for 30 min in 2X SSC and twice in 0.2X SSC at 65°C), analyzed using a Fuji phosphorimager and signals quantitated using Fuji software.
12. Signals from individual bands representing each PCR product is measured and calculated as a percentage of germline (unrearranged) DNA.

3.5.2. Rearrangement Assay Procedure

1. DNA samples are prepared using the agarose method as described in **Subheading 3.5.1.** and 1/5 of the sample is used per PCR reaction.
2. Reactions are done with a combination of an α-actin set together with one of the following:
   D-J_H4 (GGCATCGTGTTGGATTCTG and CACGAAGGAATAGCCACGC, the J_H4 primer),
   7183-J_H4 (GCAGCTGGTGGAGTCTGG and J_H4 above),
   Q52-J_H4 (TCCAGACTGAGCATCAGCAA and J_H4 above),
   J558-J_H4 (CAGGTCCAACTGCAGCAG and J_H4 above),
   V_k-J_k GGCTGCAG(G,C)TTCAGTGGCAGTG(G,A)GT(C,T)AG(G,A)G and
   TTCCAACTTTGTCCCCGAGCCG).

   The α-actin fragment serves to control for variation in efficiency of DNA preparation, PCR amplification, pipetting, and transfer. Conditions for PCR are the same as in **Subheading 3.5.1.**
3. Aliquots are withdrawn at 22 and 26 cycles for separate analysis to ensure that amplification was within the linear range.
4. PCR samples are then separated electrophoretically and blotted as in **Subheading 3.5.1.**
5. Filters are hybridized overnight (at 42°C) with random primed probes (α-actin and pJ11 or α-actin and pECk), then washed and analyzed as in **Subheading 3.5.1.**
6. Signals from individual bands representing each PCR product are measured, expressed as a ratio to the signal derived from the coamplified α-actin normalizing band, and then the ratio is expressed as a percentage of the level seen in mature B cells.

3.6. Gene Expression: RT–PCR, Cytoplasmic Staining, Western Blot

Assessment of gene expression can be done at the RNA and protein levels. Analysis of cell fractions purified by flow cytometry cell sorting will usually be sample limited, which rules out traditional Northern analysis. However, as shown in **Fig. 6**, RT–PCR can be quite reliable, particularly for documenting the very significant (i.e., 5- to 10-fold or greater) changes in gene expression that are frequently found with cells as they progress through the B lineage pathway *(14,31,32)*. Purifying total RNA from as large a sample as is reasonable (usually 10^4–10^5 cells) and then limiting the number of PCR amplification cycles may

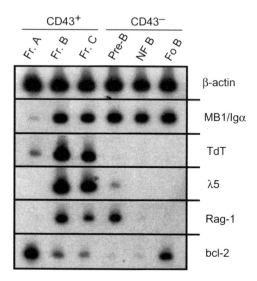

Fig. 6. RT–PCR blot showing changes in expression of several genes important in B lineage development. The β-actin gene is also included to indicate that similar amounts of RNA have been used for each analysis.

help to provide a "semiquantitative" analysis. Also, withdrawing and analyzing aliquots of the PCR sample at different numbers of cycles, particularly when combined with blotting, and detecting the amplified DNA with a radiolabeled probe, can be used to evaluate more subtle changes in RNA levels.

Analyzing protein expression when sample is limiting is a more challenging problem. Probably the simplest approach is to perform cytoplasmic staining with a fluorescent-labeled detecting antibody *(11)*. It is even possible to combine surface staining with cytoplasmic analysis, although the quality of the surface staining may suffer or background autofluorescence may increase, degrading the analysis. Western blot analysis is more definitive than cytoplasmic staining, as the molecular mass separation will help to eliminate background signal. Such analysis is quite feasible with as few as 10^6 cells and, if highly specific detecting reagents are available, even 10-fold fewer cells *(33)*.

3.6.1. RT–PCR Procedure

3.6.1.1. RNA ISOLATION

1. RNA is prepared by an acid phenol extraction procedure. 1×10^5 cells are sorted directly into 350 µL 1.5X solution D. This denatured cell lysate can be stored frozen.
2. For RNA isolation, add 500 µL acid phenol (Sigma, no. P-4682); vortex.

3. Then add 100 μL chloroform-isoamyl alcohol; vortex.
4. Incubate on ice for 15 min.
5. Centrifuge 8800*g* (10,000 rpm) at 4°C for 20 min.
6. Transfer the aqueous phase (top) to a new tube, add 1 μL glycogen (20 μg/μL) and an equal volume of isopropanol.
7. RNA precipitates overnight at –20°C.
8. Pellet RNA by centrifugation, 8800*g* (10,000 rpm) at 4°C for 30 min.
9. Wash RNA with 70% ethanol. Pellet again, 8800*g* (10,000 rpm) at 4°C for 10 min.
10. Dry RNA pellet in a SpeedVac for 10 min (no heat).
11. Resuspend RNA in 20 μL DepC H_2O and store at –70°C.

3.6.1.2. cDNA Synthesis

1. 4 μL of RNA and 4.5 μL of H_2O are mixed with 2 μL of 100 mM MeHgOH and incubated at room temperature for 5 min.
2. After addition of 2.5 μL of 0.7 M 2-ME and 1 μL of 10 U/mL random hexamers (Pharmacia, Piscataway, NJ), the reaction is incubated at 65°C for 2 min and then placed on ice.
3. The following reagents are added: 1 μL of 20 U/μL RNase inhibitor (Promega, Madison, WI); 2 μL of 10X PCR buffer; 2 μL of a 1 mM mixture of all four deoxynucleotide triphosphates and 1 μL of M-MLV reverse transcriptase at 200 U/μL (Gibco-BRL).
4. The reaction is incubated at 37°C for 60 min and heated to 95°C for 5 min, then quickly chilled on ice.

3.6.1.3. PCR Amplification

1. cDNA is amplified by PCR in a 50 μL volume containing 4 μL of cDNA sample, 1X PCR buffer, 100 μM of each four deoxynucleotide triphosphates, 2 μM of each sense and antisense primer, and 5 U of *Taq* polymerase.
2. After an initial 10-min incubation at 80°C, PCR reactions are carried out using appropriate conditions for the primers.
3. Aliquots are withdrawn at 22 and 26 cycles for separate analysis to ensure that amplification is within the linear range and PCR is completed at 30 cycles.
4. To verify that equal amounts of RNA are present in each PCR reaction, the "housekeeping gene" β-actin is also amplified in a separate sample.
5. Highly expressed genes are detected by ethidium bromide (EtBr) staining after 1.5% agarose gel electrophoresis using a UV transilluminator.
6. For semiquantitative analysis, 15 μL of PCR samples are separated by 1.5% agarose gel electrophoresis and blotted onto nylon membrane.
7. After UV crosslinking, filters are prehybridized for 1–3 h and then hybridized overnight (at 42°C) with riboprobes prepared from the PCR products (cloned into TA vector, Invitrogen).
8. Membranes are washed (twice for 30 min in 2X SSC and twice in 0.2X SSC at 65°C) and quantitated using a Fuji phosphorimager. Intensity of individual bands representing each PCR product is measured and normalized using the β-actin

signal for each sample; then the maximum signal is set to 100%, and all other values are expressed accordingly. Several independently sorted samples of each cell phenotype are analyzed.

3.6.2. Cytoplasmic Staining Procedure

1. Use the Fix and Perm Kit (*see* **Subheading 2., item 16**). Pellet sorted cells (can be as few as 50,000 or as many as 5×10^5) in a microcentrifuge (Eppendorf) tube.
2. Add 50 μL of reagent A.
3. Incubate 15 min at room temperature.
4. Wash once with 1 mL PBS.
5. Add 50 μL reagent B, together with an appropriate dilution of the staining antibody.
6. Incubate 15 min at room temperature.
7. Wash twice with 1 mL PBS.
8. Resuspend in 150 μL PBS, transfer to a FACS tube, and analyze.

3.6.3. Western Blot (Immunoblot) Procedure

1. After sorting, wash cells twice in PBS, pelleting by centrifugation.
2. Resuspend the cell pellet in 1X Robert lysis buffer ($10^6/20$–40 μL) containing protease and phosphatase inhibitors; hold on ice 15–20 min.
3. Centrifuge 8800*g* for 5 min at 4°C to clarify lysate.
4. Harvest supernatant; discard pellet.
5. Add 1/6 volume of 6X loading buffer; boil for 5 min; sample can be stored for a short time at 4°C.
6. Perform sodium dodecyl sulfate-polyacrylamide gel electrophoresis SDS-PAGE and electroblot as in conventional Western blotting, detecting using peroxidase-labeled reagent (avidin or second-step antibody) and enhanced chemoluminescence (ECL).

3.7. Cell Cycle Status: PI Staining, BrdU Incorporation

Cell cycle status can be very easily performed with limited numbers of cells by visualizing DNA content with PI *(5)*. 10^5 cells are sorted directly into lysis solution, typically allowing DNA histograms of 10,000 events to be collected. Assessment of cells that have recently synthesized DNA can be done using BrdU labeling, feeding this into the H_2O or (more reliably) injecting it ip twice daily. As with cytoplasmic staining, analysis of BrdU incorporation can be done simultaneously with surface labeling, but the requirement for exposing the DNA necessitates harsher treatment of cells, further complicating the process. Sorted samples on the order of 10^5 cells are very simply processed for BrdU staining and often yield clearer results.

3.7.1. PI Staining Procedure

1. Cells are sorted directly into 500 μL 95% ethyl alcohol (EtOH) in microcentrifuge tubes and placed in a –20°C freezer for at least 2 h.

2. Samples are vortexed and then pelleted by centrifugation for 15 min at 550*g* (2500 rpm).
3. Aspirate supernatant, leaving approx 50 µL of EtOH.
4. Resuspend pellet by vortexing and add 400 µL PI solution per sample and transfer to FACS tubes.
5. Incubate at room temperature for at least 30 min prior to analysis.
6. As with conventional cell cycle analysis, the run speed (sample flow rate) should be medium/low for optimal resolution.

3.7.2. BrdU Labeling

1. To label cells in vivo, inject 0.2 mL of a 3 mg/mL solution of BrdU in PBS ip at 9 AM and 5 PM for 4 d.
2. Prepare tissues and stain for sorting.
3. After sorting, pellet cells by centrifugation, 200*g* (1500 rpm) at 4°C for 10 min.
4. Fix and permeabilize cells using the Fix and Perm Kit.
5. Pellet cells and resuspend in 500 µL BrdU-detection denaturing solution.
6. Incubate 30 min in a 25°C water bath.
7. Wash once with PBS, and resuspend cells in 25 µL staining medium containing a titered amount of fluorescein isothiocyanate (FITC) anti-BrdU.
8. Incubate 30 min at room temperature.
9. Wash twice with PBS. Resuspend in 200 µL staining medium and analyze.

4. Notes

1. Familiarity with the general techniques of multiparameter flow cytometry is important for properly interpreting such data, and access to expert technical help will be a great benefit. This is particularly true for carrying out sorting experiments. Proper setting of compensation is crucial for obtaining reliable data, and typical setting "by eye" that often results in overcompensation can lead to erroneous results. The recommended procedure is to set the overlap compensation so that the mean in this channel for brightly stained cells is equal to that for unstained cells.
2. It is important to select staining reagents that provide sufficient signal intensity to allow discrimination of the populations of interest. For example, PE or APC-labeled AA4.1 provides much brighter staining than an FITC conjugate, and the level of staining with FITC-AA4.1 is insufficient for detecting expression on most immature B cells in spleen. Similarly, expression of IL-7 receptor is quite difficult to detect with some antibodies, so careful choice of both the monoclonal antibody clone and the fluorochrome label is important for detecting its expression. Choice of the label will depend on the available instruments and how they can be configured (laser excitation lines, emission filters, jet in air vs flow cell).
3. Staining with certain monoclonal antibodies can produce unexpected and unwanted effects, such as cell aggregation or formation of cell doublets. For example, the anti-CD24/HSA antibody 30F1 has a tendency to induce aggregation, so its use must be carefully titered to produce adequate separations without inducing alterations in the forward/side light-scatter profile (compared to unstained cells).

4. Shifts in different subsets of B lineage cells that may be induced by mutations or expression of transgenes may lead to striking differences in results from incompletely fractionated populations, such as B220+ bone marrow. If possible, thorough analyses should be done to demonstrate that this is not the case or else analyses should be done with carefully matched sorted fractions.

5. Sorting for extended periods can lead to decreased cell viability. Chilling the sample may help to reduce this. Staining with a rich medium, such as deficient RPMI may also be a benefit in maintaining functional viability. Sorting at high speed, necessitating high sheath pressure, can also lead to decreased viability or functional capacity, owing to excessive shear force as the cell traverses the nozzle or to adverse effects of cell decompression as the cell exits the nozzle.

6. Sorting purity should always be tested to verify the expected level of purity, typically on the order of 95% or greater. Assays of bulk-sorted cells will always contain some level of contaminating cells, so optimal functional assays should be performed with the minimum number of cells required, to minimize outgrowth by contaminating cell types. For example, if a differentiated precursor population is contaminated by 1% stem cells, then cell transfer of 10^6 sorted cells will provide 10^4 stem cells, likely resulting in repopulation by these cells in preference to the more differentiated population, particularly if sufficient time is allowed for their development. The same holds for in vitro assays, so clonal assays, where possible, are greatly preferred.

References

1. Bosma, G. C., Custer, R. P., and Bosma, M. J. (1983) A severe combined immunodeficiency mutation in the mouse. *Nature* **301,** 527–530.
2. Bosma, M. J. and Carroll, A. M. (1991) The SCID mouse mutant: definition, characterization and potential uses. *Annu. Rev. Immunol.* **9,** 323–350.
3. Blunt, T., Gell, D., Fox, M., et al. (1996) Identification of a nonsense mutation in the carboxyl-terminal region of DNA-dependent protein kinase catalytic subunit in the SCID mouse. *Proc. Natl. Acad. Sci. USA* **93,** 10,285–10,290.
4. Hardy, R. R., Kemp, J. D., and Hayakawa, K. (1989) Analysis of lymphoid population in SCID mice: detection of a potential B lymphocyte progenitor population present at normal levels in SCID mice by three color flow cytometry with B220 and S7. *Curr. Top. Microbiol. Immunol.* **152,** 19–25.
5. Hardy, R. R., Carmack, C. E., Shinton, S. A., Kemp, J. D., and Hayakawa, K. (1991) Resolution and characterization of pro-B and pre-pro-B cell stages in normal mouse bone marrow. *J. Exp. Med.* **173,** 1213–1225.
6. Fried, M., Hardy, R. R., and Bosma, M. J. (1989) Transgenic SCID mice with a functionally rearranged immunoglobulin heavy chain gene. *Curr. Top. Microbiol. Immunol.* **152,** 107–114.
7. Reichman-Fried, M., Bosma, M. J., and Hardy, R. R. (1993) B lineage cell in μ-transgenic SCID mice proliferate in response to IL-7 but fail to show evidence of immunoglobulin light chain rearrangement. *Int. Immunol.* **5,** 303–310.

8. Spanopoulou, E., Roman, C. A., Corcoran, L. M., et al. (1994) Functional immunoglobulin transgenes guide ordered B-cell differentiation in Rag-1-deficient mice. *Genes Dev.* **8,** 1030–1042.

9. Fang, W., Mueller, D. L., Pennell, C. A., et al. (1996) Frequent aberrant immunoglobulin gene rearrangements in pro-B cells revealed by a bcl-xL transgene. *Immunity* **4,** 291–299.

10. Bain, G., Robanus Maandag, E. C., te Riele, H. P., et al. (1997) Both E12 and E47 allow commitment to the B cell lineage. *Immunity* **6,** 145–154.

11. Li, Y. S., Wasserman, R., Hayakawa, K., and Hardy, R. R. (1996) Identification of the earliest B lineage stage in mouse bone marrow. *Immunity* **5,** 527–535.

12. Allman, D., Li, J., and Hardy, R. R. (1999) Commitment to the B lymphoid lineage occurs before DH-JH recombination. *J. Exp. Med.* **189,** 735–740.

13. Tudor, K. S., Payne, K. J., Yamashita, Y., and Kincade, P. W. (2000) Functional assessment of precursors from murine bone marrow suggests a sequence of early B lineage differentiation events. *Immunity* **12,** 335–345.

14. Li, Y.-S., Hayakawa, K., and Hardy, R. R. (1993) The regulated expression of B lineage associated genes during B cell differentiation in bone marrow and fetal liver. *J. Exp. Med.* **178,** 951–960.

15. Allman, D., Lindsley, R. C., DeMuth, W., Rudd, K., Shinton, S. A., and Hardy, R. R. (2001) Resolution of three nonproliferative immature splenic B cell subsets reveals multiple selection points during peripheral B cell maturation. *J. Immunol.* **167,** 6834–6840.

16. Kondo, M., Weissman, I. L., and Akashi, K. (1997) Identification of clonogenic common lymphoid progenitors in mouse bone marrow. *Cell* **91,** 661–672.

17. Wells, S. M., Kantor, A. B., and Stall, A. M. (1994) CD43 (S7) expression identifies peripheral B cell subsets. *J. Immunol.* **153,** 5503–5515.

18. Hardy, R. R. and Hayakawa, K. (1991) A developmental switch in B lymphopoiesis. *Proc. Natl. Acad. Sci. USA* **88,** 11,550–11,554.

19. Hardy, R. R., Shinton, S. A., and Hayakawa, K. (1992) Repopulation of SCID mice with fetal-derived B-lineage cells. *Curr. Top. Microbiol. Immunol.* **182,** 73–80.

20. Cumano, A., Dorshkind, K., Gillis, S., and Paige, C. J. (1990) The influence of S17 stromal cells and interleukin 7 on B cell development. *Eur. J. Immunol.* **20,** 2183–2189.

21. Cumano, A., Paige, C. J., Iscove, N. N., and Brady, G. (1992) Bipotential precursors of B cells and macrophages in murine fetal liver. *Nature* **356,** 612–615.

22. Carleton, M., Haks, M. C., Smeele, S. A., et al. (2002) Early growth response transcription factors are required for development of CD4(–)CD8(–) thymocytes to the CD4(+)CD8(+) stage. *J. Immunol.* **168,** 1649–1658.

23. Schmitt, T. M. and Zuniga-Pflucker, J. C. (2002) Induction of T-cell development from hematopoietic progenitor cells by delta-like-1 in vitro. *Immunity* **17,** 749–756.

24. Kawamoto, H., Ohmura, K., and Katsura, Y. (1997) Direct evidence for the commitment of hematopoietic stem cells to T, B and myeloid lineages in murine fetal liver. *Int. Immunol.* **9,** 1011–1019.

25. Dou, Y. M., Watanabe, Y., Aiba, Y., Wada, K., and Katsura, Y. (1994) A novel culture system for induction of T-cell development: modification of fetal thymus organ culture. *Thymus* **23,** 195–207.

26. Ikawa, T., Kawamoto, H., Fujimoto, S., and Katsura, Y. (1999) Commitment of common T/Natural killer (NK) progenitors to unipotent T and NK progenitors in the murine fetal thymus revealed by a single progenitor assay. *J. Exp. Med.* **190,** 1617–1626.

27. Schlissel, M. S. and Baltimore, D. (1989) Activation of immunoglobulin kappa gene rearrangement correlates with induction of germline kappa gene transcription. *Cell* **58,** 1001–1007.

28. Gu, H., Kitamura, D., and Rajewsky, K. (1991) B cell development regulated by gene rearrangement: arrest of maturation by membrane-bound D mu protein and selection of DH element reading frames. *Cell* **65,** 47–54.

29. Gu, H., Tarlinton, D., Muller, W., Rajewsky, K., and Forster, I. (1991) Most peripheral B cells in mice are ligand selected. *J. Exp. Med.* **173,** 1357–1371.

30. Kitamura, D. and Rajewsky, K. (1992) Targeted disruption of mu chain membrane exon causes loss of heavy-chain allelic exclusion. *Nature* **356,** 154–156.

31. Wasserman, R., Li, Y.-S., and Hardy, R. R. (1995) Differential expression of the blk and ret tyrosine kinases during B lineage development is dependent on Ig rearrangement. *J. Immunol.* **155,** 644–651.

32. Wasserman, R., Li, Y. S., and Hardy, R. R. (1997) Down-regulation of terminal deoxynucleotidyl transferase by Ig heavy chain in B lineage cells. *J. Immunol.* **158,** 1133–1338.

33. Wasserman, R., Li, Y. S., Shinton, S. A., et al. (1998) A novel mechanism for B cell repertoire maturation based on response by B cell precursors to pre-B receptor assembly. *J. Exp. Med.* **187,** 259–264.

2

Characterization of B-Cell Maturation in the Peripheral Immune System

Rita Carsetti

Summary

In the periphery different populations of B cells can be identified, corresponding to subsequent stages of B-cell development. Transitional 1 B cells are recent bone marrow emigrants traveling with the blood to the spleen. Here they further develop to transitional 2 and mature B cells. Marginal zone B cells are a sessile population only found in the spleen. The distinction of these cell types is only possible by three- and four-color flow cytometry, analyzing the relative expression of several developmentally regulated markers. We describe the method for the staining of the cells and the analysis of the collected data and show examples of the results obtained in normal and mutant mice.

Key Words

B cells; development; T1; T2; spleen.

1. Introduction

B cells are generated from hematopoietic stem cells (HSC) in the bone marrow during adult life. Their development can be divided into several stages, identified on the basis of their distinct phenotype and characterized by the progressive acquisition of the B-cell antigen receptor (BCR) components. The fitness of developing B cells is continuously verified. Only B cells with a signaling BCR and able to appropriately respond to growth factors derived from microenvironment of the bone marrow progress along the differentiation pathway ADDIN ENRfu *(1–6)*

B cells leave the bone marrow at the transitional 1 (T1) stage when they are still functionally immature. With the blood they reach the spleen, where they further develop to transitional 2 (T2), mature (M), and marginal zone (MZ) B cells. As in the bone marrow, this phase of development in the spleen is

From: *Methods in Molecular Biology, vol. 271: B Cell Protocols*
Edited by: H. Gu and K. Rajewsky © Humana Press Inc., Totowa, NJ

also dependent on the signaling function of the BCR, on the availability of microenvironment-derived factors, and on the adequate response of B cells to these factors ADDIN ENRfu *(7–11)*.

The accurate discrimination and quantification by flow cytometry of the B-cell populations in the periphery allows (a) the identification of alterations and blocks of B-cell development; (b) the separation of the different B-cell developmental stages, an indispensable tool for their molecular and genetic comparison both in normal and mutant mice.

2. Materials

1. Medium for single-cell suspensions: Iscove's Dulbecco's modified Eagle's medium (DMEM) (GIBCO-BRL) supplemented with 2% heat-inactivated fetal calf serum (FCS; GIBCO-BRL), 2% L-glutamine (GIBCO-BRL), $5 \times 10^{-5} M$ 2-β-mercaptoethanol (Sigma, St. Louis) and 20 mg/mL gentamycin (GIBCO-BRL).
2. Gey's solution: 14 mL sterile H_2O, 4 mL solution A, 1 mL solution B, 1 mL solution C. Solution A (for 1 L H_2O): 35 g NH_4Cl, 1.85 g KCl, 1.5 g $Na_2HPO_4 \cdot 12H_2O$, 0.119 g KHI_2PO_4, 5.0 g glucose, 25 g gelatin, 0.05 g phenol red. Autoclave at 120°C. Store at 4°C. Solution B (for 100 mL H_2O): 0.14 g $MgSO_4$ 7 H_2O, 0.42 g $MgCl_2$. 6 H_2O, 0.34 g $CaCl_2$. 2 H_2O. Autoclave at 120°C. Store in dark at 4°C. Solution C (for 100 mL H_2O): 2.25 g $NaHCO_3$. Autoclave at 120°C. Store in dark at 4°C.
3. FACS buffer: 1X phosphate-buffered saline (PBS) without Ca^{2+} and Mg^{2+}, supplemented with either 1% bovine serum albumin (BSA) or 3% dialyzed FCS and containing 0.1% NaN_3 (*see* **Notes 1** and **4**).
4. Antibodies: anti-IgM (clone 2911), anti-IgD (clone 11.26c), anti-CD21 (clones 7G6 and 7E9), anti-CD23 (clone B3B4), anti-B220 (clone RA3-6B2), and anti-heat-stable antigen (HSA; clone M1/69) were obtained from Pharmingen (San Diego, CA). Dead cells were excluded from analysis by side/forward scatter gating and/or propidium iodide (PI) staining.

3. Methods

3.1. Cell Suspensions and Basic Staining Procedure

1. Remove the organs surgically and keep them in a 15-mL plastic Petri dish containing 5 mL of cold medium for single-cell suspensions on ice. Make a single-cell suspension by pushing the organs with the piston of a syringe through a metal sift. Keep tubes on ice.
2. Remove red blood cells from samples using Gey's solution. Centrifuge single-cell suspension for 5 min at 290*g* in a refrigerated centrifuge (4°C), remove supernatant, and add 5 mL ice-cold Gey's solution. Underlay with 1 mL FCS (**Caution:** Use a 1-mL pipet or a Pasteur pipet and keep on ice for 5 min. A longer incubation time may result in reduced lymphocyte vitality.
3. Centrifuge as in **step 2** and remove supernatant. Wash with 10 mL FACS buffer and resuspend in 5 mL FACS buffer.

4. Count cells.
5. Use for each staining 5×10^5 to 3×10^6 cells per sample. The stainings can be done either in round-bottom 96-microwell plates or in FACS tubes. Washing volume is 250 µL (three times) or 2.5 mL (once), respectively.
6. After the last washing resuspend the pellet in 10 µL of antibody solution and mix gently (*see* **Notes 2** and **3**).
7. Incubate for 15 min on ice.
8. Wash (as indicated in **step 5**) and resuspend in 200–300 µL FACS buffer for acquisition.
9. For live/dead discrimination, add 10 µL PI solution (stock at 10 µg/mL). If fixing cells before analysis, do not add PI (*see* **Note 5**).

3.2. Four-Color Fluorescence Staining

Three very important things have to be kept in mind when making multiple color stainings: (a) The biotin-labeled antibody has to be added one step before the streptavidin-bound fluorochrome. If both reagents are added together, complexes are immediately formed between excess antibody and fluorochrome, and the cells will remain unstained. (b) Always use the smallest fluorochromes first to avoid steric hindrance. Fluorescein isothiocyanate (FITC), biotin, peridinin chlorophyll protein (PerCP), and Cy5 are small. PE, allophycocyanin (APC), and all tandem conjugates are big molecules. (c) Use the brightest fluorochromes for the less well-expressed markers. PE, Cy5, and most tandem conjugates are bright; PercP is very weak. Fluorochrome choice can really make a difference. For example, CD25 on pre-B cells and on regulatory T-cells can be detected with PE- or biotin-labeled antibodies, but it is almost undetectable with FITC.

1. Start from the pellet obtained in **step 5** of **Subheading 3.1.** Resuspend the cells in the antibody mix (10 µL) containing FITC and biotin-labeled antibodies.
2. Follow **steps 7** and **8**, but after centrifugation add the second-step antibodies to the pellet—for example, those labeled with Cy5 and PE.
3. Repeat **steps 7** and **8**, and after centrifugation resuspend the cells in the solution containing the streptavidin-bound fluorochrome (normally a tandem conjugate).
4. Proceed as in **step 8**.
5. To identify T1, T2, MZ, and M B cells, we normally use CD21-FITC, CD23-biotin, IgD-PE, and IgM-Cy5 plus streptavidin-cychrome. Keep in mind that CD23 is the weakest marker.

3.3. Flow-Cytometric Analysis

The identification of T1, T2, MZ, and M B cells in the spleen is only possible through the combination of several markers, which are differentially expressed during development (*see* **Table 1**). The best combination includes surface immunoglobulin IgM and IgD, the complement receptor type 2, CD21, and the low-affinity receptor for IgE, CD23 *(10)*.

Table 1

Phenotype and Location of Different B-Cell Populations in the Spleen

	T1	T2	M	MZ
sIgM	+++	+++	+	+++
sIgD	–	+++	+++	–
CD23	–	++	++	–
CD21	–	+++	++	+++
HSA	+++	+++	+	+
CD62L	–	+	++	–
493	+	+	–	–
Location	red pulp, outer PALS	follicle	follicle	marginal zone

A plot of forward scatter (FSC) vs CD23 can be used to separate CD23^{pos} and CD23^{neg} cells of lymphocyte size (small cells, low FSC) (**Fig. 1A**). IgM and IgD expression in CD23^{neg} cells discriminates non-B cells, negative for both markers, from IgM^{pos} and IgD^{neg} B cells, including both T1 and MZ B cells (**Fig. 1B**, CD23^{neg}). The relative size of these two compartments can be determined by combining IgM with CD21. T1 B cells will appear in the CD21^{neg} region, whereas MZ B cells are bright for CD21 (**Fig. 1C**, CD23^{neg}). In the spleen of adult C57BL/6 mice T1 B cells are 5–10% and MZ 3–5% of all B cells. CD23^{pos} B cells analyzed for the expression of IgD and IgM appear positive for both markers. T2 B cells, however, are bright for both IgM and IgD (IgM^{bright} IgD^{bright}), whereas M B cells express high amounts of IgD and low amounts of IgM (IgD^{bright} IgM^{pos}) (**Fig. 1B**, *right panel*). T2 can also be discriminated from M B cells based on the different levels of IgM and CD21: M B cells express moderate amounts of both markers (IgM^{pos} CD21^{pos}), whereas T2 B cells are IgM^{bright} CD21$^{pos\ bright}$ (**Fig. 1C**). T2 B cells are 15–20% of all B cells in the spleen.

The detection of IgM, IgD, and CD21 is sufficient to compare the different distribution of B cells in the spleen, blood, and lymph nodes. M B cells (IgM^{pos}, IgD^{pos}, CD21^{pos}) are present in all tissues. T2 (IgM^{bright}, IgD^{bright}, CD21^{bright}) and MZ B cells (IgM^{bright}, IgD^{neg}, CD21^{bright}) are present only in the spleen and T1 in spleen and blood (**Fig. 2A**). Because MZ B cells can only be found in the spleen, it is not indispensable to use CD23 in this type of analysis.

The frequency of T1 B cells in the blood can be, however, precisely calculated only by taking into consideration that IgM^{bright}, IgD^{neg}, and CD21^{neg} pop-

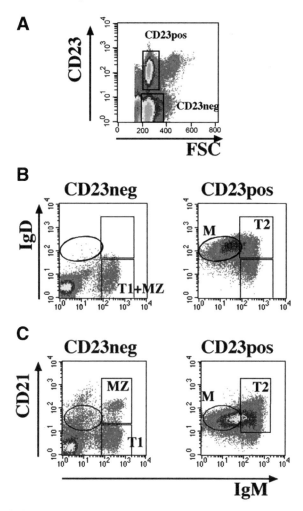

Fig. 1. Discrimination of T1, T2, M, and MZ B cells in the spleen by four-color fluorescence flow cytometry. (**A**) Cells stained for CD23, IgM, IgD, and CD21were first gated, based on size (FSC) and CD23 expression. The CD23^{neg} population is shown in the *left panels* and the CD23^{pos} in the *right panels* in (**B**) and (**C**). (**B**) The staining for IgM and IgD does not allow the discrimination of T1 and MZ B cells, which are both IgM^{bright} and IgD^{neg}. M and T2 B cells can be recognized based on the relative amount of IgM and IgD: T2 B cells are bright for both IgM and IgD (IgM ^{bright} and IgD ^{bright}), whereas M B cells express high amounts of IgD and low amounts of IgM (IgD^{bright} and IgM^{pos}). (**C**) The high level of CD21 distinguishes MZ (IgM ^{bright} and CD21^{bright}) from T1 B cells (IgM ^{bright} and CD21^{neg}). Different levels of both IgM and CD21 are present in T2 (IgM^{bright} and CD21$^{pos\ bright}$) and M B cells (IgM^{pos} and CD21^{pos}).

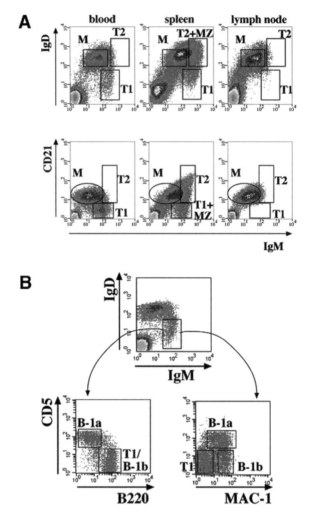

Fig. 2. Discrimination of T1, T2, M, and MZ B cells in the periphery. (**A**) Three-color fluorescence analysis of mononuclear cells isolated from blood, spleen, and lymph nodes. IgM and IgD staining is shown in the *top panels* and CD21 and IgM in the *lower panels*. (**B**) Four-color fluorescence flow cytometry of peripheral blood mononuclear cells IgD and IgM (shown in the *top panel*) in combination with either CD5 and B220 (*left panel*) or CD5 and MAC-1 (*right panel*).

ulations contain not only T1 B cells but also B-1a and B-1b B cells *(11)*. T1, B-1a, and B-1b B cells are identified by four-color fluorescence analysis using a combination of either IgD, IgM, CD5, and B220, or IgD, IgM, CD5, and Mac-1. A gate is set on IgM[bright] and IgD[neg] B cells (**Fig. 2B**, top panel). By plotting CD5 vs B220 only B-1a B cells appear as a defined population (CD5^{pos}, B220^{dull}, Fig. 2B, lower left panel), but T1 and B-1b B cells are both CD5^{neg} and B220^{bright}. CD5 and MAC-1 instead separate B-1a (CD5^{pos} and MAC-1^{pos}) from B-1b (CD5^{neg} and MAC-1^{pos}) and T1 (CD5^{neg} and MAC-1^{neg}) B cells (**Fig. 2B**, left panel).

An example of the analysis of mutant mice using this method is shown in **Fig. 3**. We studied the development of B cells in mice with genetic defects of molecules involved in the BCR signaling complex. In the spleen of mice lacking the phosphatase CD45 (CD45–/–) and of CBA/N mice lacking the natural mutant for the tyrosine kinase Btk, we observed an increase of the T2 population (**Fig. 3A** CD23^{pos}) and a reduction of the mature pool, with a normal frequency of both T1 and MZ B cells (**Fig. 3A** CD23^{neg}). We concluded that B-cell development is arrested at the T2 stage in both mouse strains and that a perfect signaling function of the BCR is important for the progression from T1 to T2 and M. This hypothesis was confirmed by the observation that B-cell development further than the T1 stage was completely abrogated in mice with a severely impaired BCR signaling function (mb1Δc/Δc–).

Several other mouse mutants have been analyzed using this method confirming the presence of important checkpoints at the T1 and T2 stage before the final differentiation of peripheral B cells. Based on the results of these studies we propose a model suggesting that the cooperation of signals from the BCR and from a growth factor produced and active in the spleen is indispensable in the late phases of B-cell development (**Fig. 4A**). Arrest of development at the T1 stage eliminates all subsequent B-cell types and can be caused either by a severe impairment of the BCR signaling function or by the lack of the B-cell activation factor (BAFF) or of its receptor (**Fig. 4B**) ADDIN ENRfu *(10,12–14)*. The block at the T2 stage is observed in mice with qualitative defects of the BCR signaling function or overexpressing BAFF (**Fig. 4C**) *(10,15–23)*. This model can be used as a framework to classify known and new mutations and will be continuously improved and modified by additional studies.

4. Notes

1. Dialysis of FCS helps to remove free biotin, which, if present in high amounts, binds streptavidin-coupled fluorochromes and impairs the detection through biotin-conjugated antibodies.
2. It is necessary to determine the proper concentration for each antibody used. For that, test antibody dilutions using the concentration of cells most often used— for example, 1×10^6. Choose the concentration that best separates positive from negative cells. For FITC-labeled antibodies, test dilutions from 1:10 to 1:80. It is possible to dilute all other antibodies much more; therefore, test dilutions up to 1:200, with the exception of PerCP-labeled antibodies, which may need to use undiluted.

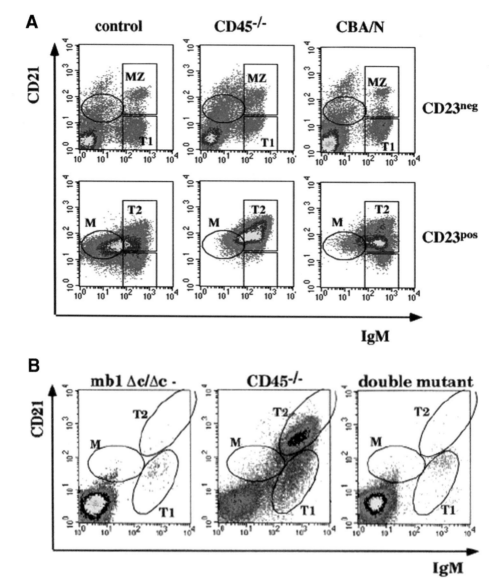

Fig. 3. Identification of blocks of B-cell development through the analysis of peripheral B cells. Four-color fluorescence flow cytometry using CD23, IgM, IgD, and CD21 in control and mutant mice. Block at the T2 stage. Block at the T1 stage.

Fig. 4. *(see facing page)* Model of normal B-cell development in the spleen. The arrest of development at the T1 stage impairs the development of T2, M, and MZ B cells. The mutations causing this phenotype are indicated on the right. The block at the T2 stage is observed in the mutant mice indicated on the right.

A NORMAL DEVELOPMENT

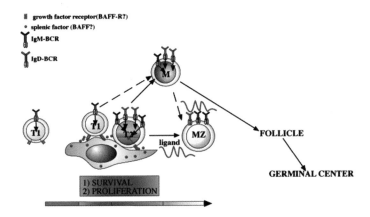

B BLOCK AT THE T1 STAGE

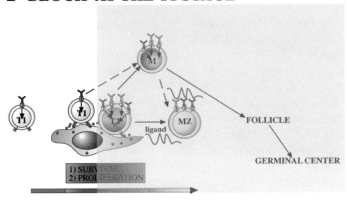

syk-/-
mb1 Δc/Δc-
Rel/RelA

BAFF-/-
A/WySnJ (BAFF-R)

C BLOCK AT THE T2 STAGE

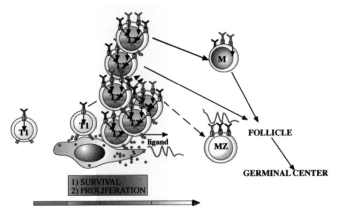

CD45 -/-
Btk-/-
Lyn-/-
Vav-/-
PI3K-/-
PLCγ-/-
BCAP-/-
BLNK-/-

BAFF transgenic

3. Antibody concentration, not its quantity, is critical. When staining with 10 μL of antibody solution added to a dry pellet, spare a lot of antibody. If two antibodies are added together, each in 10 μL, consider that the final volume will be 20 μL, and the final antibody concentration will thus be half of what is really needed.
4. The concentration of NaN$_3$ in the FACS buffer may appear high, but it is necessary to block receptor modulation. If cells have to be functional after analysis or sorting, the medium should not contain NaN$_3$, and receptor modulation may only be controlled by keeping the cells cold during the whole procedure, including sorting (use a refrigerating chamber at 4°C).
5. In a two- or three-color fluorescence analysis, PI can be used for live/dead discrimination. Add 10 μL PI solution (stock at 10 μg/mL) just before acquisition. Remember that PI can be seen either in FL2 or FL3, and when fixing cells before analysis, you cannot add PI (all cells will be positive).

References

1. Carsetti, R. (2000) The development of B cells in the bone marrow is controlled by the balance between cell-autonomous mechanisms and signals from the microenvironment. *J. Exp. Med.* **191,** 5–8.
2. Dorshkind, K. (1990) Regulation of hemopoiesis by bone marrow stromal cells and their products. *Annu. Rev. Immunol.* **8,** 111–137.
3. Hardy, R. R., Carmack, C. E., Shinton, S. A., Kemp, J. D., and Hayakawa, K. (1991) Resolution and characterization of pro-B and pre-pro-B-cell stages in normal mouse bone marrow. *J. Exp. Med.* **173,** 1213–1225.
4. Rolink, A., Haasner, D., Melchers, F., and Andersson, J. (1996) The surrogate light chain in mouse B-cell development. *Int. Rev. Immunol.* **13,** 341–356.
5. Stoddart, A., Fleming, H. E., and Paige, C. J. (2000) The role of the preBCR, the interleukin-7 receptor, and homotypic interactions during B-cell development. *Immunol. Rev.* **175,** 47–58.
6. Loffert, D., Schaal, S., Ehlich, A., et al. (1994) Early B-cell development in the mouse: insights from mutations introduced by gene targeting. *Immunol. Rev.* **137,** 135–153.
7. Carsetti, R., Kohler, G., and Lamers, M. C. (1995) Transitional B cells are the target of negative selection in the B cell compartment. *J. Exp. Med.* **181,** 2129–2140.
8. Chung, J. B., Sater, R. A., Fields, M. L., Erikson, J., and Monroe, J. G. (2002) CD23 defines two distinct subsets of immature B cells which differ in their responses to T-cell help signals. *Int. Immunol.* **14,** 157–166.
9. Chung, J. B., Silverman, M., and Monroe, J. G. (2003) Transitional B cells: step by step towards immune competence. *Trends Immunol.* **24,** 342–348.
10. Loder, F., Mutschler, B., Ray, R. J., et al. (1999) B cell development in the spleen takes place in discrete steps and is determined by the quality of B cell receptor-derived signals. *J. Exp. Med.* **190,** 75–89.
11. Wardemann, H., Boehm, T., Dear, N., and Carsetti, R. (2002) B-1a B cells that link the innate and adaptive immune responses are lacking in the absence of the spleen. *J. Exp. Med.* **195,** 771–780.

12. Turner, M., Gulbranson-Judge, A., Quinn, M. E., Walters, A. E., MacLennan, I. C., and Tybulewicz, V. L. (1997) Syk tyrosine kinase is required for the positive selection of immature B cells into the recirculating B cell pool. *J. Exp. Med.* **186,** 2013–2021.
13. Grossmann, M., O'Reilly, L. A., Gugasyan, R., Strasser, A., Adams, J. M., and Gerondakis, S. (2000) The anti-apoptotic activities of Rel and RelA required during B-cell maturation involve the regulation of Bcl-2 expression. *EMBO J.* **19,** 6351–6360.
14. Thompson, J. S., Bixler, S. A., Qian, F., et al. (2001) BAFF-R, a newly identified TNF receptor that specifically interacts with BAFF. *Science* **293,** 2108–2111.
15. Chan, V. W., Meng, F., Soriano, P., et al. (1997) Characterization of the B lymphocyte populations in Lyn-deficient mice and the role of Lyn in signal initiation and down-regulation. *Immunity* **7,** 69–81.
16. Doody, G. M., Bell, S. E., Vigorito, E., et al. (2001) Signal transduction through Vav-2 participates in humoral immune responses and B cell maturation. *Nat. Immunol.* **2,** 542–547.
17. Fu, C., Turck, C. W., Kurosaki, T., and Chan, A. C. (1998) BLNK: a central linker protein in B cell activation. *Immunity* **9,** 93–103.
18. Cornall, R. J., Cyster, J. G., Hibbs, M. L., et al. (1998) Polygenic autoimmune traits: Lyn, CD22, and SHP-1 are limiting elements of a biochemical pathway regulating BCR signaling and selection. *Immunity* **8,** 497–508.
19. Fruman, D. A., Snapper, S. B., Yballe, C. M., et al. (1999) Impaired B cell development and proliferation in absence of phosphoinositide 3-kinase p85alpha. *Science* **283,** 393–397.
20. Pappu, R., Cheng, A. M., Li, B., et al. (1999) Requirement for B cell linker protein (BLNK) in B cell development. *Science* **286,** 1949–1954.
21. Wang, D., Feng, J., Wen, R., et al. (2000) Phospholipase Cgamma2 is essential in the functions of B cell and several Fc receptors. *Immunity* **13,** 25–35.
22. Mackay, F., Woodcock, S. A., Lawton, P., et al. (1999) Mice transgenic for BAFF develop lymphocytic disorders along with autoimmune manifestations. *J. Exp. Med.* **190,** 1697–1710.
23. Mackay, F. and Browning, J. L. (2002) BAFF: a fundamental survival factor for B cells. *Nat. Rev. Immunol.* **2,** 465–475.

3

Identification of B-Cell Subsets

An Exposition of 11-Color (Hi-D) FACS Methods

**James W. Tung, David R. Parks, Wayne A. Moore,
Leonard A. Herzenberg, and Leonore A. Herzenberg**

Summary

In the last few years, the effectiveness of developmental and functional studies of individual subsets of cells has increased dramatically owing to the identification of additional subset markers and the extension of fluorescence-activated cell sorter (FACS) capabilities to simultaneously measure the expression of more markers on individual cells. For example, introduction of a 6–8 multiparameter FACS instrument resulted in significant advances in understanding B-cell development. In this chapter, we describe 11-color high-dimensional (Hi-D) FACS staining and data analysis methods that provide greater clarity in identifying the B-cell subsets in bone marrow, spleen, and peritoneal cavity. Further, we show how a single Hi-D FACS antibody reagent combination is sufficient to unambiguously identify most of the currently defined B-cell developmental subsets in the bone marrow (Hardy fractions A–F) and the functional B-cell subsets (B-1a, B-1b, B-2, and marginal zone [MZ] B cells) in the periphery.

Although we focus on murine B-cell subsets, the methods we discuss are relevant to FACS studies conducted with all types of cells and other FACS instruments. We introduce a new method for scaling axes for histograms or contour plots of FACS data. This method, which we refer to as *Logicle visualization*, is particularly useful in promoting correct interpretations of fluorescence-compensated FACS data and visual confirmation of correct compensation values. In addition, it facilitates discrimination of valid subsets. Application of Logicle visualization tools in the Hi-D FACS studies discussed here creates a strong new base for in-depth analysis of B-cell development and function.

Key Words

FACS; flow cytometry; Hi-D; B cells; B-1a; B-1b; marginal zone B cells; follicular B cells; fluorescence compensation; Fluorescence-Minus-One (FMO); Logicle visualization; FlowJo; contour plot; quantile contours.

From: *Methods in Molecular Biology, vol. 271: B Cell Protocols*
Edited by: H. Gu and K. Rajewsky © Humana Press Inc., Totowa, NJ

1. Introduction

The fluorescence-activated cell sorter (FACS) can be seen as one of the first molecular biology instruments because it was designed and developed to measure the expression of sets of genes by individual cells. Since the introduction of FACS instruments *(1)*, they have been widely used to identify and sort lymphocyte subsets defined by the selective expression of genes encoding surface proteins and other markers. Studies discussed here have made use of the latest advances in FACS technology to identify B-cell developmental and functional subsets.

The B-cell subset analyses discussed here are broadly useful in that they serve as an atlas of B-cell subsets in bone marrow and the periphery. In addition, they provide detailed examples of how we use 11-color (Hi-D FACS) methods with a new FACS data-visualization method to resolve subsets that may represent considerably less than 1% of the cells in a given sample. Thus, this chapter provides a guide both to modern Hi-D FACS methodology and to the isolation and study of the development and functions of B-cell subsets.

1.1. Using FACS to Identify B-Cell Subsets

FACS measures expression levels of proteins and other molecules in and on cells by their intrinsic fluorescence (e.g., green fluorescent protein [GFP]) or, more usually, by their ability to bind fluorochrome-conjugated antibodies or other fluorescent probes. When cells stained with fluorescent reagents/probes are passed through the FACS, the fluorochromes are excited by lasers at appropriate wavelengths and the amount of fluorescence emitted is detected by detectors filtered to accept the appropriate emitted wavelength. Over a period of roughly 30 years, FACS has evolved from a one-laser machine capable of detecting two fluorochrome colors and a size marker to a three-laser machine capable of detecting 12 (or more) fluorochromes plus indicators of cell size and granularity.

The identification of B-cell subsets has been intimately tied to the development of the FACS. Antibody-secreting (plasma) cells were the first cells to be identified and sorted with the single-laser FACS (1969) *(1)*. Later (1982), the basic B-cell subsets in spleen and lymph nodes were defined in the first immunology studies conducted with a dual-laser FACS instrument *(2–4)*. Since these early studies, a series of B-cell development subsets, including pro-B, pre-B, immature, and recirculating B-cell populations, have been identified and characterized in the bone marrow *(5–7)*. Similarly, several functionally distinct peripheral B-cell populations, including B-1a, B-1b, B-2, and marginal zone (MZ) B cells, have been recognized *(4,9–12)*.

B-cell development in bone marrow has been classified into six to seven sequential subsets designated Fractions A, B, C, C′, D, E, F by Hardy, Hayakawa, and colleagues *(5)*. Fractions A–C′, all of which express B220 and CD43, are subsets of pro-B cells actively undergoing heavy chain rearrangement. CD43 surface expression is downregulated after successful Ig heavy chain rearrangement as cells progress to fraction D (pre-B) cells.

Cells in fraction D rearrange Ig light chains and differentiate to fraction E (immature B) cells, which express surface IgM, exit the bone marrow, and continue maturation in the spleen. Bone marrow fraction F contains mature recirculating B cells that display the same phenotype as splenic and lymph node follicular (FO) B cells (relatively low levels of surface IgM and high levels of surface IgD) and probably represents a recirculating pool of these cells.

The B-cell subsets (in some cases lineages) in the periphery include immature B cells recently migrated from bone marrow to spleen, FO B cells, MZ B cells, and small numbers of B-1a and B-1b cells *(11–13)*. Collectively, these subsets/ lineages (referred to hereafter as subsets) are distinguished by differential surface expression of CD21–CD24, CD43, and CD5. The origins, anatomical locations, and functions of these subsets have been widely discussed *(8,13–18)*. Here, we focus on their phenotypes and demonstrate the Hi-D FACS methods for resolving them from one another when they are resident in the same organ.

In the examples presented in **Subheading 3.**, we show how the simultaneous staining and Hi-D FACS detection of 11 surface markers enables resolution of fractions A–F and the mature B-cell subsets in bone marrow, spleen, and in the peritoneal cavity (PerC). In addition, we show how Hi-D FACS methods enable detailed exploration of heterogeneity within the known B-cell subsets.

1.2. Fluorescence Compensation in FACS Studies

In discussing methods for resolving the B-cell subsets, we pay particular attention to problems introduced by fluorescence compensation and to related factors that influence the accuracy of conclusions drawn from FACS analyses. These considerations apply to data collection and analysis with all FACS instruments, not just Hi-D FACS, and to all FACS studies in which fluorescence compensation is necessary.

For many years, standard FACS data collection was based on analog (hardware) "compensation" to correct for overlap in the fluorescence emission spectra of fluorochromes excited by the same laser. However, these methods introduce nonlinear systematic errors into the evaluations and cannot resolve between-laser (excitation) spectral overlaps. These problems can be avoided with computed compensation based on uncompensated primary measurements.

This can be done in real time on some newer instruments and is supported for off-line analysis by most current cytometry software. In measurements using analog logarithmic amplifiers the accuracy of computed compensation is dependent on the accuracy of the logarithmic amplifier (logAMP) calibration. Therefore, the most reliable method is to use high-dynamic-range linear digital electronics with computed compensation. The Logicle data-visualization tool (developed in our laboratory, offerred commercially by FlowJo, and incorporated into other software packages; Parks, Moore, et al., in preparation), provides crucial help in evaluating compensated data and is useful in assuring that correct compensation values are being used.

Compensation effects and background (autofluorescence) levels also limit our ability to accurately delineate (gate) subsets deemed "positive" or "negative" for a given marker. Therefore, we discuss methods for choosing staining combinations that optimize gating accuracy, minimize compensation problems, and facilitate determination of background staining levels for individual subsets within an overall population. Together, the methods we describe offer a series of readily applicable techniques that can improve all multiparameter and Hi-D FACS studies, regardless of the number of fluorochromes detected and whether the studies are conducted with traditional or digital flow cytometry instruments.

2. Materials

1. Newborn calf serum (NCS); keep at –20°C until use.
2. Staining medium: deficient hRPMI, 3% NCS; optional: 1 mM ethylenediaminetetraacetic acid (EDTA) (for aggregation-prone cells). Store at 4°C.
3. 100X propidium iodide (PI): 100 µg/mL PI in 1X phosphate-buffered saline (PBS). Store at –20°C.
4. For staining: FACS tubes (5-mL polystyrene round-bottom tubes), 96-well plates, fluorochrome-conjugated antibodies (these are light sensitive so keep in dark at 4°C).
5. For data collection: We use a modified flow cytometer (Hi-D FACS) capable of detecting 11 colors with 3 lasers (488, krypton, and dye laser). The 488-nm laser excites fluorescein isothiocyanate (FITC), phycoerythrin (PE), Cy5PE/Cy-Chrome, Cy5.5PE (or Cy5.5 peridinin chlorophyll protein [PerCP]), and Cy7PE reagents. The dye laser, emitting at 598 nm, excites Texas Red/Alexa594, allophycocyanin (APC), Cy5.5APC, and Cy7APC reagents. The krypton laser (407 nm) excites Cascade blue and Cascade yellow/Alexa430 reagents. Calibration beads (Spherotech, Libertyville, IL); store at –20°C.
6. For data analysis: FlowJo software (Treestar, San Carlos, CA). Version 4.3 (includes Logicle visualization) or above is preferable.

3. Methods

The following methods include (1) staining procedure, (2) collection of 11-color data on flow cytometer, and (3) data analysis using FlowJo software.

Table 1
11-Color Stain Combination Used to Identify B-Cell Subsets

Specificity	Fluorochrome	Second step
CD21	FITC	–
B220	PE	–
CD5	Cy5PE	–
CD24	Biotin	SA-Cy5.5PerCP
IgD	Cy7PE	–
BP-1	Alexa594	–
CD43	APC	–
CD23	Cy5.5APC	–
IgM	Cy7APC	–
CD3ε, CD8α, Gr-1	CasBlu	–
Mac-1	CasYel	–

3.1. Basic Staining Protocol (see Note 1)

1. Prepare cell suspension from Balb/c spleen, peritoneal cavity, or bone marrow. Resuspend 0.5–1 million cells in 50 µL staining media for each stain combination. Add Fc block (CD16/CD32) prior to staining to prevent Fc receptor binding to antibodies.
2. Make up the stain combination: Use the recommended titer for each reagent and add an appropriate amount of the fluorochrome-conjugated antibodies to 50 µL of staining media. *See* **Table 1** for 11-color B-cell staining combination.
3. Add cells to the stain and incubate for 15 min on ice.
   If FACS tubes are used for staining, add 4 mL staining medium to wash away unbound antibodies. Centrifuge at approx 800*g* for 5 min. Aspirate and resuspend cells in 200 µL staining medium, and add PI to a final concentration of 1 µg/mL.
   If 96-well plates are used for staining, stain cells in alternate columns and rows to avoid potential contamination from adjacent wells.
   Add 150 µL staining medium, and spin down cells after the first incubation. Aspirate, add 200 µL staining medium for washing, and centrifuge. Repeat the washing step twice. Aspirate and resuspend cells in 200 µL staining medium and add PI to a final concentration of 1 µg/mL. Transfer the stained cells to FACS tubes.
4. If biotin-conjugated antibody is used in the staining cocktail, then a second step is required. Add an appropriate amount of streptavidin-fluorochrome (SA-fluorochrome) to the stained cells (stained with fluorochrome-coupled and biotin-coupled antibodies), and incubate on ice for 15 min. Wash away the unbound second-step reagent with 4 mL staining media (in FACS tubes). Centrifuge at approx 800*g* for 5 min. Aspirate and resuspend cells in 200 µL staining medium and add PI to a final concentration of 1 µg/mL.

5. The stained cells can be covered and kept at 4°C for 3–4 h until analysis.
6. A compensation sample (cells or antibody-capturing beads) must be made fresh for each fluorochrome used in the staining combination. A compensation sample consists of cells or capturing (antirat Ig or antimouse Ig) beads and only one staining reagent. Generally, a single-stain sample should be included for each reagent in an experiment. In particular, tandem dye lots differ in spectrum so a separate single stain sample is almost always necessary for each tandem reagent.
7. Fluorescence-minus-one (FMO) samples are made by staining cells with a reagent cocktail from which a selected reagent is omitted. FMO-stained cells are necessary for determining the effective background signal distribution in the fluorescence channel of the omitted reagent. FMO methodology is discussed in **Note 2**.

3.2. Data Collection on 11-Color Flow Cytometer (Hi-D FACS)

3.2.1. Instrumentation Standardization

Before collecting data, the three lasers in the Hi-D FACS are carefully aligned. If an automated standardization system is available, the instrument is standardized with stable fluorescent microspheres ("beads"). If automated standardization is not available (most flow instruments), the beads are used to manually standardize the instrument by tuning each photomultiplier tube (PMT) voltage in turn until the beads in each channel are at the standard location assigned for the channel.

Except under very special circumstances, no changes are made to the PMT settings after standardization is completed. If the instrument develops a nozzle clog or other mechanical problem, it must be restandardized before data collection is resumed. After the standardization is completed, it is wise to collect data for a calibration bead sample (10,000 events will do) and to store this standardization data along with the data collected for the samples in the experiment.

In our experience, the multiple dye-containing rainbow beads obtained from Spherotech (*see* **Subheading 2., step 5**) provide a very convenient, uniform, and stable reference for standardization.

3.2.2. Fluorescence Compensation

The emission spectrum of each fluorochrome is not restricted to the wavelength band collected for that fluorochrome. In most cases, some fluorescence is detectable at wavelengths collected for other fluorochromes. Because the amount of the spectral overlap fluorescence is proportional to the emission in the primary channel, the signal owing to the overlap fluorescence can be estimated and subtracted from the overall fluorescence collected in a given channel. This process, through which signals owing to overlap fluorescence are removed, is called *fluorescence compensation*. Fluorescence compensation is necessary whenever a dye generates signal on a detector other than its primary detector.

Our method of setting fluorescence compensation parameters has changed recently, owing to introduction of software that provides good user support in specifying correct compensation values and owing to increased appreciation of the pitfalls and irreversibility of data manipulation by analog compensation. Previously, instrument hardware was used to subtract the amount of overlap fluorescence and adjust the net signal to match background/autofluorescence. However, this subtraction introduces significant artifacts for two reasons:

Hardware subtraction with peak detection becomes biased as the net corrected signal approaches zero.

Logarithmic scales do not contain a zero point and cannot display negative numbers. However the single stain data used to set compensation values yield populations of events distributed around zero and a mean value close to zero. Logarithmic displays of such populations give views that are not suited to selection of correct compensation values. In addition, measurements of compensated signals using logarithmic amplification distort results by reporting all very low and negative values at the minimum of the logarithmic scale.

Therefore, good practice dictates that the hardware compensation utility on FACS instruments be turned off and that uncompensated data be collected. In this case, data must also be collected from compensation samples and stored so that it can be used later to do the fluorescence compensation off-line. The method for doing this is discussed in the next section.

Analog hardware compensation must still be used for most multicolor sorting on instruments that are without real-time software-computed compensation. In all cases data on control samples should be recorded with the same settings as the test samples.

3.2.3. Data Collection on Multicolor Experimental Samples

After standardizing the FACS instrument and collecting the data on calibration beads, adjust the FACS instrument forward scatter gain to bring cell samples to the proper scale. Select number of cells to be analyzed. Collect data on individual samples. Adjust the sample flow rate to a proper rate as the accuracy of measurements and fluorescence compensations are degraded if the sample flow rate (μL/s more than cells/s) is too high. Minimize changes in sample flow rate between samples.

3.3. Data Analysis Using FlowJo Software

1. Create a new FlowJo workspace with the data that is collected.
2. Live cell gate: double-click on a sample to view the collected data in a new window. Dead cells (PI positive) will show up as very bright cells on the Cy5PE (488 nm excitation, approx 665 nm emission) channel. Set a gate that includes all cells except the dead cells. Click on down arrow to bring up a new window showing all cells within the gate. It is preferable to view the analysis on contour plots

or color density plots instead of the dot plots because the structure of the data in dense regions is obscured in dot plots owing to contiguous placement and over-writing of dots (**Fig. 1**). Both contour plots (with outliers, **Fig. 1**) and color den-sity plots with outliers (not shown) in FlowJo maintain resolution regardless of cell numbers and are thus better suited for viewing populations.

3. Compensation gates: each fluorescent reagent must have a compensation sample associated with it. Gate on fluorochrome-positive cells (or antibody-capturing beads) for each of the 11 colors. However, very bright cells (at the upper boundary of the plot) must be excluded, even when they are part of a continuous population.

   a. To define compensation matrix, go to *Platform → Compensate sample → Define matrix*. Using the populations defined in the previous step, place the gate name for each of the fluorochrome-positive cell samples in the appropri-ate boxes on the menu. Use unstained cells for fluorochrome-negative cells. Use gated unstained beads if capturing beads are used for compensation. Click "compute" to generate a compensation matrix for the staining combination.

      One special caution: Different lots of tandem fluorochromes (i.e., Cy5PE, Cy5.5PE, Cy7PE, Cy5.5APC, and Cy7APC) will have different emission spec-tra. Similarly, different conjugates with the same or different APC lots may have different emission spectra. Therefore, it is necessary to generate separate compensation matrices if different lots of tandem fluorochromes are used in different stain combinations (**Fig. 2**).

   b. Apply the compensation matrix to the cells stained with 11 colors (*Platform → Compensate Samples → Comp Matrix*). Gate on live cells before analyzing each sample.

3.4. Logicle Visualization

The Logicle data display addresses problems introduced by the absence of a zero point and negative values on the logarithmic axes traditionally used to dis-play FACS data (David Parks, Wayne Moore, et al., manuscript in preparation) (*see* **Fig. 3**). Logarithmic plots "pile up" at baseline all events with values at or below that level and often show a peak above the true center of a population with a low-to-negative mean (**Fig. 3A,** *upper right panel*). Logicle axes, in con-trast, enable display of the data on mathematically defined axes that are essen-tially linear in the region around zero and essentially logarithmic at high data values.

Note that the Logicle transformation does not change the data. It only provides a method for visualizing data that cannot be displayed well on a logarithmic scale.

Figure 3A illustrates the usefulness of Logicle display of single stain con-trol data in assessing the accuracy of fluorescence compensation. In the loga-rithmic view the data is correctly compensated (as confirmed by statistical evaluation), but there is no way to determine that from the visual display. The

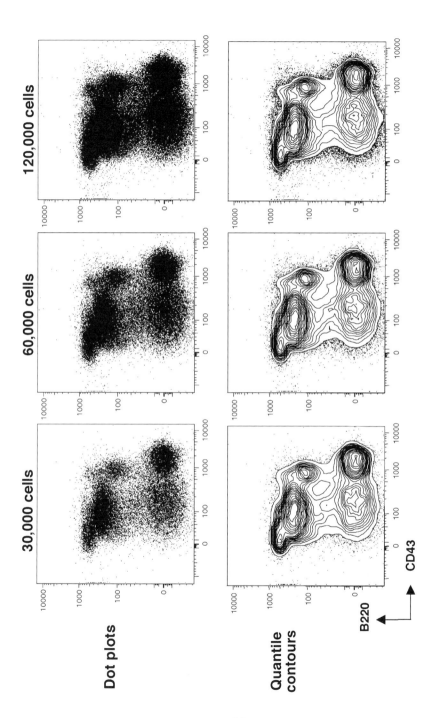

Fig. 1. Dot plots are sensitive to cell numbers. Dot plots tend to lose resolution when cell numbers are too high or too low. Contour plots (with outliers) maintain the resolution of subsets regardless of cell numbers.

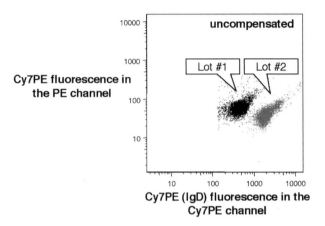

Fig. 2. Different lots of tandem dyes have different fluorescence properties. IgD conjugated to two different lots of Cy7PE (shown in *gray* and *black*) gives different Cy7PE to PE fluorescence ratios.

Logicle plot, however, provides clear visual confirmation that the FITC high and FITC low/negative populations are matched in terms of their average PE channel signal.

Deviation from this pattern will indicate errors in compensation (**Fig. 3B**).

Logicle visualization tool has recently been established as the default in our laboratory. It has only been added to FlowJo in version 4.3, March 2003. Therefore, in future versions the method for accessing it may be different than that described here.

In essence, a Logicle transformation is defined to match a user-selected, compensated, and gated data set and then applied to related compensated data sets. Live, size-gated populations work well as the selected data set. Using cells from the same source as those to which the transformation will be applied is advisable. Logicle is currently accessed in FlowJo by turning on the *Allow custom visualization* option in the FlowJo Preferences. Once Logicle is turned on and compensation has been done, select the gated data set (fully stained) intended as the Logicle base (i.e., live, size-gated lymphocytes), then select *Platform → Compensate Sample → Define Transformation*. FlowJo will then apply the transformation to all samples compensated with the compensation matrix used for the initial transformation sample.

Note that to use different transformation base populations for samples compensated with the same matrix, it is necessary to copy and rename the matrix and use the copy to compensate the target subgroup. Using different base populations may be useful because the Logicle transformation is determined by the selected base population.

A
Log visualization

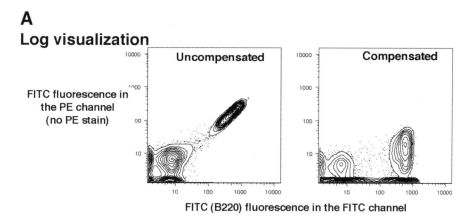

visualization

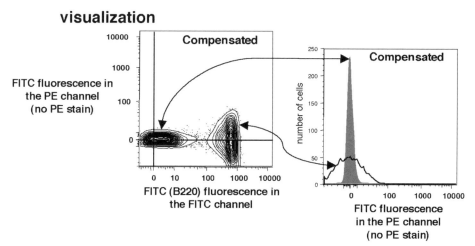

Fig. 3A. Visualization methods. Spleen cells are singly stained with B220-FITC and shown in two-dimensional plots (FITC vs PE) on top panels. *Top left*: uncompensated data. *Top right*: compensated data with standard logarithmic scale. Note that while the computed median in the PE dimension of the FITC-positive population is near the baseline of this plot, the apparent center of the population is well above this. *Bottom left*: compensated data with custom visualization. *Bottom right*: Histograms of compensated PE fluorescence detected in FITC[+] and FITC[-] cells. Note that both FITC[+] and FITC[-] cells are distributed symmetrically around a common center near zero.

3.5. Gating and Visualizing B-Cell Subsets

Figs. 4–6 show data for bone marrow, spleen, and peritoneal cavity samples stained with an 11-color staining combination designed to distinguish subsets of developing and mature B cells. We point out the advantages of using this single

B
Logicle visualization

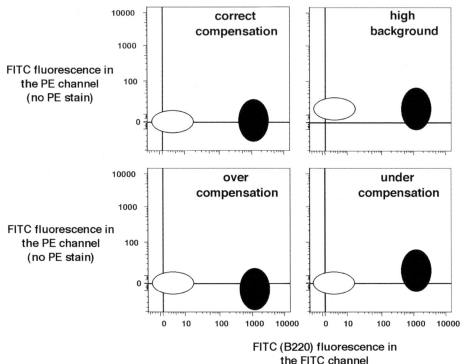

Fig. 3B. Visualization methods *(continued)*. Deviation from proper compensation can be visualized on Logicle axes. *Upper left panel*: cells that are properly compensated. *Upper right panel*: high background or autofluorescence level on the PE channel. *Lower left panel*: FITC⁺ cells that are overcompensated (too much fluorescence subtraction). *Lower right panel*: FITC⁺ cells that are undercompensated (too little fluorescence subtraction).

Hi-D FACS stain, as opposed to using several combinations with fewer colors in each, as we move through the gating methods.

1. In adult Balb/cN bone marrow (**Fig. 4A**), live cells are gated from total cells (in FSC and Cy5PE channels). Within the live cells, a lymphocyte gate is applied to gate out the large, nonlymphocytic cells. Within the live, lymphocyte gate, a B220⁺ dump (CD3ε⁻, CD8α⁻, GR-1⁻) gate is applied to gate on B cells. This live, lymphocyte, B220⁺ gating scheme is also used for bone marrow, splenic, and peritoneal B cells.

A

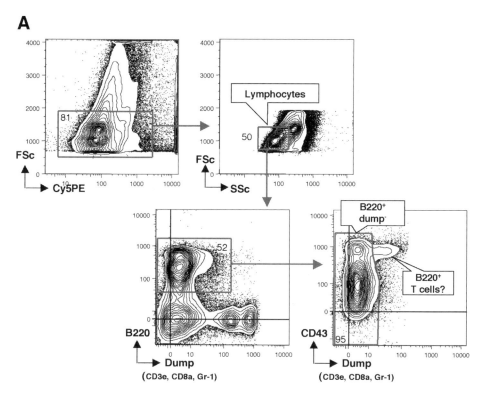

Fig. 4A. Identification of bone marrow B-cell populations. Bone marrow cells are first gated on live cells (*top left panel*), on lymphocyte size (*top right panel*), followed by eliminating contaminating non-B cells by gating on B220 and the dump channel (CD3ε⁻, CD8α⁻, and Gr-1⁻) (*bottom panels*).

The classic gating scheme for identifying the B-cell developmental subsets is shown in **Fig. 4B**. In this figure, CD43 and B220 (*upper panel*) jointly identify two major B-cell developmental subsets within the live, lymphocyte, B220⁺ population: Hardy fractions A–C′, which contain the earliest developing B cells; and Hardy fractions D–F, which contain pre-B, immature and recirculating B cells.

Fraction A–C is then further separated into fractions A, B, C, and C′ by CD24 and BP-1 (**Fig. 4B,** *lower right panel*). Fractions D–F (**Fig. 4B,** *lower left panel*) is separated into fractions D, E, and F by staining for IgM and IgD. Further gating (data not shown) demonstrates that fractions D (pre-B) and E (immature B) continue to express CD24 and do not express CD5, CD21, and CD23. Fraction F (recirculating B) cells express CD23 and low levels of CD21.

Fig. 4C presents an alternate gating scheme that better resolves fractions A–C′ and reveals a "new" subset akin to fraction D but expressing higher levels of

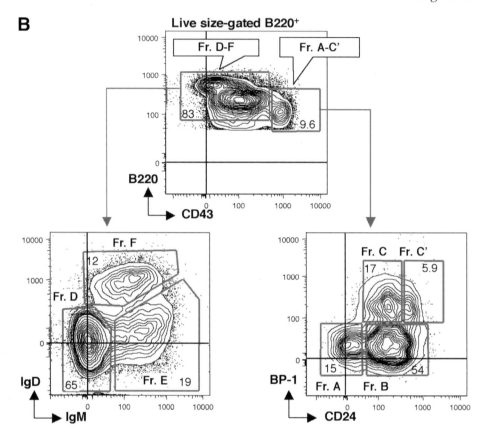

Fig. 4B. Identification of bone marrow B-cell populations *(continued).* In live, size-gated, B220⁺ dump⁻ bone marrow cells, fractions A–C are identified as CD43⁺ B220⁺, whereas fractions D–F are CD43⁻ B220⁺. Fractions A–C can be separated into individual subsets by CD24 and BP-1. Fractions D–F can be separated based on the IgM and IgD surface expressions.

CD43 and CD24. In this scheme, fractions E and F (B220⁺ cells expressing IgM and IgD) are sequentially gated out. The remaining cells are gated on CD24 and CD43 and are resolved into four subsets that can then be further gated to resolve fractions A–D and reveal a previously unresolved subset labeled "D" in the *third panel* down on the right. In addition to revealing this "new subset," this gating scheme has the advantage of cleanly resolving fraction C from fraction C′.

Fig. 4C. *(see facing page)* Identification of bone marrow B-cell populations *(continued).* An alternative way of separating fractions A–C′ based on CD24 and CD43.

C

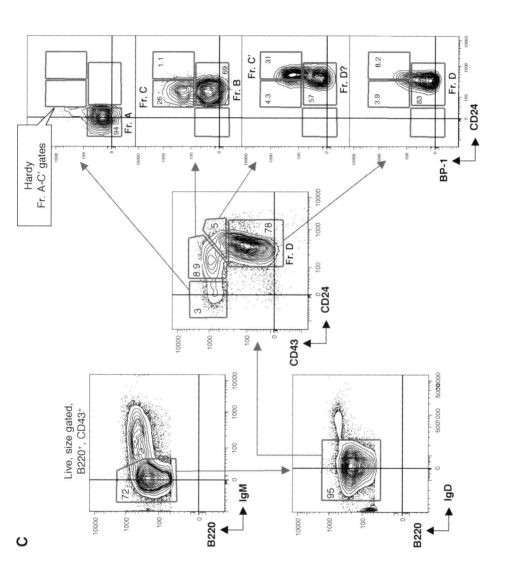

Live, size gated,
B220+, CD43+

Hardy
Fr. A–C' gates

51

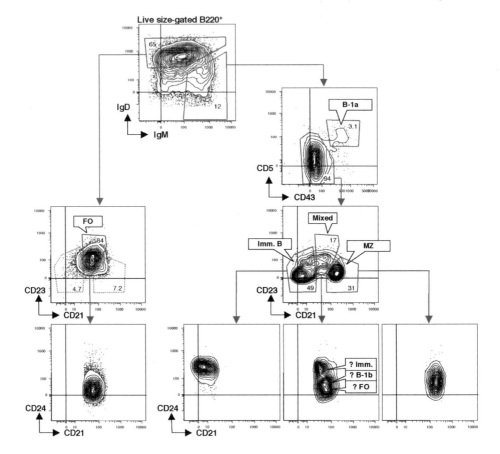

Fig. 5. Identification of splenic B-cell populations. Live, size-gated, B220⁺ dump⁻ splenic B cells are separated based on IgM and IgD expression. The IgM^{lo} IgD^{hi} cells are mainly FO B (B-2) cells. The IgM^{hi} IgD^{lo} cells can be further separated into MZ, B-1a, and immature B populations.

2. **Figure 5** presents the gating scheme for mature B-cell subsets in adult Balb/cN spleen. Live B220⁺ cells are selected using the gating scheme described in the previous section and then resolved according to surface Ig expression into two major subsets.

   a. FO B cells (*left panels*) are a fairly homogeneous IgM^{lo} IgD^{hi} subset. They express intermediate levels of CD21, little or no CD24 (*lower left panel*), high levels of CD23, and they do not express detectable levels of CD5, BP-1, or Mac-1 (data not shown).

   b. The IgM^{hi} IgD^{lo} subset (*right panels*) contains two subsets:

      (i) B-1a cells, which express CD5 and CD43, are readily resolved by visualizing CD5 and CD43 expression.

(ii) The remaining cells, which express neither of these markers (*center right panel*). CD21 and CD23 expression resolves this latter subset into three smaller subsets (*lower panels on right*).

(a) Immature B cells, which do not express either CD21 or CD23 but express high levels of CD24.

(b) A small subset that expresses intermediate levels of CD21 and CD23 and appears to be a mixture of immature B, B-1b, and perhaps some follicular B cells.

(c) Marginal zone (MZ) B cells, which express high levels of CD21 but little or no CD23 and CD24.

3. **Figure 6** presents a gating scheme for resolving B-cell subsets in the Balb/cN peritoneal cavity. The Ig$^+$ population that constitutes the vast majority of the B220$^+$, size-gated live cells is resolved into three subsets by B220 and CD5 expression levels (*top right panel*).

a. B-1a cells, which express low levels of CD5, high levels of CD43, low to intermediate levels of CD24, high levels of IgM, low levels of IgD, and little or no CD21 or CD23 (although a few cells may express intermediate CD23 levels).

b. B-1b cells, which do not express detectable levels of CD5 but are otherwise similar in phenotype to B-1a for the markers tested.

c. A subset with the same phenotype as FO B cells in the spleen.

4. Notes

1. Considerations in designing reagent combinations: FACS data collection consists basically of using a laser to excite a particular fluorochrome and collection of the emitted fluorescence via a fluorescence detector. However, as the number of lasers and detectors increase, the complexity of the data collection and the analysis also increases because of the spectral overlap that occurs between fluorochromes. Designing staining combinations for multiparameter and Hi-D FACS experiments therefore requires careful choice of reagent–fluorochrome combinations to minimize data uncertainty owing to high background and fluorescence compensation limitations.

When the staining cocktail is being designed to detect two or more determinants of the same cells, reagent–fluorochrome pairs should be chosen to minimize the need to subtract large fluorescence compensation values. Ideally, the fluorochromes associated with reagents that detect determinants on the same cells should be excited by separate lasers (and hence the fluorescence signals are collected into different detectors at different times). Initial studies can be done with pairwise stains to determine the spectral overlap between the two reagents when excited by the separate lasers. For example, in **Fig. 3B**, because PE and APC have minimal spectral overlap, B220-PE and CD43-APC are chosen to identify fractions A–C and D–F in the bone marrow so that overlap from the bright B220 (488 nm laser) signal will not have to be subtracted from the CD43 (dye laser) signal, which may be dull or negative on some subsets (fractions D–F).

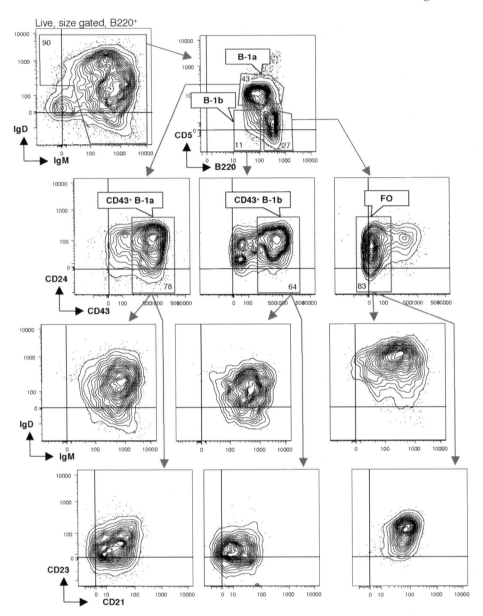

Fig. 6. Identification of peritoneal B-cell populations. Live, size-gated, B220⁺ dump⁻ peritoneal cells are separated based on B220 and CD5 expression. These B220⁺ cells can be separated into FO B, B-1a, and B-1b cells.

If the only reagents available to stain a pair of markers on a given cell type must be excited by the same laser, choice of which fluorochrome is used to mark the reagent detecting the duller of the two markers is quite important. In general, efforts should be made to assign markers giving high levels of staining to dyes that do not have strong spectral overlap onto several other detectors. Conversely, markers expected to yield low or negative staining on cells of interest can be preferentially assigned to dyes that show strong spectral overlaps.

In addition, it is important to titrate the antibody reagents before staining cells with a reagent cocktail. Generally, staining concentrations are selected to stain the cell type of interest at saturation for the target determinant. Minimization of background staining is also taken into account. Titrations also enable estimation of the amount of target determinant expressed on subsets of interest and thus provide key information for designing efficient reagent cocktails.

2. Background signal estimation and fluorescence-minus-one (FMO) controls: it is always difficult to distinguish cells that do not express a given determinant from cells that express low levels of the determinant because cells may have different levels of intrinsic autofluorescence or nonspecifically bound fluorochrome-conjugated reagents. For estimating background owing to nonspecific binding of fluorochrome-coupled reagents, staining samples with a nonreactive antibody that otherwise mimics the staining reagent (i.e., an "isotype control") can be useful (although interpretation can be confounded by differences in "stickiness" of the two putatively similar antibodies). However, background staining tends to be minimal when appropriate Fc and other blocking antibodies are used before staining, and if staining reagents are prepared and stored properly and centrifuged when necessary to remove aggregates.

Autofluorescence contributions to background can be estimated by collecting data for an unstained cell sample. However, in multiparameter and Hi-D FACS studies, overall background estimation is more complicated because spectral overlap and the statistics of fluorescence compensation differentially affect subsets of cells. For example, consider a sample that contains two subsets of cells, one of which expresses a determinant detected by a FITC-coupled reagent and the other does not. If no PE-conjugated reagent is included in the cocktail, both subsets would *a priori* be expected to show the same level of background fluorescence in the PE channel. However, because FITC reagents give a substantial signal in the PE channel, the FITC-positive population will have a higher fluorescence in this channel.

On average, fluorescence compensation reduces the corrected fluorescence in the PE channel to the background level for both subsets in this example. Thus, both subsets will be distributed symmetrically around the center of the negative population, but because of the statistics of the compensation process, the FITC-negative population will have a substantially smaller spread than the FITC-positive subset (*see* **Fig. 3A**).

The compensation effects shown in **Fig. 3A** can be readily visualized because the cells in this example were only stained with B220-FITC. Therefore, the amount of FITC fluorescence represents the total fluorescence on each subset

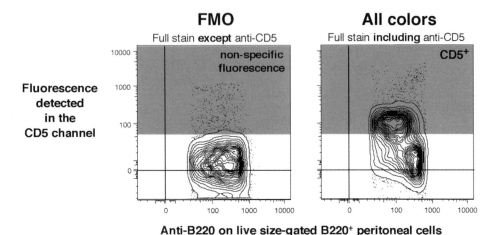

FMO

Full stain **except** anti-CD5

All colors

Full stain **including** anti-CD5

Fluorescence detected in the CD5 channel

non-specific fluorescence

CD5⁺

Anti-B220 on live size-gated B220⁺ peritoneal cells

Fig. 7. Fluorescence-minus-one (FMO). An FMO stain is performed with peritoneal cells to determine background level in the CD5 (Cy5PE) channel. The standard live, size-gated, B220⁺" gating is applied to both 11-color and FMO-stained cells. The B220⁺ subsets in each stain are then compared to determine background levels.

above the unstained autofluorescence. The FITC primary signal can be shown on the *x*-axis, and the compensated values can be shown on the *y*-axis (PE channel). In multiparameter and Hi-D FACS experiments, however, the expression of multiple markers in individual subsets often means that several stains may contribute to signals obtained for a particular subset in a given channel, necessitating compensation for several fluorescence signals in that channel. For this reason, the boundary between no detectable expression and low expression of a particular determinant must be determined directly.

FMO staining, that is, staining the cells with all reagents in the cocktail except the one that registers in the channel of particular interest, is critical for this threshold determination *(19,20)*. In the example shown in **Fig. 7**, FMO staining is used to determine CD5 expression in peritoneal B cells (live, size-gated B220⁺) stained with the 11-color staining cocktail shown in **Table 1**. The upper bound of the B220⁺ cells in the FMO stain (no CD5 reagent) is 55 fluorescence units. Therefore, when the CD5 reagent is present (the full cocktail), B cells that show more that 55 fluorescence units can be taken as CD5⁺. The CD5 threshold for B220⁻ cells (not shown in this example) could be considerably lower if they do not express any other determinant that contributes signals for which compensation is necessary. In all studies shown in the figures in this publication, FMO stains were done to evaluate thresholds for positivity whenever they were in question.

Acknowledgment

We thank Ometa Herman, Sue Sheppard, and John Mantovani for their technical expertise and help. Work from this publication was supported in part by NIH Grant #EB00231.

References

1. Hulett, H. R., Bonner, W. A., Barrett, J., and Herzenberg, L. A. (1969) Cell sorting: automated separation of mammalian cells as a function of intracellular fluorescence. *Science* **166,** 747–749.
2. Hardy, R. R., Hayakawa, K., Haaijman, J., and Herzenberg, L. A. (1982) B-cell subpopulations identifiable by two-color fluorescence analysis using a dual-laser FACS. *Ann. NY Acad. Sci.* **399,** 112–121.
3. Hardy, R. R., Hayakawa, K., Parks, D. R., and Herzenberg, L. A. (1983) Demonstration of B-cell maturation in X-linked immunodeficient mice by simultaneous three-colour immunofluorescence. *Nature* **306,** 270–272.
4. Hayakawa, K., Hardy, R. R., Parks, D. R., and Herzenberg, L. A. (1983) The "Ly-1 B" cell subpopulation in normal immunodefective, and autoimmune mice. *J. Exp. Med.* **157,** 202–218.
5. Hardy, R. R., Carmack, C. E., Shinton, S. A., Kemp, J. D., and Hayakawa, K. (1991) Resolution and characterization of pro-B and pre-pro-B cell stages in normal mouse bone marrow. *J. Exp. Med.* **173,** 1213–1225.
6. Li, Y. S., Wasserman, R., Hayakawa, K., and Hardy, R. R. (1996) Identification of the earliest B lineage stage in mouse bone marrow. *Immunity* **5,** 527–535.
7. Allman, D., Li, J., and Hardy, R. R. (1999) Commitment to the B lymphoid lineage occurs before DH-JH recombination. *J. Exp. Med.* **189,** 735–740.
8. Kantor, A. B. and Herzenberg, L. A. (1993) Origin of murine B cell lineages. *Annu. Rev. Immunol.* **11,** 501–538.
9. Kantor, A. B., Stall, A. M., Adams, S., and Herzenberg, L. A. (1992) Differential development of progenitor activity for three B-cell lineages. *Proc. Natl. Acad. Sci. USA* **89,** 3320–3324.
10. Kantor, A. B., Stall, A. M., Adams, S., and Herzenberg, L. A. (1992) Adoptive transfer of murine B-cell lineages. *Ann. NY Acad. Sci.* **651,** 168, 169.
11. Martin, F. and Kearney, J. F. (1999) CD21high IgMhigh splenic B cells enriched in the marginal zone: distinct phenotypes and functions. *Curr. Top. Microbiol. Immunol.* **246,** 45–50; discussion 51–52.
12. Martin, F., Oliver, A. M., and Kearney, J. F. (2001) Marginal zone and B1 B cells unite in the early response against T-independent blood-borne particulate antigens. *Immunity* **14,** 617–629.
13. Hardy, R. R. and Hayakawa, K. (2001) B cell development pathways. *Annu. Rev. Immunol.* **19,** 595–621.
14. Herzenberg, L. A., Baumgarth, N., and Wilshire, J. A. (2000) B-1 cell origins and VH repertoire determination. *Curr. Top. Microbiol. Immunol.* **252,** 3–13.

15. Herzenberg, L. A. and Kantor, A. B. (1992) Layered evolution in the immune system: a model for the ontogeny and development of multiple lymphocyte lineages. *Ann. NY Acad. Sci.* **651,** 1–9.
16. Haughton, G., Arnold, L. W., Whitmore, A. C., and Clarke, S. H. (1993) B-1 cells are made, not born. *Immunol. Today* **14,** 84–87; discussion 87–91.
17. Berland, R. and Wortis, H. H. (2002). Origins and functions of B-1 cells with notes on the role of CD5. *Annu. Rev. Immunol.* **20,** 253–300.
18. Hardy, R. R., Li, Y. S., Allman, D., Asano, M., Gui, M., and Hayakawa, K. (2000) B-cell commitment, development and selection. *Immunol. Rev.* **175,** 23–32.
19. Roederer, M. (2001) Spectral compensation for flow cytometry: visualization artifacts, limitations, and caveats. *Cytometry* **45,** 194–205.
20. Baumgarth, N. and Roederer, M. (2000) A practical approach to multicolor flow cytometry for immunophenotyping. *J. Immunol. Methods* **243,** 77–97.

4

Analysis of B-Cell Life-Span and Homeostasis

Irmgard Förster

Summary

In the past, the life-span of B cells in rodents has been determined by a variety of methods, leading to conflicting results. Among the various techniques employed, labeling of dividing cells with the thymidine analog 5-bromo-2'-deoxyuridine (BrdU) has turned out to be a versatile and reliable procedure. Labeling of the cells can be easily performed in vivo by feeding BrdU in the drinking water for extended periods of time or by an ip injection of BrdU for short-term labeling experiments. Using the protocol described, it is possible to combine flow cytometric detection of incorporated BrdU simultaneously with fluorescence staining of various cell surface markers.

Key Words

BrdU; B cell; life-span; flow cytometry; proliferation.

1. Introduction

1.1. Determination of B-Lymphocyte Life-Span In Vivo

The life-span of a cell is generally defined as the time between the generation of the cell and its death or entry into cell division. Measurements of B-cell life-span in animal models have been performed by many researchers in the past but have led to conflicting results depending on the experimental approach used (1–3). The earliest experiments were performed using incorporation of radioactively labeled nucleotides (^{3}H-thymidine) into the deoxyribonucleic acid (DNA) of dividing cells in vivo. For this purpose, animals were regularly injected with ^{3}H-thymidine for defined periods of time and quantification of radiolabeled cells was performed by autoradiography (4–6). These experiments have led to the conclusion that most peripheral B cells are long-lived with a life-span exceeding several weeks. However, it could not be excluded that the irradiation resulted in premature death of labeled cells and thus to an underestimation of the number of dividing cells. In a different approach, a compre-

From: *Methods in Molecular Biology, vol. 271: B Cell Protocols*
Edited by: H. Gu and K. Rajewsky © Humana Press Inc., Totowa, NJ

hensive quantification of cell division during B-cell development in the bone marrow was performed by induction of metaphase arrest and detection of metaphases in histological sections (1,7). From this data it is known that in mice $1–2 \times 10^7$ B cells are released from the bone marrow per day, accounting for approx 1/10 of the total peripheral B-cell pool. Only a small fraction of these cells appear to be integrated into the stable peripheral B-cell pool, whereas the majority of these cells rapidly disappear from the circulation. Other experiments on determination of B-cell life-span were based on the selective depletion of proliferating cells by application of hydroxyurea or ganciclovir (8,9). In this case, up to 80% of peripheral B cells were depleted within 1 wk of treatment, indicating that the majority of peripheral B cells were short-lived, in contrast to the results obtained by incorporation of ³H-thymidine. However, this approach may also be influenced by the substantial toxicity of hydrox-yurea such that it cannot be excluded that nonproliferating cells are also affected by the drug. For further discussion of this discrepancy see **refs. 1–3**.

1.2. Use of BrdU for Analysis of Cellular Life-Span

Because of the potential toxicity of the various drugs or radioactive nucleotides introduced into the animals, a different technique was established that appears to be less harmful in vivo. This method is based on the incorpo-ration of the thymidine analog 5-bromo-2′-deoxyuridine (BrdU) into the cellu-lar DNA and detection of incorporated BrdU with BrdU-specific monoclonal antibodies, requiring denaturation of the genomic DNA (10–15; **Fig. 1**). Although BrdU may also exert toxic effects on cells because of its potential mutagenicity (16) and interference with cellular differentiation processes (17), careful choice of the experimental conditions (concentration of BrdU, time of labeling, type of mouse strain) and appropriate controls of toxicity have proved the absence of alteration of B-cell life-span through incorporation of BrdU in both rats and mice (18–20).

Over the last decade, determination of lymphocyte life-span by BrdU-incorporation has been widely employed and proven as a reliable technique that can be combined with detection of various differentiation markers on the cell surfaces using flow cytometry (21–24). Essentially, BrdU-labeling exper-iments have confirmed the initial observation that the peripheral B-cell pool is mainly composed of long-lived B cells, whereas a minor fraction of 20–30% of all B cells are constantly renewed from the bone marrow (19).

1.3. Different Methods of Detection of Incorporated BrdU by Monoclonal Antibodies

As mentioned previously, binding of monoclonal antibodies to incorporated BrdU requires denaturation of the genomic DNA for optimal accessibility. Ini-

Thymidine **5-Bromo-2'-deoxyuridine**

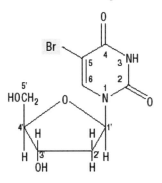

Fig. 1. Structure of thymidine and the thymidine analog BrdU.

tial experiments were performed by treating the cells with 3N HCl, resulting in efficient denaturation of the DNA but irreversible destruction of most cell surface markers *(11,14)*. A milder treatment with 1N HCl containing 0.5% Tween-20 resulting in a partial denaturation was shown to preserve the ability to detect prior staining of surface markers with biotinylated or FITC-labeled antibodies on cells fixed in 70% methanol *(11,12,14,19)*. A significant disadvantage of the HCl-based denaturation was the failure to use fluorescent proteins (e.g., phycoerythrin) as fluorochromes for cell surface staining, thereby restricting the possibility of multiparameter labeling of the cells. The optimized protocol described in **Subheading 3** is based on a modified protocol in which cells are fixed and permeabilized in 1% paraformaldehyde containing Tween-20, and accessibility of the genomic DNA is achieved by treatment of the cells with DNaseI *(13,14)*. Using this protocol surface staining can be readily performed with phycocyanins in addition to biotinylated antibodies prior to fixation of the cells and detection of incorporated BrdU. An example of an anti-BrdU staining is depicted in **Fig. 2**.

2. Materials

2.1. In Vivo BrdU Labeling

1. BrdU (Sigma B5002).
2. Optional: sucrose (Sigma S-1888) (*see* **Note 1**).
3. Phosphate buffered saline (PBS).
4. H_2O.

2.2. Fluorescence Staining and Detection of Incorporated BrdU

1. PBS.
2. PBS/1% bovine serum albumin (BSA)/0.03% NaN_3, sterile filtered.

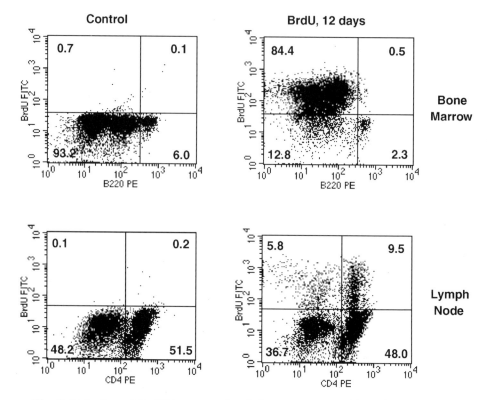

Fig. 2. Detection of BrdU incorporation in bone marrow and lymph node cells after 12 d of BrdU feeding in the drinking water. Note that B220^{negative} and B220^{low} cells in bone marrow undergo rapid turnover, whereas B220^{high} cells are predominantly long-lived cells. Mice were of the C3HeB/FeJ strain and were 8 wk old.

3. 70% Ethanol.
4. 2% Paraformaldehyde/0.02% Tween-20.
5. Deoxyribonuclease I (DNaseI) solution: 150 mM NaCl, 5 mM MgCl$_2$, 10 µM HCl, 300 µg/mL DNaseI (500 Kunitz units/mL).
6. Anti-BrdU fluorescein isothiocyanate (FITC)-coupled monoclonal antibody (Becton-Dickinson, cat. no. 347583).
7. Antibodies required for surface staining (depending on the experiment).
8. 1.5-mL Microtubes (Eppendorf).

3. Methods

3.1. In Vivo BrdU-Labeling

The methods described here have been tested on mice exclusively. Similar approaches for treatment of rats are described in the literature *(18)*.

1. Prepare a stock solution of BrdU at 50 mg/mL in H_2O. If you have problems getting the powder into solution adjust pH to neutral. Store stock solution at 4°C in the dark, or freeze aliquots at –20°C.
2. For feeding, dilute stock solution to 1 mg/mL in H_2O and add 1% sucrose (*see* **Note 1**). Use normal tap water for dilution because this will be used as drinking water. If necessary, sterilize by filtration. Keep storage bottles protected from light at 4°C. Under these conditions diluted BrdU is stable for about 1 mo. For feeding, water bottles have to be protected from light (best method is to paint the bottles dark), and BrdU–water should be replaced at least every 3rd day. We calculate 10 mL of drinking water per mouse per day. When handling the solution and cages, please keep in mind that BrdU is a potential carcinogen (*see* **Note 2**).
3. For short-term labeling (1 d or less), it is advisable to inject BrdU ip. Use 1 mg BrdU per injection (dilute BrdU to 5 mg/mL in PBS); two injections 6 h apart are optimal. For long-term treatment, feeding BrdU in the drinking water is the method of choice because repeated injections are too stressful for the mice.

3.2. Staining Protocol

We use 1.5-mL microfuge tubes for all staining steps. A flowchart of the staining protocol is given in **Fig. 3**.

1. Prepare single-cell suspensions of spleen, lymph nodes, thymus, or bone marrow according to standard procedures (*see* **Note 3**). Because of the various fixation steps required, it is advisable to start with a higher number of cells than for a normal surface staining. Approximately $2–4 \times 10^6$ cells per sample are optimal.
2. Perform surface staining as normal using biotinylated antibodies in the first step and phycoerythrin (PE)-conjugated antibodies as well as streptavidin-conjugates in the second step. The choice of the fluorescence label of streptavidin depends on the fluorescence-activated cell sorter (FACS) analyzer and lasers available (CyChrome, Texas Red, PE-Cy5, or allophycocyanin [APC] can be used). Use $PBS/BSA/NaN_3$ as staining buffer.
3. Wash cells in $PBS/BSA/NaN_3$ and resuspend in 200 µL PBS without BSA.
4. Using a P200 pipetman, inject the stained cells into 1 mL ice-cold 70% ethanol. This injection prevents cell clumping and works better than dropwise addition of ethanol. Fix the cells for 30 min on ice, protected from light.
5. Spin cells down at 2000 rpm in a swing-out rotor (Varifuge) for 5 min (in all further centrifugation steps use 1200*g*). Resuspend cells in 500 µL PBS.
6. Add an equal volume of 500 µL 2% paraformaldehyde/0.02% Tween-20 in PBS giving a final concentration of 1% paraformaldehyde/0.01% Tween-20. Incubate on ice for at least 1 h. Incubation overnight improves the BrdU-staining sensitivity. Protect from light (*see* **Note 4**).
7. Wash once in PBS.
8. Resuspend in 1 mL DNaseI buffer (*see* **Subheading 2.2., item 5**). DNaseI should be added to the buffer directly before use. Incubate for 10 min at room temperature.
9. Wash once with $PBS/BSA/NaN_3$.

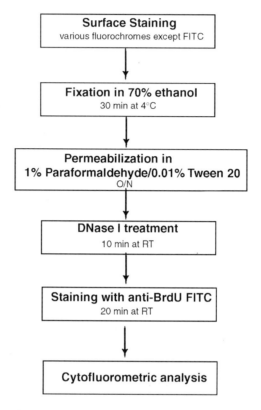

Fig. 3. Flowchart of the BrdU staining protocol.

10. Stain with anti-BrdU FITC for 20 min at room temperature. The concentration of anti-BrdU antibody should be titrated (*see* **Note 5**).
11. Wash once in PBS/BSA/NaN$_3$.
12. Resuspend cells in 300 µL PBS/BSA/NaN$_3$ for analysis (*see* **Notes 6** and **7**).

4. Notes

1. When kept in the dark, BrdU stock solution is stable for 3–6 mo at –20°C or 4 wk at 4°C. BrdU-containing drinking water can be stored for up to 4 wk at 4°C in the dark. Water bottles with BrdU-containing drinking water in the animal cages should be protected from light and replaced every 2–3 d. Sucrose may be added to the drinking water to improve taste and to increase the amount of BrdU taken up by the animals.
2. For most mouse strains the concentration of BrdU in the drinking water is optimal at 1 mg/mL. C57BL/6 mice are particularly sensitive to BrdU labeling and tolerate this dose for only 1–2 wk without signs of toxicity. In this case the dose may be reduced to 0.5 mg/mL. First signs of toxicity are most readily apparent in the

thymus, that is, through reduction of the proportion of rapidly proliferating double-positive thymocytes.

3. As a positive control for efficient uptake of BrdU and for the staining procedure, it is useful to include a bone marrow sample from BrdU-treated mice in each analysis. B precursor cells undergo a rapid turnover in the bone marrow such that more than 80% of bone marrow cells (gated on small lymphocytes) are labeled within 3 d. As a negative control it is recommended to include samples from a mouse that has not been treated with BrdU (*see* **Fig. 2**).

4. The sensitivity of BrdU detection is higher when treatment with paraformaldehyde/ Tween-20 is extended overnight *(14)*. Cells can also be kept for several days at this step when protected from light.

5. For staining with anti-BrdU antibodies after DNaseI treatment of the DNA, it is advisable to use FITC-coupled antibodies only. Apparently the accessibility of the DNA is not good enough for staining with PE-conjugated antibodies because of the large size of the fluorochrome. This problem does not occur after HCl treatment of the DNA.

6. In general, the HCl method *(12)* of DNA denaturation appears to be more sensitive for BrdU detection than the DNaseI method but has the significant disadvantage of restricted use of fluorochromes for surface staining. In our laboratory it has been observed that the B220 surface marker resists treatment with 1N HCl and can be stained together with anti-BrdU using phycoerythrin-labeled antibodies (clone RA3-3B2) after denaturation of the DNA.

7. In principle anti-BrdU staining can be combined with DNA-staining (e.g., by 7-AAD) for cell cycle analysis *(11)* or with intracellular cytokine staining. In the latter case cells should be permeabilized using saponin as a detergent.

References

1. Osmond, D. G. (1993) The turnover of B-cell populations. *Immunol. Today* **14,** 34–37.
2. Freitas, A. A. and Rocha, B. B. (1993) Lymphocyte lifespans: homeostasis, selection and competition. *Immunol. Today* **14,** 25–29.
3. Rajewsky, K. (1993) B-cell lifespans in the mouse—why to debate what? *Immunol. Today* **14,** 40, 41; discussion 41–43.
4. Sprent, J. and Basten, A. (1973) Circulating T and B lymphocytes of the mouse, II: lifespan. *Cell Immunol.* **7,** 40–59.
5. Röpke, C., Hougen, H. P., and Everett, N. B. (1975) Long-lived T and B lymphocytes in the bone marrow and thoracic duct lymph of the mouse. *Cell Immunol.* **15,** 82–93.
6. Röpke, C. and Everett, N. B. (1975) Life span of small lymphocytes in the thymolymphatic tissues of normal and thymus-deprived BALB/C mice. *Anat. Rec.* **183,** 83–94.
7. Opstelten, D. and Osmond, D. G. (1983) Pre-B cells in mouse bone marrow: immunofluorescence stathmokinetic studies of the proliferation of cytoplasmic μ-chain-bearing cells in normal mice. *J. Immunol.* **131,** 2635–2640.

8. Freitas, A. A., Rocha, B., Forni, L., and Coutinho, A. (1982) Population dynamics of B lymphocytes and their precursors: demonstration of high turnover in the central and peripheral lymphoid organs. *J. Immunol.* **128,** 54–60.

9. Heyman, R. A., Borrelli, E., Lesley, J., et al. (1989) Thymidine kinase obliteration: creation of transgenic mice with controlled immune deficiency. *Proc. Natl. Acad. Sci. USA* **86,** 2698–2702.

10. Gratzner, H. G. (1982) Monoclonal antibody to 5-bromo- and 5-iododeoxyuridine: a new reagent for detection of DNA replication. *Science* **218,** 474, 475.

11. Houck, D. W. and Loken, M. R. (1985) Simultaneous analysis of cell surface antigens, bromodeoxyuridine incorporation and DNA content. *Cytometry* **6,** 531–538.

12. Förster, I., Vieira, P., and Rajewsky, K. (1989) Flow cytometric analysis of cell proliferation dynamics in the B cell compartment of the mouse. *Int. Immunol.* **1,** 321–331.

13. Carayon, P. and Bord, A. (1992) Identification of DNA-replicating lymphocyte subsets using a new method to label the bromo-deoxyuridine incorporated into the DNA. *J. Immunol. Methods* **147,** 225–230.

14. Penit, C. and Vasseur, F. (1993) Phenotype analysis of cycling and postcycling thymocytes: evaluation of detection methods for BrdUrd and surface proteins. *Cytometry* **14,** 757–763.

15. Tough, D. F. and Sprent, J. (1994) Turnover of naive- and memory-phenotype T cells. *J. Exp. Med.* **179,** 1127–1135.

16. Kaufman, E. R. and Davidson, R. L. (1978) Bromodeoxyuridine mutagenesis in mammalian cells: mutagenesis is independent of the amount of bromouracil in DNA. *Proc. Natl. Acad. Sci. USA* **75,** 4982–4986.

17. Tapscott, S. J., Lassar, A. B., Davis, R. L., and Weintraub, H. (1989) 5-bromo-2′-deoxyuridine blocks myogenesis by extinguishing expression of MyoD1. *Science* **245,** 532–536.

18. Gray, D. (1988) Population kinetics of rat peripheral B cells. *J. Exp. Med.* **167,** 805–816.

19. Förster, I. and Rajewsky, K. (1990) The bulk of the peripheral B-cell pool in mice is stable and not rapidly renewed from the bone marrow. *Proc. Natl. Acad. Sci. USA* **87,** 4781–4784.

20. Schittek, B., Rajewsky, K., and Förster, I. (1991) Dividing cells in bone marrow and spleen incorporate bromodeoxyuridine with high efficiency. *Eur. J. Immunol.* **21,** 235–238.

21. Schittek, B. and Rajewsky, K. (1990) Maintenance of B-cell memory by long-lived cells generated from proliferating precursors. *Nature* **346,** 749–751.

22. Tough, D. F. and Sprent, J. (1995) Lifespan of lymphocytes. *Immunol. Res.* **14,** 1–12.

23. Penit, C. and Vasseur, F. (1997) Expansion of mature thymocyte subsets before emigration to the periphery. *J. Immunol.* **159,** 4848–4856.

24. Kamath, A. T., Henri, S., Battye, F., Tough, D. F., and Shortman, K. (2002) Developmental kinetics and lifespan of dendritic cells in mouse lymphoid organs. *Blood* **100,** 1734–1741.

5

Differentiation of B Lymphocytes From Hematopoietic Stem Cells

Paulo Vieira and Ana Cumano

Summary

Differentiation of B lymphocytes can be efficiently obtained when multipotent hematopoietic precursors are cocultured with stromal cell lines and soluble growth factors. Stromal cell lines provide yet-undefined signals required for the expansion of the precursor population and/or lineage commitment and soluble mediators. In consequence, the type of the exogenously added interleukins depends on the stromal support used in the assay. In contrast to S17 and OP9 stroma, the fibroblast line NIH3T3 does not support B-cell precursor expansion of CD19+ fetal liver cells; neither does it induce B-lineage differentiation from embryonic multipotent progenitors, in the absence of added cytokines. Under these conditions *c-kit* ligand, interleukin-7 (IL-7), and Flt3 ligand (Flt3-L) are added to the cultures to ensure optimal B-cell differentiation. Another cytokine, stroma-derived lymphopoietin, can also be used instead of IL-7 in embryonic but not adult hematopoietic precursors.

Key Words

Stromal cells; lymphocytes; cytokines.

1. Introduction

Lymphocytes are ultimately generated from hematopoietic stem cells (HSCs), which are able to sustain blood cell production throughout life owing to two main properties: self-renewal and multilineage differentiation. Common lymphoid progenitors (CLPs) *(1)*, which are more advanced than HSC in the hematopoietic hierarchy, are also capable of multilineage differentiation: they can give rise to B, T, and NK cells, although they lack erythromyeloid differentiation potential. However, CLPs lack self-renewal capacity and can only originate a wave of mature cell production, in vivo. Both cell types are found in the adult bone marrow *(1)* where they differentiate in direct contact with the local cytokine-producing stromal compartment.

From: *Methods in Molecular Biology, vol. 271: B Cell Protocols*
Edited by: H. Gu and K. Rajewsky © Humana Press Inc., Totowa, NJ

HSCs are generated once in the lifetime of mammals—during embryonic development—and colonize the hematopoietic organs where they expand and differentiate. Two populations of hematopoietic precursors coexist and are independently generated in the mouse embryo *(2)*. The first to be detected is generated in the yolk sac (YS) at 7.5 d postcoitum (dpc) and the second in the intraembryonic mesoderm around the aorta rudiment, starting at 9 dpc. These populations differ in the two main aspects of HSCs: The former does not differentiate into lymphocytes and does not support long-term blood production in irradiated hosts. Therefore they are neither multipotent nor self-renewable. The latter population of hematopoietic precursors colonizes adult recipients and contributes for more than 6 mo to all hematopoietic lineages of recipient mice *(3)*. The intraembryonic-derived hematopoietic precursors therefore qualify as HSC.

Both embryonic and adult HSCs as well as CLPs can differentiate into multiple hematopoietic lineages in vitro. This development is strictly dependent on the presence in the cultures of soluble growth factors.

In contrast to what is observed in vivo, T lymphocytes are conspicuously absent in these cultures, reflecting requirements for differentiation and/or survival that remained, for long, elusive. T cells can, however, be obtained when HSCs or CLPs are assayed in fetal thymic organ cultures (FTOCs) *(4)*, where precursors are allowed to differentiate in a fetal thymic lobe depleted of hematopoietic components. T-cells obtained under these conditions closely resemble thymocytes, and the system is highly efficient in that single cells are capable of generating large numbers of T lymphocytes. We will not discuss here the FTOC cultures that have been extensively described elsewhere.

For many years, the attempt to maintain unmanipulated HSCs in vitro for long periods of time has been pursued, and different culture systems were devised, most dependent on the presence of bone marrow-derived stromal cells. So far no culture system allowed self-renewal and expansion of either embryonic or adult HSCs. Multipotent precursors with multilineage long-term reconstituting capacity cannot be maintained in vitro and rapidly differentiate. This constraint creates serious restrictions on the genetic manipulation required for autologous transplantation after gene transfer. The understanding of the requirements for self-renewal of stem cell populations remains therefore a major focus of research in hematopoiesis.

2. Materials

2.1. Dissection of Embryo

1. 70% ethanol.
2. Phosphate-buffered saline (PBS) with calcium and magnesium.

2.2. Organ Culture In Toto and Analysis of Hematopoietic Potential

1. 24-Well plate (TPP).
2. OptiMem (Invitrogen).
3. Fetal calf serum (FCS from ICN).
4. Penicillin–streptomycin (Invitrogen).
5. β-Mercaptoethanol (Invitrogen).
6. OP9 stromal cells (T. Nakano, S.-I. Nishikawa, Kyoto, Japan). S17 cells (from K. Dorshkind), NIH-3T3.
7. Trypsin–ethylenediaminetetraacetic acid (EDTA) (Invitrogen).
8. 96-Well plates (TPP).
9. Hank's balanced salt solution (HBSS from Invitrogen).
10. Trypan blue (Invitrogen).
11. Irradiator (X-ray or cesium source).
12. Interleukin-7 (IL-7) supernatant. We use the supernatant of J558 cells transfected with the complementary deoxyribonucleic acid (cDNA) encoding IL-7, a kind gift from F. Melchers (Basel, Switzerland). Cells are grown to confluency and the supernatant is collected. Titration is done in 2E8 cells, a kind gift from P. Kincade (Oklahoma, USA), dependent on IL-7 for growth. The supernatant is serially diluted on the 2E8 cells, and growth is scored 3 d later either by counting the cells or by thymidine incorporation. The highest dilution that gives maximum proliferation is chosen. Murine thymic stroma-derived lymphopoietin (TSLP) can be purchased from R&D.
13. Flt3 ligand (Flt3-L) supernatant: we use the supernatant of the Sp2.0 myeloma transfected with the cDNA encoding Flt3-L and Baf3 cells transfected with the cDNA encoding Flt3, kind gifts from R. Rottapel (Toronto, Canada). Supernatant collection and titration is done as in **step 12**.
14. A *c-kit*-ligand (kL) supernatant: we use the supernatant of CHO cells transfected with the cDNA encoding kL, a kind gift from Genetics Institute (Boston, MA). Titration is done as in **step 12** on freshly prepared bone marrow mast cells cultured for 1 wk in kL.

2.3. Flow Cytometry Analyses

2.3.1. Monoclonal Antibodies From Pharmingen (Becton-Dickinson)

1. CD19–fluorescein isothiocyanate (FITC; clone 1D3).
2. Ter119–phycoerythrin (PE).
3. Mac-1– allophycocyanin (APC) (clone M1/70).
4. CD4-PE.
5. CD8-APC.
6. Ly5.1–FITC.

2.3.2. Other Products

1. 1X PBS.
2. Round-bottom 96-well plate (CML).
3. Propidium iodide (PI) (Sigma).

3. Methods

3.1. Culture System

3.1.1. Stromal Cell Lines

Multipotent hematopoietic progenitors do not efficiently give rise to B lymphocytes in suspension culture without the presence of stromal cells, reflecting the requirement for yet-undefined survival and/or proliferation signals. We have successfully used two different mouse stromal lines obtained from the bone marrow. S17 is capable of supporting B lymphocyte, myeloid, and erythroid cell differentiation from single HSCs. More recently we have been using the OP9 stromal cell line, derived from bone marrow of the *op/op* mouse. This line is therefore deficient in macrophage colony-stimulating factor (M-CSF) production and does not sustain macrophage overgrowth that can impair lymphocyte expansion. In contrast, OP9 produces IL-15, strictly required for NK cell development. Consequently, the development of hematopoietic multipotent cells in OP9 stroma leads to the differentiation of large numbers of B-cell precursors and of NK-cells (*see* **Fig. 1**). OP9 stroma is 30% more efficient than S17 stroma in B-cell development from single HSC cultures (*5,6*).

The interaction between committed B-cell precursor and the surrounding stroma does not require a specialized cell type as shown by our finding that a fibroblast line, NIH-3T3, can replace classic stromal lines in these cultures (*7*). The use of fibroblasts affords a better control of the cytokines in the culture because 3T3 cells produce no factors able to support differentiation of B lineage cells. Indeed committed B-cell precursors from either fetal liver or bone marrow do not proliferate on NIH-3T3 feeder layers in the absence of exogenous growth factors.

All cell lines can be amplified and maintained in OptiMem, 15% fetal calf serum (FCS), 1% antibiotics, and 5×10^{-5} M β-mercaptoethanol, at 37°C, 5% CO_2. For the preparation of the culture plates coated with stromal cells, cells are trypsinized by adding 4 mL of trypsin–EDTA to the culture flask that has been preemptied of culture medium. Trypsin is allowed to act for 15 min at room temperature. Cells are then collected and washed once in HBBS containing 5% of FCS. The pellet is recovered in 1 mL complete medium (OptiMem, 10% FCS, 1% antibiotics). Stromal cells are numbered, after dilution in trypan blue, and plated at a concentration of 5×10^4 living cells per mL of complete medium for

---

Fig. 1. *(see facing page)* Phenotypic analysis of B-cell precursors derived from single multipotent embryonic precursors, in culture with OP9 cells, IL-7, and Flt3-L. Cells were stained with anti-NK1.1 and CD19. Gaited populations are shown in the *right panels.*

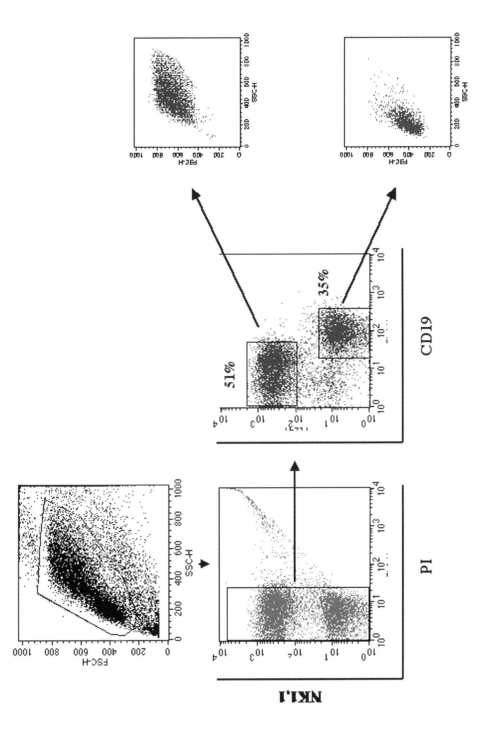

both S17 and OP9, or 10^5 per mL in the case of NIH-3T3. 50 µL of stromal cell suspension are dispensed per well of flat-bottom 96-well plates or 500 µL of 24-well plates. Stromal cells are left overnight at 37°C. Plates seeded with stromal cells are prepared 1 d before the culture starts *(8)*.

3.1.2. Culture Conditions for B-Cell Differentiation

Before plating hematopoietic cells in culture, stromal-cell-coated plates are irradiated to arrest cell proliferation. A 10-Gy dose of irradiation is necessary. Hematopoietic precursors are suspended at the required concentrations in complete medium; 50 µL of this suspension is then deposited in each well onto the stromal cell layer. Each well now contains 100 µL (stromal cells + progenitors). Finally, 100 µL of complete medium supplemented with growth factors is added to a final volume of 200 µL.

Although IL-7 is an important requirement for B-cell development in vitro, Flt3-L and kL were shown to potentiate B lineage cell recovery. Stromal cells lines, like fresh stroma, secrete cytokines important for hematopoietic development. However, individual cell lines are heterogeneous in cytokine production reflecting either diversity in the stromal bone marrow compartment or the consequence of epigenetic modifications during the establishment of the line. S17 stromal cells produce undetectable levels of IL-7 when expanded in vitro, although they are able to support the expansion and differentiation of precursor B cells, in the absence of added factors, albeit at low frequency. In contrast OP9 cells do not produce M-CSF and Flt3-L, but express kL and IL-7, though in limited amounts. As mentioned in **Subheading 3.1.1.**, NIH-3T3 cells alone are unable to support the expansion of committed B-cell precursors *(7)*.

We presently supplement all cultures with Flt3-L, IL-7, and kL to provide an appropriate environment for proliferation and differentiation. Complete medium supplemented with growth factors (each of those twice concentrated) is prepared and 100 µL added to each well. Under those conditions, multipotent hematopoietic precursors can differentiate toward lymphoid, myeloid, and erythroid lineages (if the erythroid lineage differentiation is to be studied, erythropoietin should be added to the cytokine cocktail at a concentration of 8 U/mL). It is noteworthy that OP9 stromal cells are less efficient in sustaining myeloid cell development than S17.

The culture lasts 12–14 d, at 37°C, 5% CO_2. During the culture period, half of the medium is removed on d 5 and replaced by 100–150 µL of fresh medium in each well (carefully remove the medium with a multichannel preferentially in the surface, so as not to take any developing hematopoietic cells) supplemented with Flt3-L and IL-7 only. An excess of kL favors the development of mast cells that can become predominant in the culture. By the end of the culture period, the well content can be analyzed by flow cytometry. Recently, TSLP, has been

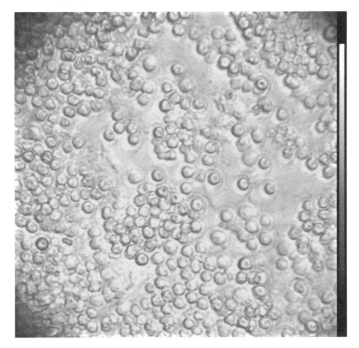

Fig. 2. Phase-contrast view of a clone of B-lineage cells developed from single micromanipulated multipotent precursors from embryonic origin.

characterized as another B-cell growth factor. Although less well studied it apparently targets more mature subsets of B-cell precursors than IL-7 (*see* **Subheading 3.1.3.**).

3.1.3. B-Cell Development In Vitro

B-lineage precursors expand extensively and if obtained from embryonic tissues can give rise to large clones (up to 10^7 cells) of small, round lymphocytes (*see* **Fig. 2**). If the starting population consists of HSCs, the first CD19⁺ cells appear around d 10, whereas CLPs give rise to CD19⁺ cells within 5 d of culture. The detection of CD19⁺ cells is followed by differentiation to IgM⁺ cells that rapidly die thereafter. During this period, cells can respond to lipopolysaccharide (LPS). After an additional 8–10 d, IgM is detectable in the culture supernatant by enzyme-linked immunosorbent assay (ELISA). Mature B cells do not develop easily in vitro, and most keep an immature phenotype IgM⁺IgD⁻ (*see* Chapter 7). Cultures of BM-derived B-cell precursors can be maintained for long periods of time by the expansion of CD19⁺IgM⁻CD43⁺ cells. This phenomenon was well characterized in cultures derived from AGM HSCs *(9)*.

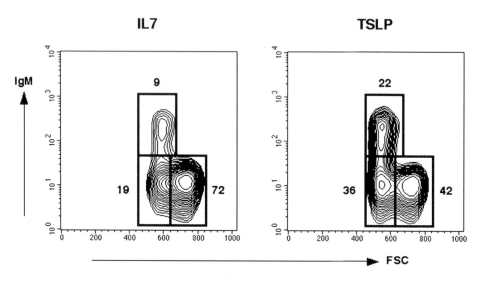

Fig. 3. Phenotypic analysis of BM-derived B-cell precursors, after 5 d in culture with NIH-3T3 in the presence of exogenous IL-7 (*left panel*) or TSLP (*right panel*). The cells were stained for surface IgM and analyzed in a FACS calibur. FSC, forward light scatter.

The earliest subsets of B lineage cells to respond in vitro to IL-7 are BM fraction A/B. IL-7 is, however, not the only characterized growth factor for B-cell precursors. As previously mentioned, TSLP has been shown to promote the expansion of B-cell precursors from mouse BM. The target population is a later subset than fraction A/B *(10)* because TSLP cultures do not last as long as IL-7 cultures, and the phenotypic analysis of the cells shows a predominance of small IgM+ cells compared to cultures in IL-7 (**Fig. 3**). This result, which has been interpreted as evidence that TSLP is a differentiation rather than a growth factor, instead indicates that TSLP promotes expansion only of a more mature BM fraction, leading consequently to the absence of proliferating pro-B cells in the culture *(9)*. When FL precursors are used, however, IL-7 and TSLP have nearly indistinguishable activities. Both factors support proliferation of FL pro-B cells, and in this case, TSLP does not lead to the predominance of small B cells in the culture *(7)*.

We have analyzed cultures starting with single micromanipulated embryonic HSCs that lasted for up to 40 d in the presence of OP9 and exogenous kL, Flt3L, and IL-7. Although CD19 expression was first detected after 10 d of culture, IgM+ cells could only be detected between days 19 and 30 *(9)*. As cultures further developed, IgM+ cells could no longer be detected. At this point a phenotypic analysis reveals an increasing fraction of pro-B cells that express

Table 1

		Cultures of AGM Cells	
	Spleen	Day 10	Day 21
Productive/nonproductive VDJ rearrangements	5	0.5	0.7
Ratio of VDJ/DJ rearrangements	1.25	0.5	0.3

B cells developed in vitro have an increased ration of productive/nonproductive VDJ rearrangements and a decreased ratio of VDJ/DJ rearrangements. (For experimental details *see* **ref. 9**.)

CD19 and CD43, and the cells do not respond to LPS. B-cell precursors that were expanded for more than 4 wk in culture show a ratio of $D–J_H$ vs $V_H–D–J_H$ rearrangements fivefold higher than in IgM^- bone marrow cells, and most $V_H–D–J_H$ rearrangements are out of frame (*see* **Table 1**). Taken together these results indicate the accumulation of cells that cannot progress in differentiation to a pre-B cell and later B cell stages in these cultures. These are apparently blocked in the pro-B-cell stage either by downregulation of the proteins required for immunoglobulin loci rearrangement or by the survival of cells that underwent nonproductive rearrangements. These cells respond well to IL-7, and both would account for the results described above. They can eventually grow indefinitely, and after a crisis period they can be cloned and generate IL-7-dependent cell lines.

3.2. Phenotypic Analysis of Hematopoietic Progeny

3.2.1. Flow Cytometric Analysis

1. After a 2-wk culture, cells from each well are analyzed by flow cytometry for the presence of lineage-specific surface markers: CD19 for lymphoid B cells, Mac-1 for myeloid cells, and Ter119 for erythrocytes.
2. Transfer cells from each culture well into a round-bottom 96-well plate. Spin down the plate 2–3 min at 1300 rpm, 4°C.
3. Excess of medium in the wells is flicked off by inverting the plate vigorously onto a paper towel.
4. Staining mix is prepared in 1X PBS, 3% FCS with 10 *mM* sodium azide (FACS medium). We use CD19–FITC and NK1.1–PE, at the proper dilutions, previously estimated. Resuspend each cell pellet in 25 µL of staining mix. Incubate 15–20 min at 4°C, in the dark.
5. By the end of the incubation period wash the cells in FACS medium, by adding 200 µL per well, then spin 2–3 min at 1300 rpm, 4°C. During the spin, FACS

medium with PI is prepared (0.5 µg/mL PI). PI is a DNA intercalator, entering passively in the nucleus of dead cells. Interestingly, PI gives equal fluorescence in the flow cytometer on filters detecting PE and Cy5–PE conjugates. Therefore it allows the elimination of dead cells that appear in the diagonal of the plot of both parameters (**Fig. 1**). After washing stained cells, resuspend the pellets in FACS medium–PI and analyze the cells. Use the SSC parameter in linear mode so that you can easily distinguish between mononuclear cells (low SSC) and granulocytes (high SSC).

Acknowledgments

This work was supported by the Ligue Contre le Cancer to the Lymphocyte Development Unit as an associated laboratory.

References

1. Kondo, M., Weissman, I. L., and Akashi, K. (1997). Identification of clonogenic common lymphoid progenitors in the mouse bone marrow. *Cell* **91,** 661–672.
2. Cumano, A., Dieterlen-Lievre, F., and Godin, I. (1996) Lymphoid potential probed before circulation in mouse, is restricted to caudal intraembryonic splanchnopleura. *Cell* **86,** 907–916.
3. Cumano, A., Ferraz, J. C., Klaine, M., DiSanto, J. P., and Godin, I. (2001) Intraembryonic but not yolk sac hematopoietic precursors, isolated before circulation, provide long-term multilineage reconstitution. *Immunity* **15,** 477–485.
4. Jenkinson, E. J., Anderson, G., and Owen, J. J. (1992) Studies on T cell maturation on defined thymic stromal cell populations. *J. Exp. Med.* **176,** 845–853.
5. Nakano, T., Kodama, H., and Honjo, T. (1994) Generation of lymphohematopoietic cells from embryonic stem cells in culture. *Science* **265,** 1098–1101.
6. Cho, S. K., Webber, T. D., Carlyle, J. R., Nakano, T., Lewis, S. M., and Zuniga-Pflucker, J. C. (1999) Functional characterization of B lymphocytes generated in vitro from embryonic stem cells. *Proc. Natl. Acad. Sci. USA* **96,** 9797–9802.
7. Voßhenrich, C. A. J., Cumano, A., Müller, W., Di Santo, J. P., and Vieira, P. (2003) Thymic stroma-derived lymphopoietin distinguishes fetal from adult B cell development. *Nat. Immunol.* **4,** 773–779.
8. Godin, I., Dieterlen-Lievre, F., and Cumano, A. (1995) Emergence of multipotent hemopoietic cells in the yolk sac and paraaortic splanchnopleura in mouse embryos, beginning at 8.5 d postcoitus. *Proc. Natl. Acad. Sci. USA* **92,** 773–777.
9. Nourrit, F., Doyen, N., Kourilsky, P., Rougeon, F., and Cumano, A. (1998) Extensive junctional diversity of Ig heavy chain rearrangements generated in the progeny of single fetal multipotent hematopoietic cells in the absence of selection. *J. Immunol.* **160,** 4254–4261.
10. Hardy, R. R., Carmack, C. E., Shinton, S. A., Kemp, J. D., and Hayakawa, K. (1991). Resolution and characterization of pro-B and pre-pro-B-cell stages in normal mouse bone marrow. *J. Exp. Med.* **173,** 1213–1225.

6

Analysis of Lymphocyte Development and Function Using the RAG-Deficient Blastocyst Complementation System

Faith M. Young, Carl A. Pinkert, and Andrea Bottaro

Summary

The RAG-deficient blastocyst complementation system (RBCS) represents a flexible and rapid method for the genetic analysis of lymphocyte function using a gene-targeting approach. In chimeras derived from manipulated embryonic stem cells injected into VDJ recombination-incapable, RAG-deficient blastocysts, any lymphoid cells past the prolymphocytic stage will be embryonic stem cell-derived. This approach can therefore bypass pitfalls such as pleiotropy and embryonic lethality to allow the analysis of targeted gene mutations with respect to lymphocyte development and function in a genetically uniform cell population. Thanks to recent advances in targeting techniques and in mouse embryo manipulation, this remarkably efficient technique has become a highly feasible and useful addition to any immunology research program. In this review, we discuss the technical aspects of the procedure, as well as its advantages and drawbacks compared to alternative approaches, and our practical experience in establishing the system at the University of Rochester.

Key Words

Lymphocyte, gene targeting, knockout, RAG, embryonic stem cells, chimera, genetic analysis, mutation, B cell, T cell.

1. Introduction

Ten years after its original publication, the RAG-deficient blastocyst complementation system (RBCS) has established itself as a unique and efficient approach for the genetic analysis of lymphoid cell function, with applications reported so far in over 100 research papers. Several advantages set the RBCS apart from alternative approaches, including cost containment, speed, efficiency, and flexibility. Nevertheless, with the exception of a few laboratories, the RBCS

From: *Methods in Molecular Biology, vol. 271: B Cell Protocols*
Edited by: H. Gu and K. Rajewsky © Humana Press Inc., Totowa, NJ

has yet to become entirely commonplace for most institutions, in part because, to our knowledge, no comprehensive technical discussion of the system and its methods has yet appeared in the literature.

In this review, we aim to supply the necessary information for any established laboratory or transgenic facility with experience in gene-targeting technology to successfully adopt the RBCS as a routine approach for their experimental goals. First, we provide a general discussion of the RBCS, its rationale, advantages, and limitations. Rather than describing in exhaustive detail all techniques required for the RBCS, we focus specifically on the key steps of the process, highlighting unique requirements of the system compared to alternative approaches. More important, all the protocols described are those used within the University of Rochester Transgenic Facility and Vivarium, and though hopefully useful as a set of general guidelines, they will likely require specific adaptations in other settings. Finally, we describe expected yields and known pitfalls of the system, from our direct experience.

The RBCS is based on the ability to generate genetically homogeneous embryonic stem (ES)-cell-derived, physiologically developed mature lymphocytes, within the context of otherwise chimeric mice with tissues of mixed genetic origin. The feasibility of reconstituting the mature lymphoid compartment of an immunodeficient animal in somatic ES cell chimeras was originally shown by Roes and Rajewsky in 1991 *(1)*. Using blastocysts from severe combined immunodeficient (SCID) mice and ES cells bearing a heterozygous mutation in the IgH Cδ gene, Roes and Rajewsky showed successful generation of ES-cell-derived mature lymphocytes, capable of responding to immunization with T-dependent antigens. However, the interpretation of the results from complemented SCID mice could theoretically be affected by the known leakiness of the SCID mutation *(2)*. Following the finding that RAG2-deficient mice, unlike SCID, display an absolute block in lymphocyte development at the very early, pro-B/T-cell stage *(3)*, Jianzhu Chen and collaborators in Fred Alt's laboratory demonstrated the significant potential of this mouse strain as a system for complementation studies *(4)*. More important, they also showed in their original report that mutant ES cells could be extensively manipulated to generate complex genotypes (in their case, homozygous JH mutant ES cells, complemented with a rearranged immunoglobulin heavy chain transgene), without losing the ability to efficiently reconstitute the peripheral lymphoid compartment in chimeric mice *(4)*. These observations set the stage for the exploitation of the technique in a number of different settings (reviewed in **ref. 5**).

1.1. Advantages and Applications

There are several notable advantages of the RBCS over its conventional counterparts. The first, and arguably the most significant one is the shortening of the experimental timeline. Because the RBCS bypasses the requirement for germline transmission of any genetic alteration introduced into ES cells, as well as the necessity in most cases for breeding to homozygosity, the study can rapidly proceed to the phenotypic analysis.

A second advantage, also related to the lack of germline transmission requirement, is that high levels of ES cell totipotency are not essential for experimental success; indeed, as discussed later, even low levels of chimerism can result in efficient reconstitution of the lymphoid compartment. This represents a major benefit in cases in which poor targeting efficiency generates only a limited number of candidate ES cell lines for analysis *(6–11)*.

At the time of its development, the RBCS approach also provided the only available alternative strategy for a common pitfall of gene-targeting technology; that is, embryonic lethality owing to early developmental defects. Indeed, several of the first articles to apply the technique were specifically addressed to evaluate the effect on lymphocyte function of embryonic lethal mutations, such as those of the *Rb*, *c-jun*, or *GATA-2* genes, to only mention a few *(12–14)*. Analogously, the RBCS can find application in cases in which the gene mutation is not necessarily lethal but results in extensive pleiotropic effects in other cell lineages, which may obscure or complicate the analysis of a lymphoid-specific phenotype; this is the case for instance of the calcineurin A α-deficient mice *(15)*. Currently, the availability of conditional and tissue-specific gene-targeting techniques *(16)* can efficiently and elegantly bypass most of these hurdles, although the RBCS remains an option in cases where the possibile use of of recombinase-dependent somatic gene deletion systems was not considered in the original experimental design.

The reduction in cost and effort related to the long-term maintenance of targeted mouse strains also has to be considered as a significant benefit of the RBCS, particularly in experiments that require introduction of multiple different transgenes in a gene-targeted background, or extensive backcrossing. For similar reasons, the RBCS has been used extensively to analyze the function of immunoglobulin and T-cell receptor gene regulatory elements *(17–23)*. These studies can be very effectively focused on a few predefined, specific functional properties of the targeted region (e.g., transcription levels, recombination efficiency, etc.). Thus, turnaround time and the possibility of analyzing multiple mutations in parallel are often key, whereas the generation of germline mutant mouse strains is in most cases of secondary importance. For these reasons, the RBCS arguably constitutes the ideal system for mechanistic studies of this kind.

Overall, the main practical advantages of the RBCS are its remarkable flexibility, speed, and efficiency. By allowing the time- and cost-effective phenotypic analysis of genetically modified lymphocytes, the RBCS usefully complements conventional targeting techniques; in addition, it probably should constitute the system of choice in cases in which a rapid turnaround is critical, when there is a requirement to perform pilot analysis of certain phenotypic effect, or when the testing of multiple independent mutations is needed.

2. Materials
1. Normal and mutant ES cell lines and relative culture media and supplements.
2. RAG-deficient mouse colony.
3. Pregnant mare's serum gonadotropin (PMSG) (Calbiochem, San Diego, CA, or A. F. Parlow, Harbor-UCLA Medical Center, Torrance, CA).
4. Human chorionic gonadotropin (hCG) (ICN Biomedical, Cleveland, OH).
5. 1-mL syringes.
6. 26-gage $\frac{1}{2}$-in needles.
7. 7.5% Povidone-iodine (PVD-I) solution.
8. HEPES-buffered M16 or BMOC3 medium.
9. Two sets of sterile surgical instruments.
10. 35- and 60-mm Tissue-culture dishes.
11. Microscope slides.
12. Trypsin 0.25% w/v, 1 mM ethylenediaminetetraacetic acid (EDTA) solution.
13. 15-mL conical tubes.
14. Microinjection equipment.
15. Pseudopregnant (24-h postcoitum) female mice (ICR or other outbred Swiss strains).
16. Heparinized microhematocrit capillary tubes (VWR 15401-628).
17. Heparin solution (100 USP U/mL).
18. Eppendorf tubes.
19. Avertin 2.5% solution.
20. Fetal bovine serum (FBS).
21. Red blood cell lysis buffer: 1.55 M NH$_4$Cl, 10 mM Tris-HCl, pH 7.5.
22. Phosphate-buffered saline (PBS), 5% FBS (fluorescence-activated cell sorting [FACS] buffer).
23. 50% rat serum in PBS.
24. Fluorochrome-labeled antibodies to mouse B- and T-lymphocyte markers (mouse CD19-, B220-, IgM-, and CD3ε-, TCRβ-, CD4-, or CD8-specific).
25. Flow cytometer and analysis software.

3. Methods
3.1. Mutant ES Cell Generation and Culture

Culture of wild-type and genetically modified ES cells should be performed according to standard protocols (*11,24,25*). If necessary, homozygous mutants can be generated by successive targeting or by selection of heterozygotes under increased G418 concentration (*26*). In the latter case, the mutant, low-efficiency form of the neomycin-resistance (*neo^R*) gene (*27*) must be used in the original targeting construct (the wild-type form of the *neo^R* gene provides high resistance levels that cannot be further selected). As discussed, whereas care must be taken in preserving ES cell totipotency during culture, the RBCS is known to work effectively even with relatively poor-quality ES cell lines. In our protocols, all ES cell lines are tested for the presence of known murine pathogens including *Mycoplasma* species contamination prior to the time of injection; although chimera complementation can be obtained with contaminated cell lines, overall efficiencies are generally lower and transmission of the pathogen to the animal colony is a significant risk.

3.2. Maintenance of the RAG2-Deficient Mouse Colony

The RBCS has been applied successfully using either mixed-background (e.g., C57BL/6x129xCBA) RAG2-deficient mice, or C57BL/6-backcrossed RAG1-deficient mice (available from the Jackson Laboratory) as blastocyst donors. Here we provide details for the maintenance of a small colony of donors on a mixed background, sufficient to generate cohorts of females for superovulation and embryo harvest on a biweekly basis.

As for every immunodeficient mouse strain, strict specific-pathogen-free practices must be observed; limited access to the colony, as well as specific procedures to avoid transfer of pathogens from normal colonies to the RAG-deficient colony are highly recommended.

The production colony is composed of about 12–15 actively breeding pairs, usually aged 2–6 mo. Newly established pairs tend to produce a first litter within 3–4 wk, and most pairs produce new litters regularly every 3 wk. Litters average 6 to 8 pups and are weaned at 3 wk. Mixed-background RAG2-deficient females are good mothers and can handle large, close litters effectively; culling of large litters is not necessary. Pairs are replaced on rotation at age 6–8 mo, or earlier if litter yield declines. In addition, 30–40 studs (also aged 2–8 mo) are maintained in separate cages to mate with superovulated females.

3.3. Expected Yields

A colony of the size just described generates cohorts of about 20–30 females for embryo harvest every 2 wk plus the necessary mice to replenish the production colony and the stud group on rotation.

3.4. Superovulation

Young females are superovulated at age 3–4 wk according to the following regimen, which was empirically determined to provide the highest embryo yields in the RAG2-deficient strain, based on earlier studies illustrating strain-specific differences in responses to gonadotropins and various embryo transfer procedures in mice (*see* **refs. 28–30**).

Day 1: 5 U PMSG in a volume of 200 µL (25 U/mL), delivered by intraperitoneal injection using a 26-gage ½-in needle.

Day 3: 46 to 48 h after PMSG treatment, 5 U hCG (also 200 µL of a 25-U/mL solution, ip). Females are placed with studs immediately after injection.

Day 4: Check for copulatory plugs.

Day 7: Plugged females are humanely euthanized exactly 92 h post-hCG for embryo harvest.

Although not preferred because of diminished yields, nonplugged females can be used for a second round of superovulation, following >12 d with no sign of pregnancy. In this case, the dose of both gonadotropins (PMSG and hCG) is raised to 7.5 U (300 µL).

3.5. Embryo Harvest and Injection With ES Cells

3.5.1. Embryo Harvest

1. Approximately 4–6 h before scheduled injections, humanely euthanize female mice (we use rapid anesthesia with 2.5% avertin ip and cervical dislocation).
2. Liberally swab abdomen with a 7.5% PVD-I solution. A midventral abdominal incision is then made, exposing the abdominal wall.
3. Using a separate set of surgical instruments, the abdominal wall is opened. The cervix is then visualized and grasped with blunt forceps. The cervix is severed posterior to the forceps, allowing removal of the uterus and oviduct or the entire tract (by severing the fat pad between the corresponding ovary and kidney). The dissections from donor females continue as rapidly as possible to remove all donor tracts, but care is taken not to excessively manipulate the tissues, thus causing possible loss of or damage to ova.
4. Place the intact uterus and oviducts in a 35- or 60-mm tissue-culture (or Petri) dish containing HEPES-buffered medium. Removal of much of the extraneous tissue (blood vessels, fat, etc.) facilitates uterine flushing and subsequent embryo collection.
5. Transfer uteri to a second dish with fresh medium, then flush with HEPES-buffered medium to collect embryos.
6. After collection, transfer embryos to a third dish and wash in either modified BMOC3 or M16 medium; maintain embryos in microdrop cultures at 37°C in an atmosphere of 5% CO_2/5% O_2/90% N_2 until injected with ES cell stocks.

3.5.2. ES Cell Preparation

1. Cultures of rapidly growing ES cells (50% of split density—usually from a 16–24-h-old 1:4–1:5 split culture) are trypsinized (5 min at 37°C, 0.25% trypsin, 1 mM EDTA) about 1–2 h before scheduled injection time. Cells are resuspended by vigorous pipetting; complete single-cell suspension is verified by visual examination before harvesting.

2. Harvest cells in 4 mL ES cell medium, and plate onto fresh 6-cm plates. Feeder fibroblasts are adhered by incubation for 30–45 min in tissue-culture incubator (37°C, 5% CO_2).
3. Nonadherent cells (almost exclusively ES cells, by visual examination) are collected and transferred into 15-mL conical tubes, counted with a hemocytometer, and stored on ice until injection.

3.5.3. Blastocyst Microinjection

1. A 1-mL medium containing 1×10^6 ES cells is prepared. Before injection of ES cells, blastocysts are placed in either M16 or BMOC3 medium supplemented with HEPES, then transferred to a microdrop of medium overlaid with silicone oil on a microscope slide for injection. We perform injections under phase-contrast microscopy using a holding pipet (diameter of the opening is approx 20–40 μM) and an injection pipet with a sharp beveled edge.
2. A blastocyst is selected with the holding pipet, rotated to where the inner cell mass (ICM, area of greatest cell density in the embryo) is identified and held in place against the holding pipet; approx 12 small, round ES cells are retrieved with the injection pipet and injected into the blastocele cavity in apposition to the ICM. Injections of ES cells may be augmented using a pressure-controlled mechanical or hydraulic delivery system (e.g., PLI-100 Pico-Injector, Medical Systems Corp.; Eppendorf 5221, Brinkmann; Narishige IM-200).
3. After all microinjections are completed, the injected blastocysts are transferred to the ampullae of pseudopregnant (day 1 postmating) mice, by either oviductal or uterine transfer, 10–15 blastocysts/transfer *(11,28)*. Based on our observations suggesting a greater fragility of the RAG2 –/– embryos (R. L. Howell, C. L. Donegan, AB, FY, and CAP unpublished data), use of one or two "carrier" embryos (generally, earlier stage or "spare" embryos, including zygote to morula stage, as well as uninjected/collapsed blastocysts) was found to facilitate pregnancy maintenance *(28)*.

3.5.4. Expected Yields

In our hands, cycles performed according to the protocols described in **Subheadings 3.5.1–3.5.3.** yield on average about 50 high-quality, injectable embryos/session. However, because as many as 100 or more embryos can occasionally be recovered, generation of an excess of pseudopregnant foster mothers is recommended.

3.6. Chimera Screening and Analysis

Pups from the injections described in **Subheadings 3.5.1–3.5.3.** are screened at age 3 wk for successful reconstitution of the lymphoid compartment. Screening can be performed by either enzyme-linked immunosorbent assay (ELISA) for serum IgM, or flow cytometry on peripheral blood mononuclear cells. The latter system, described in **Subheading 3.6.1.**, is preferable, as it provides more quantitative information on the level of chimerism and, if desired, can be simultaneously used for more specific phenotypic characterizations. In some cases, depending on the strain of origin of the ES cell injected and the coat color of the RAG-deficient mice (in the case of our mixed background RAG2-deficient colony, almost all mice are agouti), it is possible to identify highly chimeric mice by coat color examination. However, it is highly recommended that immunological screening also be performed on apparently low- or noncontributing chimeras, as reconstitution of the lymphoid compartment is highly efficient and can occur in these mice as well (*see* **Subheading 3.6.1.**).

3.6.1. Peripheral Blood Collection and Preparation

1. Mice are anesthetized with injection of 2.5% avertin solution, ip Harvest blood from the retro-orbital sinus by gently inserting one end of a heparinized capillary glass tube behind the eye, keeping the other end inside a 1.5-mL Eppendorf tube containing 50–100 µL of heparin solution.
2. Collect about 100–200 µL of blood (max.), gently extract capillary, cap the tube, and quickly mix by inversion. Keep at room temperature until all samples are collected.
3. Spin samples in a microfuge at 500g, 5 min at 4°C. Remove heparinized plasma; if desired, this may be saved for further qualitative analysis (strictly quantitative serological assays are not possible, owing to uneven harvest and dilution of the original blood samples).
4. Resuspend cell pellet in 1 mL of red blood cell lysis buffer, and incubate 5 min at room temperature.
5. Overlay lysate on 0.5 mL of FBS, and spin again in refrigerated microfuge at 500g for 5 min. Expect a small white to slightly pink pellet; if excessive red blood cells persist (very red pellet), the lysis step may be repeated, though yields will decrease.
6. Aspirate supernatant, resuspend pellet in either FACS buffer (50 µL), or block with 20 µL 50% rat serum 20 min on ice, followed by addition of 30 µL of FACS buffer (we found that blocking significantly reduces staining background).

3.6.2. Staining and Analysis

1. Take one half (25 µL) from each sample from **Subheading 3.6.1.**, **step 6**; transfer to round-bottom 12- × 75-mm tubes. Add 25 µL of stain mix (FACS buffer containing 0.25 µg/25µL of each labeled antibody); we routinely use antibodies to mouse CD19, CD45R/B220, or IgM, and to CD3ε, but any combination of antibodies to mature lymphocyte markers can also be employed. Additional antibodies for preliminary phenotypic analysis can also be added to the staining mix, depending on analysis requirements and flow cytometer channel availability. Note that the second half of the sample can also be used in parallel for additional staining, or saved in FACS buffer at 4°C up to overnight for further analysis.

2. Mix cell suspension and stain solution; incubate 20 min on ice.
3. Wash by adding 3 mL of FACS buffer, spin in tissue-culture centrifuge (200g, 5 min at 4°C); aspirate supernatant.
4. Resuspend pellet in 300–500 µL of FACS buffer or, if analysis is not performed immediately, in FACS buffer supplemented with 1% formaldehyde.
5. Analyze by flow cytometry according to standard protocols.

3.6.3. Expected Yields

Although successful complementation ultimately depends on the quality of the ES cell stock, in the majority of cases significant fractions of circulating mature B and T cells can be detected in 30–50% of analyzed chimeras. Of these, up to 50% usually have significant reconstitution of the peripheral lymphoid compartment, often fully comparable to normal. An example of screening results from RAG-complemented chimeras from a manipulated ES cell line is shown in **Fig. 1**.

4. Notes

Fully complemented chimeric mice generated by the RBCS display peripheral immunophenotypes that are essentially indistinguishable from those of normal mice: total spleen and lymph node cells, fractions of B and T cells, and of lymphoid subpopulations are also usually normal, and splenic architecture is also conserved. Lymphocytes in these mice display normal functional phenotypes, as established both in vitro in a number of activation assays, and in vivo during antigen-specific immune responses. In the latter case, however, the possibility of specific differences in responses owing to the genetic heterogeneity of immune cells (especially in the mixed-background RAG2-deficient complemented chimeras) cannot be overlooked.

More important, good complementation is often observed in animals with little ES cell contribution to other somatic tissues, such as skin, liver, kidney, muscle (as established by coat color, or Southern blot analysis of tissue genomic deoxyribonucleic acid [DNA]). In our experience, as little as a 10% contribution to somatic tissues is often sufficient for extensive, often full complementation of the peripheral lymphoid compartment.

Another important observation relates to some specific alterations often observed in only poorly complemented mice, of which investigators must be aware to avoid erroneous interpretation of results. Most notably, it appears that complementation is significantly more efficient for the T-cell compartment than for B cells. Thus, animals can be found with detectable thymic precursor and peripheral mature T-cell repopulation, but significant reduction or even complete absence of peripheral B cells (*see* **Fig. 2** for an example). This artifact is particularly significant for chimeras derived from ES cell lines with degraded

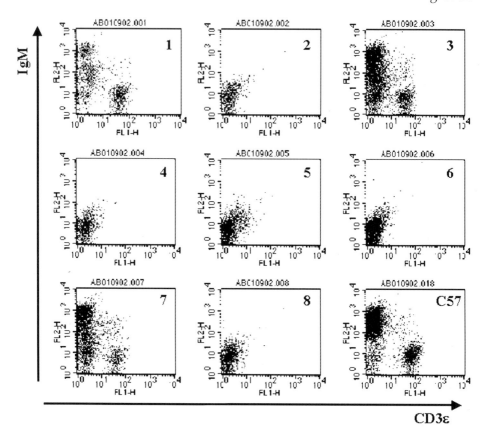

Fig. 1. Screening of chimeras by flow cytometry on peripheral blood cells. Peripheral blood leukocytes from retro-orbital bleeds were prepared as described in the text from 8 presumptive chimeric mice (nos. 1–8) and a positive C57BL/6 control (C57), stained with antibodies to IgM and CD3ε, and analyzed by flow cytometry. Plots display staining profiles of cells in the small lymphocyte gate. Three of the 8 pups displayed various levels of reconstitution of the peripheral B- and T-cell compartments, with nos. 3 and 7 having abundant B and T cells, and no. 1 having lower numbers. All other animals show a lack of circulating mature lymphocytes, as expected from noncomplemented RAG-deficient mice.

totipotency, as all these mice may display this aberrant phenotype, which can then be mistaken for a *bona fide* effect of the introduced mutation.

Similarly, differential distribution of B-cell subpopulations in the bone marrow in poor chimeras is also observed, reflecting the ability of B cells with successfully rearranged antigen receptors to rapidly exit the bone marrow. It is important to note that in most cases significant contribution to the pro-B cell (B220+CD43+IgM−) compartment will be of RAG−/− blastocyst origin. In cases

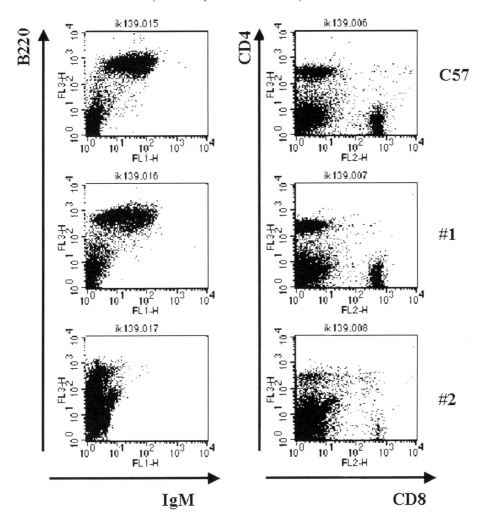

Fig. 2. Differential effects of partial chimerism on peripheral B- and T-cell complementation. Splenic cells from a C57BL/6 mouse and two chimeras with different levels of complementation were stained for B-cell markers (B220 and IgM, *left panels*) and T-cell markers (CD4 and CD8, *right panels*). In chimera #1, normal levels of B and T cells are observed, whereas chimera #2, although displaying detectable numbers of CD4$^+$ and CD8$^+$ T cells, completely lacks B220^{hi}, IgM$^+$ mature B cells. Instead, the splenic B-lymphoid compartment in this chimera is constituted exclusively by RAG-deficient, blastocyst-derived, B220^{lo}, IgM$^-$ cells.

in which differentiation of ES- from blastocyst-derived precursor B lymphocytes is necessary, strategies based on serological polymorphism can be applied; for instance, our RAG2-deficient colony consists entirely of Ly5.2

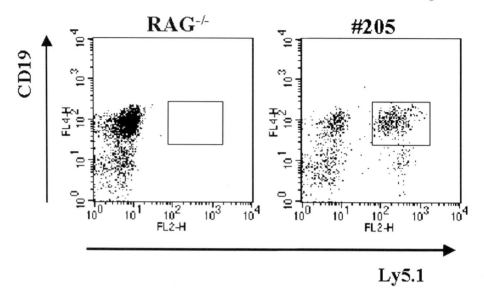

Fig. 3. Identification of ES-cell derived pro-B cells in complemented RAG-deficient chimeras. FACS analysis was performed on bone marrow samples from a RAG2-deficient mouse (*left panel*), homozygous for the Ly5.2 allele, and from a chimera derived from injection of a RAG-deficient ES cell line bearing the Ly5.1 allele (a gift from Fred Alt, HHMI, the Children's Hospital and the Center for Blood Research, Boston, MA). ES cell-derived, Ly5.1+ cells are detectable in both the B lymphoid (CD19+) and nonlymphoid (CD19−) compartments in the chimera, allowing for analysis of early B-cell precursors using the RBCS.

homozygotes, allowing differentiation of ES-cell-derived leukocytes on the basis of Ly5.1 expression (**Fig. 3**).

Finally, slight but reproducible phenotypic differences between complemented chimeras and the equivalent germline mutated mouse strains, although rare, have occasionally been observed (*1, 31, 32*, and T. Bushnell, A. Eyssallenne, AB, and FMY, unpublished); it is unclear whether this is attributable specifically to the chimeric system or, more likely, to the differential manifestation of the effects of mutations and transgenes in the context of heterogenous genetic backgrounds.

Acknowledgments

We are grateful to Igor Kuzin, Tim Bushnell, Vicki Houseknecht, and Rob Howell for comments and sharing of data and information. This work was supported in part by J. P. Wilmot Foundation and Leukemia-Lymphoma Society of America Foundation grants to FMY, NIH grants 1 RO1 AI45012 to AB,

R21RR16286 and N01DE12634 to CAP, and by funds from the University of Rochester Medical Center.

References

1. Roes, J. and Rajewsky, K. (1991) Cell autonomous expression of IgD is not essential for the maturation of conventional B cells. *Int. Immunol.* **3,** 1367–1371.
2. Bosma, M. J. and Carroll, A. M. (1991) The SCID mouse mutant: definition, characterization, and potential uses. *Annu. Rev. Immunol.* **9,** 323–350.
3. Shinkai, Y., Rathbun, G., Lam, K.-P., et al. (1992) Rag-2 deficient mice lack mature lymphocytes owing to inability to initiate V(D)J rearrangement. *Cell* **68,** 855–867.
4. Chen, J., Lansford, R., Stewart, V., Young, F., and Alt, F. W. (1993) Rag-2-deficient blastocyst complementation: an assay of gene function in lymphocyte development. *Proc. Natl. Acad. Sci. USA* **90,** 4528–4532.
5. Chen, J. (1996) Analysis of gene function in lymphocytes by Rag-2-deficient blastocyst complementation. *Adv. Immunol.* **62,** 31–59.
6. Thomas, K. R. and Capecchi, M. R. (1987) Site-directed mutagenesis by gene targeting in mouse embryo-derived stem cells. *Cell* **51,** 503–512.
7. Mansour, S. L., Thomas, K. R., and Capecchi, M. R. (1988) Disruption of the proto-oncogene int-2 in mouse embryo-derived stem cells: a general strategy for targeting mutations to non-selectable genes. *Nature* **336,** 348–352.
8. Capecchi, M. R. (1989) The new mouse genetics: altering the genome by gene targeting. *Trends Genet.* **5,** 70–76.
9. Deng, C. and Capecchi, M. R. (1992) Reexamination of gene targeting frequency as a function of the extent of homology between the targeting vector and the target locus. *Mol. Cell. Biol.* **12,** 3365–3371.
10. Doetschman, T. (1994) Gene transfer in embryonic stem cells. In Pinkert, C. A., ed., *Transgenic Animal Technology: A Laboratory Handbook,* Academic, New York, pp. 115–146.
11. Nagy, A., Gertsenstein, M., Vintersten, K., and Behringer, R. (2003) *Manipulating the Mouse Embryo: A Laboratory Manual,* 3rd ed., Cold Spring Harbor Laboratory Press, Cold Spring Harbor, New York.
12. Chen, J., Gorman, J. R., Stewart, V., Williams, B., Jacks, T., and Alt, F. W. (1993) Generation of normal lymphocytes populations by Rb-deficient embryonic stem cells. *Curr. Biol.* **3,** 405–413.
13. Chen, J., Stewart, V., Spyrou, G., Hilberg, F., Wagner, E. F., and Alt, F. W. (1994) Generation of normal T and B lymphocytes by c-jun deficient embryonic stem cells. *Immunity* **1,** 65–72.
14. Tsai, F. Y., Keller, G., Kuo, F. C., et al. (1994) An early haematopoietic defect in mice lacking the transcription factor GATA-2. *Nature* **371,** 221–226.
15. Zhang, B. W., Zimmer, G., Chen, J., et al. (1996) T cell responses in calcineurin A α-deficient mice. *J. Exp. Med.* **183,** 413–420.
16. Le, Y. and Sauer, B. (2000) Conditional gene knockout using cre recombinase. *Methods Mol. Biol.* **136,** 477–485.

17. Chen, J., Young, F., Bottaro, A., Stewart, V., Smith, R. K., and Alt, F. W. (1993) Mutations of the intronic IgH enhancer and its flanking sequences differentially affect accessibility of the JH locus. *EMBO J.* **12**, 4635–4645.

18. Zhang, J., Bottaro, A., Li, S., Stewart, V., and Alt, F. W. (1993) A selective defect in IgG2b switching as a result of targeted mutation of the Iλ2b promoter and exon. *EMBO J.* **12**, 3529–3537.

19. Bottaro, A., Lansford, R., Xu, L., Zhang, J., Rothman, P., and Alt, F. W. (1994) S region transcription per se promotes basal IgE class switch recombination but additional factors regulate the efficiency of the process. *EMBO J.* **13**, 665–674.

20. Cogne, M., Lansford, R., Bottaro, A., et al. (1994) A class switch control region at the 3′ end of the IgH locus. *Cell* **77**, 737–747.

21. Sleckman, B. P., Bardon, C. G., Ferrini, R., Davidson, L., and Alt, F. W. (1997) Function of the TCR alpha enhancer in alphabeta and gammadelta T cells. *Immunity* **7**, 505–515.

22. Manis, J. P., van der Stoep, N., Tian, M., et al. (1998) Class switching in B cells lacking 3′ immunoglobulin heavy chain enhancers. *J. Exp. Med.* **188**, 1421–1431.

23. Sakai, E., Bottaro, A., Davidson, L., Sleckman, B. P., and Alt, F. W. (1999) Recombination and transcription of the endogenous IgH locus is effected by the IgH intronic enhancer core region in the absence of the matrix attachment regions. *Proc. Natl. Acad. Sci. USA* **96**, 1526–1531.

24. Zambrowicz, B. P., Hasty, E. P., and Sands, A. T. (1998) *Mouse Mutagenesis,* Lexicon Genetics, The Woodlands, TX.

25. Doetschman, T. (2002) Gene transfer in embryonic stem cells, I: history and methodology. In: Pinkert, C. A., ed., *Transgenic Animal Technology: A Laboratory Handbook,* 2nd ed., Academic, New York, pp. 113–141.

26. Mortensen, R. M., Conner, D. A., Chao, S., Geisterfer-Lowrance, A. A. T., and Seidman, J. G. (1992) Production of homozygous mutant ES cells with a single targeting construct. *Mol. Cell. Biol.* **12**, 2391–2395.

27. Yenofsky, R. L., Fine, M., and Pellow, J. W. (1990) A mutant neomycin phosphotransferase II gene reduces the resistance of transformants to antibiotic selection pressure. *Proc. Natl. Acad. Sci. USA* **87**, 3435–3439.

28. Johnson, L. W., Moffatt, R. J., Bartol, F. F., and Pinkert, C. A. (1996) Optimization of embryo transfer protocols for mice. *Theriogenology* **46**, 1267–1276.

29. Vergara, G. J., Irwin, M. H., Moffatt, R. J., and Pinkert, C. A. (1997) In vitro fertilization in mice: strain differences in response to superovulation protocols and effect of cumulus cell removal. *Theriogenology* **47**, 1245–1252.

30. Polites, H. G. and Pinkert, C. A. (2002) DNA microinjection and transgenic animal production. In Pinkert, C. A., ed., *Transgenic Animal Technology: A Laboratory Handbook,* 2nd ed. Academic, New York, pp. 15–70.

31. Muthusamy, N., Barton, K., and Leiden, J. M. (1995) Defective activation and survival of T cells lacking the Ets-1 transcription factor. *Nature* **377**, 639–642.

32. Barton, K., Muthusamy, N., Fischer, C., et al. (1998) The Ets-1 transcription factor is required for the development of natural killer cells in mice. *Immunity* **9**, 555–563.

7

Conditional Gene Mutagenesis in B-Lineage Cells

Stefano Casola

Summary

Since its first application in mice almost 10 yr ago, the Cre/loxP has become the system of choice to study gene function in vivo in a cell-type, stage-specific, and inducible manner. This chapter provides a set of updated protocols that will help the reader to construct a vector for conditional gene targeting, to in vitro manipulate embryonic stem (ES) cells and to rapidly identify successfully targeted ES colonies. It also provides an updated list of Cre strains currently used to assess gene function at defined stages of B-cell development and guidelines to generate single-copy, knock-in transgenes regulated in a Cre-dependent manner.

Key Words

Embryonic stem cells; Cre; loxP; Flp; frt; B cells; conditional targeting; knock-in; transgene; targeting vector; pEASY-FLOX; pEASY-FLIRT.

1. Introduction

Gene targeting in embryonic stem (ES) cells is dependent on homologous recombination, a process by which a fragment of deoxyribonucleic acid (DNA) is replaced with another carrying a homologous sequence. Mutations introduced in ES cells by gene-targeting range from single-point mutations to deletions of large genomic regions (50–100 kb). Insertion of specific DNA sequences within a target site (without deleting any endogenous sequence) can also be obtained by gene targeting. However, conventional gene targeting presents numerous limitations to the study of gene function in adult mice, such as: the germline mutation leads to embryonic lethality; the expression of the mutant gene in multiple cell types leads to a complex phenotype; the impossibility of generating mouse models of human diseases, such as cancer, in which somatic mutations affecting gene function are present in the affected cells only and, finally, the function of the mutant gene may be relevant in the same cell type at mul-

From: *Methods in Molecular Biology, vol. 271: B Cell Protocols*
Edited by: H. Gu and K. Rajewsky © Humana Press Inc., Totowa, NJ

Cre/loxP system

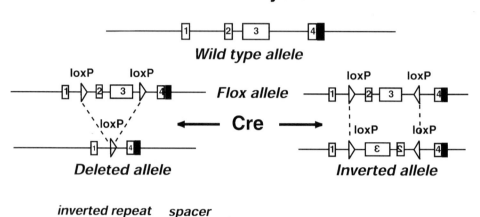

Fig. 1. Depending on the orientation of the loxP sites (shown as *triangles*) a floxed gene segment can be either deleted or inverted after Cre recombination. The loxP site consists of two 13-bp inverted repeats separated by an asymmetric 8-bp spacer region that defines the orientation of the loxP site.

tiple stages of differentiation. Conditional gene targeting overcomes all these limitations and makes it the system of choice to study the function of a gene in adult mice in a cell-type, stage-specific, and temporal manner. One of the most popular approaches for conditional targeting of endogenous genes is based on the Cre–loxP system *(1)*. The latter relies on the activity of the site-specific Cre recombinase promoting recombination between two loxP sites *(2)*. Depending on the orientation of the loxP sites it is possible to achieve deletion (loxP sites placed in the same orientation) or inversion (loxP sites placed in opposite orientation) of a loxP-flanked ("floxed") DNA segment (*see* **Fig. 1**). In a conditional gene-targeting experiment using the Cre–loxP system, the end result is the insertion into the mouse genome of two loxP sites flanking a gene or gene segment of interest. Correctly targeted ES cells are injected into 3.5-d-old blastocysts (or aggregated to 2.5-d-old morulas) to generate chimeric mice. Conditional deletion/inversion of the floxed gene in B cells (as well as in any other cell type) requires successful transmission through the germline of the conditional allele followed by the breeding of floxed mice to transgenic animals expressing the Cre recombinase in a tissue, stage-specific, or inducible manner.

Table 1
List of Cre Transgenic Mice Used for Conditional Gene-Targeting Experiments in Specific Stages of B-Cell Development

Cre Strain	Type of Transgene	Expression	Chromosome Localization	Background	Ref.
CD19-Cre	Knock-in into the CD19 locus	Constitutive— all B cells starting from B-cell progenitors	Chr 7	C57BL/6	*(16)*
CD21-Cre	BAC transgene	Constitutive— fraction of immature B cells and all mature B cells	ND	Backcross no. 4 to C57BL/6 and to Balb/c	Alimzhanov, M. and Rajewsky, K., manuscript in preparation.
Cγ1-Cre	Knock-in into the Cγ1 (IgG1) constant region locus	Constitutive— class-switched germinal center B cells, IgG1$^+$ memory B cells and plasma cells	Chr 12	C57BL/6 and Balb/c	Casola, S. and Rajewsky, K., manuscript in preparation.
Mx-Cre	Transgene	Inducible by IFNα efficient deletion tested in B-cell progenitors and mature B cells.	ND	C57BL/6	*(17)*
Rosa26-CreER^{T2}	Knock-in into the Rosa26 locus	Inducible by tamoxifen treatment	Chr 6	C57BL/6	*(18)*

BAC, bacterial artificial chromosome; ND, not determined.

A list of Cre transgenic mice currently used to study gene function in B cells is shown in **Table 1**.

1.1. Planning a Conditional Gene-Targeting Experiment

A typical vector for conditional gene targeting is schematically shown in **Fig. 2**. It comprises genomic sequences required to promote homologous recombination (arms of homology), the gene segment flanked by loxP sites that will be deleted in vivo, positive (neomycin resistance [*Neo^R*] gene) and negative (thymidine kinase or diphtheria toxin gene) selection markers. In our experience, homology arms of 2–7 kb in length and floxed fragments ranging from 0.1 to 3 kb have given good homologous recombination frequencies. Larger sizes of the floxed gene segment may reduce the efficiency of in vivo

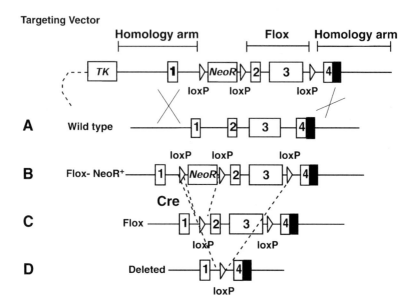

Fig. 2. Scheme of a conditional gene-targeting experiment. Depicted are the targeting vector, the wild-type allele (**A**), the floxed allele carrying the selection marker (**B**), the floxed allele after Cre-mediated deletion of the selection marker (**C**), and the deleted allele as a result of in vivo (or in vitro) Cre-mediated recombination (**D**).

Cre-mediated recombination especially in cases in which expression of the recombinase is temporally restricted. Screening of homologous recombinants is performed either by Southern blotting analysis or by polymerase chain reaction (PCR). For Southern blotting screening of targeted ES cells a few rules should be followed:

1. Probe(s) chosen to screen for a homologous recombination event should anneal in a genomic region external to the homology arms used to construct the targeting vector (TV). External probes should be obtained to detect both the 5′ and 3′ end of the targeted locus.
2. The size of the band corresponding to the targeted allele should be smaller than that of the wild-type allele. This will avoid problems in identifying correctly targeted ES colonies in cases in which incomplete digestion of genomic DNA occurs. This is achieved by inserting unique restriction sites in the TV or taking advantage of specific restriction sites present in the *Neo*^{R} gene.
3. Restriction enzymes used for Southern analysis should be tested in advance. Because a targeting experiment may require the analysis of several hundred ES colonies, it is a good practice to plan the screening of ES cells using an inexpen-

Table 2
List of ES Cell Lines Commonly Used in Our Laboratory
for Gene-Targeting Experiments

ES Cell Line	Origin	Genetic Markers	Electroporation Condition	Ref.
Bruce-4	C57BL/6	a, C, H-2^b; IgH^b	230 V	*(14)*
IB10	129/Ola	A^w, c^{ch}, H-2x^b; IgH^a	240 V	*(20)*
Balb/c-I	Balb/c	A, c, H-2^d, IgH^a	230 V	*(21)*
HM-1	129/Ola	A^w, c^{ch}, H2^b, IgH^a, hprt$^-$	240 V	*(7)*

sive and efficient restriction enzyme. *Eco*RI, *Bam*HI, *Hind*III, *Eco*RV, *Pst*I, *Xba*I, and *Stu*I are examples of low-cost enzymes efficiently digesting genomic DNA.

1.2. Constructing a Targeting Vector for Conditional Gene Targeting

1.2.1. Choosing the ES Cell Line for the Targeting Experiment

It is important to decide which ES cell line to use for the targeting experiment before planning the construction of the TV. The frequency of homologous recombination increases, indeed, if isogenic DNA is used for the preparation of the TV. The choice of which ES cell line to use is also dependent on the previous success of germline transmission. ES cell lines at earlier passages have usually a higher chance of germline transmission. A list of ES cell lines currently used in our laboratory (all proved to have a good rate of germline transmission) is shown in **Table 2**. For our conditional gene-targeting experiments aimed at addressing gene function in B cells, either C57BL/6 or Balb/c-derived ES cells are preferred. Inbred mice generated from targeted ES cells will require breeding to tissue and/or stage-specific Cre strains. Thus, it is essential to maintain the Cre mouse strain(s) that will be used for the final experiments on the same genetic background as that of the floxed mice. Most Cre strains used to study gene function in B cells are on a C57BL/6 and/or Balb/c background (*see* **Table 1**).

1.2.2. Source of Genomic DNA to Construct the Targeting Vector

Bacterial artificial chromosomes (BACs) carrying large fragments of the mouse genome (200–250 kb) represent the most convenient source of genomic DNA to construct a TV for conditional gene targeting (a depository of BAC libraries can be found at http://bacpac.chori.org/home.html). Two alternative approaches can be followed if the available mouse BAC genomic libraries do not match with the genetic background of the ES cell line that will be used for

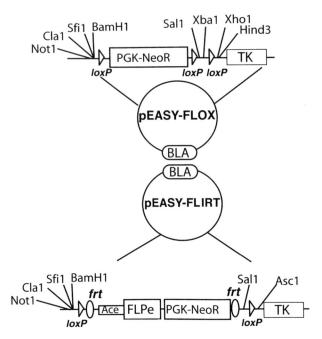

Fig. 3. Features of pEASY-FLOX and pEASY-FLIRT vectors. Unique restriction sites that are used for cloning purposes and for the linearization of the TV vector are indicated. Both vectors are linearized on the opposite site of the TK gene. pEASY-FLIRT contains an additional testis-specific Flp expression cassette (Ace-Flpe) aimed at deleting the frt-flanked *Neo*R gene in male gonads of chimeric mice. BLA, β-lactamase gene.

targeting: (a) use nonisogenic DNA to clone the arms of homology of the TV. In our experience, this has led, in many instances, to successful homologous recombination events; (b) amplify the homology arms by PCR using as DNA substrate, genomic DNA from the ES cell line that will be used for the targeting experiment. This procedure has the additional advantage of limiting the number of cloning steps required to build the TV because unique restriction sites can be inserted in the primers used for the PCR amplification of the homology arms and the floxed gene segment.

1.2.3. pEASY-FLOX and pEASY-FLIRT: Ready-to-Use Vectors for Conditional Gene Targeting

We have developed two universal plasmids that allow the rapid construction of a vector for conditional gene targeting (*see* **Fig. 3**). The main difference between the vectors consists in flanking the *Neo^R* gene with either *loxP* sites or

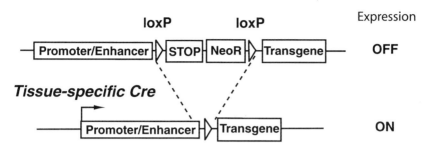

Fig. 4. Cre-dependent activation of gene expression. The STOP cassette depicted was originally described by Lakso et al. *(3)*.

frt sites (recognized by the site-specific Flp recombinase). Three cloning steps are sufficient to insert in each of the two vectors the two arms of homology and the floxed gene segment, respectively.

1.3. Conditional Activation of Gene Expression Using Single-Copy, Knock-In Transgenes

Another way to study gene function in vivo is to activate expression of target genes in a tissue-specific and temporal manner. To achieve this goal several approaches can be adopted. Cloning transgenes under the control of tissue-specific (and stage-specific) promoters offers one possibility to activate gene expression in a target tissue. Alternatively, it is possible to use the Cre–loxP system (or any analogous site-specific recombination system) as described for the first time by Lakso et al. *(3)*. Specifically, as shown in **Fig. 4**, a transcription and translation termination sequence (STOP cassette) flanked by loxP sites is cloned between the promoter of choice (that can be active in a broad range of tissues) and the target gene (or complementary DNA [cDNA]). This configuration leads to an inactive transgene. Opon tissue-specific (and stage-specific) Cre-mediated deletion of the STOP cassette, expression of the target gene is induced (*see* **Fig. 4**). Several versions of the loxP-flanked STOP cassette have been described *(3,4)*. Conditional transgenes can be inserted in the mouse genome randomly and usually in multiple copies (injecting DNA transgenic constructs directly into fertilized eggs) or in a site-directed fashion by homologous recombination, targeting DNA constructs in vitro into ES cells. Recently, single-copy, tissue-specific transgenes have also been obtained by injecting fertilized eggs with BACs modified in recombination-proficient bacterial strains through homologous recombination. *Rosa26* and hypoxanthine phosphoribosyl transferase *(Hprt)* genomic loci are frequently used to target single-copy conditional transgenes in a site-specific manner *(4,5)*. The *Rosa26* locus offers several advantages over other loci for the insertion of single-copy transgenes, including high frequency of homologous recombination (as high as 1 homologous recombinant for every 10 G418R colonies); broad tissue expression (including pro-B, mature, germinal center, and memory B cells [Casola, S. and Rajewsky, K. unpublished results; Alimzhanov, M. and Rajewsky, K. unpublished results]); and ability to test expression of the transgene directly in ES cells after transient transfection of targeted ES clones with a Cre expression vector. The *Hprt* gene represents an alternative to *Rosa26*. Taking advantage of the existence of an Hprt-deficient ES cell line (HM-1), it is possible, using a replacement vector, to restore the integrity of the *Hprt* locus by homologous recombination in these cells, inserting a conditional transgene at the same time upstream of it *(5,7)*. Selection of homologous recombinants is done growing ES cells in hypoxhantine-aminopterin-thymidine (HAT) selection medium that allows the exclusive growth of Hprt-proficient ES cells. A critical condition for successful expression of a transgene inserted into the Hprt locus is represented by the usage of a well-characterized tissue- and stage-specific promoter/ enhancer combination controlling expression of the transgene *(8)*.

2. Materials

2.1. Plasmids

1. pMMNeoFlox8: contains a loxP-flanked Neo^R cassette *(9)*.
2. pGEM-frt2NeoR: contains an frt-flanked Neo^R cassette *(10)*.
3. pMP8: targeting vector to insert single-copy transgenes into the Hprt locus *(5)*.
4. pRosa26-1: targeting vector to insert single-copy transgenes into the Rosa26 locus *(4)*.
5. pBS-loxP$_2$-STOP: loxP-flanked STOP cassette *(10)*.
6. pgk-Cre: expression vector for Cre recombinase (K. Fellenberg, Institute of Genetics, University of Cologne).
7. pCAGGS-FLPe: expression vector for FLP recombinase *(11)*.
8. pEASY-FLOX and pEASY-FLIRT: universal TV for conditional gene targeting (*see* **Subheading 1.2.3.**).

2.2. In Vitro Culture of Mouse Embryonic Fibroblasts

Mouse embryonic fibroblasts (MEF) are isolated from d-14 DR-4 embryos expressing four drug-selectable marker genes *(12)*.

1. MEFs are grown in EF medium lacking antibiotics: Dulbecco's modified Eagle's medium (DMEM) (high glucose without sodium pyruvate; Gibco-BRL, cat. no. 11960-044) plus 10% fetal calf serum (FCS) (Gemini Bioproducts, cat. no. 900-108), 1X sodium pyruvate (Gibco-BRL, 100X, cat. no. 11360-070), 200 mM 1X L-glutamine (Gibco-BRL, 100 X, cat. no. 15400-054). Store complete medium at 4°C.

2. Gelatin (cell culture grade from porcine, Sigma 2%, cat. no. G1393). MEFs are plated on 0.1% gelatin (diluted in 1X phosphate-buffered saline [PBS] without Mg^+/Ca^+)-coated tissue-culture plates. Add enough 0.1% gelatin solution to cover the tissue-culture dish. Incubate at room temperature for 10–15 min. Aspirate gelatin and plate MEFs.

3. PBS 1X without Mg^+/Ca^+ (Cellgro, cat. no. 21-040-CV). Store at room temperature.

4. Trypsin–ethylenediaminetetraacetic acid (EDTA) (Gibco-BRL 2.5%, 10X, cat no. 15090-046). Store stock solution at $-20°C$. For trypsinization of cells dilute to 1X with PBS. Diluted 1X trypsin–EDTA solution can be kept for up to 1 wk at $4°C$.

5. Mitomycin C (MMC) (Sigma, cat no. M-0503): Dissolve powder (2 mg/vial) in 2.0 mL of 1X PBS and consider it 100X concentrated. Aliquot in freezing tubes and store at $-70°C$. MMC is light sensitive.

6. Tissue-culture dishes: 9-cm dishes (Falcon, cat. no. 353003), flat-bottom 96-well plates (Corning, cat. no. 3598), and U-bottom 96-well plates (Corning, cat. no. 3799).

7. 50-mL (Falcon, cat. no. 352098) and 15-mL (Falcon, cat. no. 352097) conical tubes.

8. Cryogenic vials (Nalgene, cat. no. 5012-0020).

9. CO_2 incubator ($37°C$, 5% CO_2 and maximal humidity).

10. Freezing medium: 10% dymethylsulfoxide (DMSO; Sigma, cat. no. D2650) in FCS.

2.3. In Vitro Culture and Targeting of ES Cells

1. ES cells are grown in antibiotic-free ES medium: DMEM (high glucose without sodium pyruvate; Gibco-BRL, cat. no. 11960-044) plus 15% FCS (ES-tested Hyclone, cat. no. SH30071), 1X sodium pyruvate (Gibco-BRL, 100X, cat. no. 11360-070; store at $4°C$), 200 mM 1X L-glutamine (Gibco-BRL, 100X, cat. no. 15400-054; store at $-20°C$), 10 mM 1X nonessential amino acids (Gibco-BRL, 100X, cat. no. 11140-050; store at $4°C$), 100 μM 2-β-mercaptoethanol (Gibco-BRL, 50 mM 2000X, cat. no. 21985-023; store at $4°C$), 1000 U/mL leukemia inhibitory factor (LIF; 1:1000 dilution of LIF present in 720 LIF-D cell-conditioned media. CHO-transfected 720 LIF-D cells contain an expression vector for the secreted form of LIF encoded by the LIF-D splicing variant mRNA (Genetics Institute Cambridge, MA). Store aliquots of LIF-containing supernatant at $-70°C$).

2. 1X trypsin–EDTA solution (diluted in 1X PBS) is completed with the addition of 1% chicken serum (Gibco-BRL, cat. no. 16110082) when working with C57BL/6 or Balb/c-derived ES cell lines.

3. ES transfection is done using a Bio-Rad electroporator at 230–240 V, 480 μF.

4. Electroporation cuvets: 4-mm Gene pulser cuvets (Bio-Rad, cat. no. 1652088).

5. ES transfection buffer: RPMI-1640 without phenol red (Gibco-BRL, cat. no. 11835-030).

6. G418-sulfate (neomycin or geneticin, Gibco-BRL 1 g, cat. no. 11811-023). Resuspend in sterile water at a concentration of 100 mg/mL. Check specific activity

when calculating the amount of G418 to give to ES cells. Store stock solution at −70°C. G418-containing ES medium is kept at 4°C.

7. Gancyclovir (GANC; Syntex Cymeven): Measure 4.3 mg GANC sodium salt and add 80 μL of sterile ddH$_2$O (2×10^{-1} M). Dilute 10 μL of this 2×10^{-1} M concentrated solution in 1 mL ES complete medium (2×10^{-3} M). Filter-sterilize the solution (with 0.2 μm filters) and dilute 10 μL of this 2×10^{-3} M concentrated solution in 10 mL ES medium containing G418 (final 2 μM). Prepare solution fresh every day.

8. Trypsinization, expansion, and daily feeding of picked ES colonies in 96-well plates are performed using a 12-well multipipettor (Finnpipette 4531X20 and 4531X30) or 8-well multistepper pipet (Finnpipette 4531X50). Medium is aspirated from 96-well plates using a multiwell plate washer manifold (Sigma M-2656).

9. Freezing medium: 10% dimethyl sulfoxide (DMSO) in ES-tested FCS.

10. ES cell lysis buffer: 10 mM NaCl, 10 mM Tris-HCl, pH 7.5, 10 mM EDTA, 0.5% sarcosyl and 0.4 mg/mL freshly added proteinase K (Roche, cat. no. 1092766; 1 g is resuspended in water at a final concentration of 10 mg/mL. Stock solution is stored at −20°C).

11. Restriction enzyme mixture: 1 mM dithiothreitol (DTT; Invitrogen 0.1 M, cat. no. 18064-014), 1 mM spermidine (Sigma, cat. no. S0266), 100 μg/mL bovine serum albumin (BSA), 50 μg/mL RNase A and 20 U of restriction enzyme per sample.

12. 0.15 N HCl.

13. Transfer solution for Southern blotting: 0.4 N NaOH/0.6 M NaCl.

14. Hybridization buffer: 1 M NaCl, 50 mM Tris-HCl, pH 7.5, 10% dextran sulfate (Sigma, cat. no. 31403), 1% sodium dodecyl sulfate (SDS) and 250 μg/mL sonicated fish sperm (Roche, 10 mg/mL, cat. no. 1467140).

15. 2 M Sodium acetate, pH 6.5.

16. Absolute and 70% ethanol.

17. CO$_2$ incubator (37°C, 5% CO$_2$, and maximal humidity).

3. Methods

3.1. Preparing MEFs for the Transfection of ES Cells

1. *D 0:* Rapidly thaw 1 vial (3×10^6 cells) of MEFs and transfer cells into 15-mL conical tubes containing 10 mL of EF medium.

2. Spin at 195g for 10 min at 4°C.

3. Resuspend pellet in 5 mL and plate cells in six 9-cm gelatin-coated tissue-culture dishes (Passage 1: MEF-1). Incubate in a humidified incubator at 37°C. Feed cells daily with 10 mL EF medium.

4. *D 3:* Four confluent MEF-1 plates are trypsinized; **steps 5–14** (*see* **Note 1**).

5. Wash MEF-1 twice with 10 mL PBS.

6. Add 3 mL of 1X trypsin–EDTA solution and incubate for 3–5 min at 37°C.

7. Pipet cells gently until a single-cell suspension is obtained (do not incubate MEF in trypsin–EDTA solution for longer than 7–8 min).

8. Neutralize trypsin by adding 5 mL of EF medium.
9. Count the number of viable cells by the trypan blue exclusion method and make aliquots of 3×10^6 cells/15-mL conical tube.
10. Spin cells at 195g for 10 min at room temperature.
11. Resuspend pellet in 1 mL of freezing medium (kept at 4°C).
12. Rapidly transfer cells into labeled cryogenic vials.
13. Place freezing vials in appropriate Styropore boxes, and transfer cells overnight to a –70°C freezer.
14. Next day allocate freezing vials into a liquid N_2 tank where they can be kept for an indefinite time.
15. *D 3:* One MEF-1 9-cm tissue-culture plate is treated with MMC; **steps 16–22** (*see* **Note 2**). These cells will represent the feeder layer for ES cells that will be thawed for the transfection experiment.
16. Aspirate medium and wash cells twice with PBS.
17. Add to cells 4 mL of EF medium containing 40 µL of a 100X MMC solution.
18. Incubate cells with MMC for 3–6 h in the incubator at 37°C.
19. Wash cells three times with PBS.
20. Trypsinize plate following the method in **steps 4–14**.
21. Resuspend pellet in 10 mL EF medium and plate on one 9-cm tissue-culture dish.
22. *D 3:* One MEF-1 plate is trypsinized and cells are expanded to three 9-cm tissue-culture dishes (MEF-2).

3.2. Preparing ES Cells for Transfection

1. *D 4:* Thaw rapidly a vial of frozen ES cells ($2–3 \times 10^6$ cells), dilute cells in 10 mL ES cell medium, and spin at 195g for 10 min at room temperature.
2. Resuspend ES cell pellet in 10 mL ES medium and plate cells onto MMC-treated MEF-1 (as described in **Subheading 3.1., steps 15–21**).
3. *D 4:* Thaw 2 vials of MEF-1 (described in **Subheading 3.1., steps 1–3**) and plate cells on 10 gelatin-coated 9-cm tissue-culture dishes (MEF-2).
4. *D 6:* Treat three MEF-2 plates with MMC (described in **Subheading 3.1., step 22**), trypsinize cells, and plate in three 9-cm tissue-culture dishes.
5. *D 7:* Trypsinize ES cells as described in **Subheading 3.1., steps 4–14** (*see* **Note 3**), count viable cells (*see* **Note 4**), and plate 1×10 ES cells on each of three MMC-treated MEF-2 plates as described in **step 4**, this subheading (*see* **Note 5**).
6. *D 8:* Treat MEF-2 (10 plates) for 3 h with MMC (described in **step 4**, this subheading); trypsinize and count cells.
7. Plate 3×10^6 MEF-2 cells in each 9-cm gelatin-coated tissue-culture dish (8 plates). Six plates will contain MMC-treated MEF-2 that will be used to plate ES cells from two electroporation experiments. The rest of MMC-treated EF cells are plated onto two 9-cm tissue-culture dishes and used for the following transfection control plates:
   a. Plate 1×10^6 transfected ES cells treated with G418 but without GANC. This control gives an estimate of the enrichment in the selection process given by the usage of the negative selection marker thymidine kinase (tk).

b. Plate 0.5×10^6 untransfected cells. This control plate will indicate the stringency of the G418 selection. Never start picking ES cells before this control plate is free of viable ES colonies.

3.3. Transfection of ES Cells

The following protocol is for one transfection experiment.

1. *D8:* Linearize targeting vector:
   a. Digest in a microfuge tube 30 µg of high-quality DNA for 3–5 h at the appropriate temperature (*see* **Note 6**).
   b. Phenol (saturated in Tris-HCl, pH 8.0)/chloroform (50% v/v) extraction of the linearized targeting vector.
   c. Precipitate vector in 1/10 vol sodium acetate (2 *M*, pH 6.5) and 2.5 vol of absolute ethanol.
   d. Precipitate DNA at –20°C for at least 20 min.
   e. Spin at 15,700*g* in a microcentrifuge for 20 min at 4°C.
   f. Discard supernatant and wash DNA pellet twice with cold (–20°C) 75% ethanol.
   g. Discard supernatant carefully under the cell culture hood.
   h. Air-dry pellet for 30–40 min.
   i. Resuspend pellet in 100 µL of transfection buffer.
2. *D9:* Feed ES plates with fresh ES medium 2 h before transfection.
3. *D9:* Trypsinize ES cells (*see* **Note 7**).
   a. Count cells and aliquot 1×10^7 ES cells in a 50-mL conical tube.
   b. Spin ES cells at 195*g*, room temperature, for 10 min.
   c. Wash ES cell pellet once in 45 mL of PBS.
   d. Spin ES cells at 195*g*, room temperature, for 10 min.
4. Resuspend ES cell pellet in 700 µL of transfection buffer.
5. Add DNA solution (100 µL) to the ES cell suspension. Mix well, avoiding bubbles.
6. Transfer cells to an electroporation cuvet.
7. *D9:* Electroporate ES cells at 230–240 V (depending on the ES cell line chosen; *see* **Table 2** and **Note 8**).
8. Incubate cells for 5–10 min at room temperature.
9. Transfer electroporated ES cells to a 50-mL conical tube containing 30 mL of ES medium.
10. Plate 10 mL of the transfected ES cell suspension into each of three 9-cm tissue-culture dishes containing MMC-treated MEF-2 cells. Transfer cells to a 37°C incubator.
11. Feed ES cells daily throughout the entire transfection experiment with 10 mL of fresh ES medium.
12. Two days after transfection (d 11), start feeding transfected ES cells with G418-containing ES medium (*see* **Note 9**).

13. If the thymidine kinase gene is included in the targeting vector, start 5 d after transfection (d 14) to select against random integrants adding GANC to the G418-containing ES medium (*see* **Note 10**).
14. Six days after transfection (d 15) thaw 2 vials of MEF-2 and plate cells onto 10 gelatin-coated 9-cm tissue-culture dishes (MEF-3) (*see* **Note 11**).
15. Feed MEF-3 daily with EF medium.
16. By d 7–8 posttransfection (d 16–17) distinct G418, GANC double-resistant colonies, should be visible only in the plates containing transfected ES cells.
17. The day before picking resistant ES colonies (d 17), treat MEF-3 (10 plates described in **step 14**) with MMC, trypsinize, and count cells (*see* **Note 12**).
18. Transfer MMC-treated MEF-3 cells from 10 confluent 9-cm plates to 20 flat-bottom 96-well plates (*see* **Notes 13** and **14**).

3.4. Picking Double-Resistant ES Colonies

Start picking at d 9 posttransfection (d 18) and continue until d 11 (d 20). The following protocol is to pick 96 colonies in one 96-well plate.

1. Feed transfected ES plates 2 h before picking.
2. Prepare one 96-well plate (U-bottom) containing in each well 20 μL of a 2X trypsin–EDTA solution (plus 1% chicken serum if picking C57BL/6 or Balb/c-derived-ES colonies).
3. Wash one 9-cm tissue-culture plate containing double-resistant ES colonies twice with 10 mL PBS.
4. After the second wash aspirate PBS and add 10 mL of PBS to the plate.
5. Pick colonies under a tissue-culture hood, with the help of a stereomicroscope, using a 20-μL micropipettor, setting the pipet to 20 μL and using sterile disposable tips.
6. Transfer each picked colony into 1 well of the U-bottom 96-well plate containing 20 μL of 2X trypsin–EDTA.
7. Pick from one ES plate for 30–40 min. Check the status of picked colonies under the microscope. If cells are not yet dissociated, place the plate for 5 min in the incubator at 37°C.
8. Add 110 μL of ES medium to each well. Dissociate cells using a multi-channel pipettor and sterile tips (change tips for each set of wells), pipetting vigorously.
9. With a multichannel pipettor, transfer cells (150 μL) of each well to the respective well of a 96-well plate (flat bottom) containing MMC–MEF-3 cells.
10. If additional colonies need to be picked from the same 9-cm tissue-culture plate, wash twice with PBS and feed plate with ES medium (+G418). Picking from the same plate can be done later in the same day or during the following days.
11. To increase the chance to identify homologous recombinants, ES colony picking is performed over a 2–d time period (*see* **Note 15**).
12. Feed picked colonies for the following 2–3 d using a multistepper pipettor, adding 150 μL of ES medium (with G418, without GANC) to each well.

13. When cells reach confluence, wash 96-well plates twice with PBS and split 1:3.
    a. Trypsinize colonies adding 50 µL of 1X trypsin–EDTA to each well.
    b. Place 96-well plate in the incubator for 5 min at 37°C.
    c. Dissociate cells by pipetting them vigorously.
    d. Neutralize trypsin–EDTA adding100 µL of ES medium to each well.
    e. Divide the 150 µL-cell suspension in three parts and transfer cells to the corresponding well (50 µL each) of three 96-well plates containing MEF-3 (each well having already 150 µL of G418-containing ES medium).
14. Feed 96-well plates daily with ES medium (containing G418).
15. Three or 4 d later freeze down one of the three 96-well replica plates (*see* **Subheading 3.5., steps 1–6**).
16. The second replica plate is frozen down the day after.
17. The third replica plate is trypsinized as described in **step 13**. Cells are plated to three 96-well plates coated with gelatin and fed daily with ES medium. From these plates, as soon as ES clones are grown to confluence, DNA is extracted and screened for the occurrence of homologous recombination.

3.5. Freezing ES Cells in 96-Well Plates

1. Grow ES cells to subconfluence.
2. Aspirate medium, wash each well twice with 100 µL PBS, using a multistepper pipet, and trypsinize cells (as described in **Subheading 3.4., step 13**).
3. Neutralize trypsin by adding 50 µL of 2X freezing medium to each well.
4. Dissociate cells vigorously by pipetting them several times with a multichannel pipettor.
5. Overlay 100 µL of sterile mineral oil in each well.
6. Seal 96-well plates with Parafilm, and store at –70°C until the screening is completed.

3.6. Rapid Screening of Double-Resistant ES Colonies

The following is a modification of the protocol described in **ref. *13***.
1. Grow ES colonies on gelatin-coated 96-well plates until they reach confluence.
2. Wash each well twice with 100 µL PBS.
3. After the last wash, aspirate PBS and add to each well 50 µL of ES lysis buffer.
4. Wrap 96-well plate with Parafilm, and lyse cells at 56°C overnight (*see* **Note 16**).
5. Next day, cool the humidified box, allowing it to stand for 1 h at room temperature.
6. Precipitate DNA, adding to each well 100 µL of absolute ethanol and 1.5 µL of 5 *M* NaCl.
7. Incubate plate on the bench for 2–3 h on a shaker.
8. Discard the supernatant, carefully inverting the plate.
9. Wash 96-well plate, adding to each well 100 µL of ice-cold 70% ethanol.
10. Repeat **step 9** two more times, inverting plates after each wash.
11. After the last ethanol wash, plates are left on the bench for 20 min to air-dry (*see* **Note 17**).
12. Apply to each well 35 µL of the restriction digest mix.

13. Wrap plate in saran wrap and incubate overnight at the appropriate restriction-enzyme temperature.
14. Next day, block digestion by adding 9 μL of 5X loading buffer (*see* **Note 18**).
15. Digested DNA from each well is loaded with a multichannel pipettor onto a 0.8–1.1% agarose gel (1X TAE) prepared using gel chambers accommodating up to four sets of gel combs (with each gel comb set allowing the loading of 50 DNA samples). Gel is run overnight at 40 V (*see* **Note 19**).
16. Denature DNA, incubating the agarose gel for 20 min in 0.15 *N* HCl and subsequently for 1 h in transfer solution.
17. Transfer DNA onto positively charged nylon membranes (in transfer solution) following conventional upward alkaline capillary transfer protocols (completed in 16–24 h), or by performing a 1-h downward alkaline capillary transfer *(14)*.
18. Crosslink DNA, baking membranes at 80°C for 1 hr, or by ultraviolet (UV) irradiation (254 nm wavelength) using a Stratagene Stratalinker.
19. Prehybridize membrane (for 4 h to overnight) and hybridize overnight in hybridization buffer using either a 5′ or 3′-labeled external probe.
20. Wash membrane (*see* **Note 20**) and, finally, expose overnight in an autoradiography cassette at –70°C.
21. Develop film.

3.7. Thawing ES Cells From 96-Well Plates

1. Prepare MMC-treated MEF-3 and plate onto 48-well plates.
2. Warm up sterile distilled water to 37°C, and pour it into a sterile Pyrex dish under the cell-culture hood.
3. Remove one frozen 96-well plate from the –70°C freezer and place it on the surface of the water until its content is thawed.
4. Transfer correctly targeted ES clones into 15-mL conical tubes containing 5 mL of ES medium using a 1-mL disposable pipet (*see* **Note 21**).
5. Spin ES cells 10 min at 195*g* at room temperature.
6. Aspirate supernatant and resuspend cell pellet in 0.5 mL of ES medium containing G418. Plate ES cells into 1 well of a 48-well plate.
7. Feed cells daily with fresh ES medium containing G418.
8. Passage ES cells when they reach subconfluence.

3.8. Deleting the Selection Marker From the Targeted Locus

Once targeted ES clones are identified successfully, the positive selection marker is removed from the targeted allele. Depending on which type of TV is used, targeted ES cells are transiently transfected with either a Cre or Flp expression vector. In case the frt/FLP recombination system is used, removal of the selection marker can also be achieved in vivo breeding mouse chimeras to the Flp deleter strain *(15)*.

1. Thaw a vial of targeted ES cells and expand cells as described in *see* **Subheading 3.2.**

2. Feed cells daily with ES medium (+G418) until they reach a density of 1×10^7 cells.

3. The day before transfection, precipitate 30 μg of Cre or Flp expression vectors in a microfuge tube with 1/10 vol Na acetate (2 *M*, pH 6.5) and 2.5 vol absolute ethanol, for at least 40 min at –20°C.

4. Spin DNA in a microcentrifuge at 15,700*g* for 20 min at 4°C.

5. Wash pellet twice with 75% ice-cold ethanol.

6. Open microfuge tube under the hood and air-dry pellet for 20 min.

7. Resuspend DNA using sterile disposable tips in 100 μL of transfection buffer.

8. On the day of transfection, trypsinize, count, and aliquot 5×10^6 ES cells in a 50-mL conical tube.

9. Perform electroporation of ES cells as described in **Subheading 3.3., steps 5–9**.

10. Plate 5×10^6 transfected ES cells in one 9-cm tissue-culture plate containing MMC-treated MEF-2 or MEF-3. ES cells are fed with ES medium lacking G418.

11. As soon as ES colonies are visible (usually 2 d after transfection), trypsinize ES cells and plate 1×10^3 cells each onto three 9-cm tissue-culture plates containing MMC-treated MEF-2 or MEF-3.

12. Feed ES cells daily with fresh ES medium (without G418).

13. Three or 4 d after plating, ES colonies are picked following the same protocol described in **Subheading 3.4.**, with one exception: after trypsinization each ES colony is directly divided into two flat-bottom 96-well replica plates containing MMC-treated MEF-3 cells (*see* **Note 22**).

14. After 2 d, one 96-well plate is fed with ES medium containing G418, and the second plate is fed with only ES medium (*see* **Note 23**).

15. Feed the ES colonies present in replica plates daily with the appropriate ES medium.

16. Expand ES colonies from the 96-well plate fed with ES medium alone, choosing those clones that are clearly dying in the same well of the replica plate fed with G418-containing medium.

17. Grow ES clones, extract DNA, and confirm by Southern analysis the successful deletion of the selection marker (*see* **Note 24**).

4. Notes

1. A plate of confluent EF cells can contain from 4 to 6×10^6 cells.

2. Handle with care. Mitomycin is a chemotherapeutic agent inhibiting DNA synthesis and crosslinking DNA.

3. Trypsinization of C57BL/6 and Balb/c-derived ES cells requires the addition of chicken serum (1% final concentration) to the 1X trypsin–EDTA solution to protect cells during the trypsinization procedure (chicken serum lacks trypsin inhibitors).

4. To calculate the number of ES cells in a 9-cm tissue-culture plate, subtract from the total number of counted cells, 4×10^6 that should correspond to the approximate number of feeder cells present in a confluent 9-cm tissue-culture dish.

5. Freeze down the rest of ES cells (2×10^6 ES cells per vial in 1 mL freezing medium).

6. For each transfection use 20 µg of linearized vector.
7. Follow ES cells under the microscope while incubating in 1X trypsin–EDTA. Large cell clumps should be avoided because this will reduce significantly the efficiency of transfection.
8. Good transfection efficiencies are associated with a time constancy of 6–9 ms.
9. The concentration of G418 to add to transfected ES cells should be first titrated on wild-type ES cells. In our experience 180–220 µg/mL (100% activity) of G418 is sufficient by d 8 posttransfection to kill all nontransfected ES colonies present in the control plate.
10. GANC solution (final concentration 2 µ*M*) is prepared fresh every day and added to the ES medium. GANC should not be added to the control plate described in **Subheading 3.2., step 7a**.
11. At confluence, 10 plates of MEF-3 should be sufficient to prepare a feeder layer for picking and expanding 500 double-resistant ES colonies.
12. Count approximately how many double-resistant ES colonies are present to estimate how many feeders are required for picking.
13. 1.7×10^6 MEF-3 cells are sufficient for one 96-well plate.
14. To pick and expand 480 double-resistant ES colonies (plated in five 96-well plates), twenty 96-well plates are required.
15. ES clones carrying properly targeted loci may sometime grow slower and, thus, picking ES colonies on different days increases the chance of identifying the correct ones.
16. Place 96-well plates into sealed containers in which a humidified atmosphere is achieved by adding wet paper towels.
17. Do not allow plates to over-dry, because this will affect the efficiency of DNA digestion.
18. At this point, samples can be either run on an agarose gel or frozen down at –20°C.
19. Depending on the size of the expected bands, gel can also be run over the day at a maximum of 80 V.
20. Stringency of washing conditions varies from probe to probe. Usually start washing membranes in 2X SSC, 0.1% SDS, at 65°C for 30–40 min.
21. Thaw also ES clones present in wells preceding and following the well containing the targeted clone. Sometimes screening by Southern blotting of a large number of ES clones may accidentally lead to mistakes in loading samples onto the agarose gel.
22. The number of ES colonies to pick depends on the efficiency of transfection and of Cre (or Flp) recombination. In case the Cre–loxP system is used to delete the selection marker, usually 50% of Cre-transfected ES cells become G418 sensitive. Thus, picking 200 ES colonies will be enough to obtain at least 10 G418-sensitive colonies (with a 10% efficiency of transfection). Please note that among the G418-sensitive colonies only some will carry the deletion of just the selection marker. If the Flp–frt recombination system is employed to delete the neomycin resistance gene, a significantly higher number of ES colonies should be picked (300–400)

because the efficiency of recombination is around 5–10% of transfected cells. To enrich for Flp-transfected ES cells it is possible to cotransfect targeted ES cells with a puromycin expression vector, followed 12 h later by selection of ES cells for 48 h with puromycin.

23. In our experience increasing by 25% the amount of G418 given to the ES cells makes it easier to detect sensitive (dying) colonies.

24. Internal probes can also be used for this purpose. In this case choose for the Southern analysis a restriction enzyme that cuts once in the targeted region and once outside.

Acknowledgments

S. Casola was supported by the Human Frontiers Science Program, the Cancer Research Institute, and in part by the CBR Institute for Biomedical Research. Dedicated to L. Casola.

References

1. Rajewsky, K., Gu, H., Kuhn, R., et al. (1996) Conditional gene targeting. *J. Clin. Invest.* **98,** 600–603.
2. Sternberg, N. and Hamilton, D. (1981) Bacteriophage P1 site-specific recombination, I: recombination between loxP sites. *J. Mol. Biol.* **150,** 467–486.
3. Lakso, M., Sauer, B., Mosinger, B., Jr., et al. (1992) Targeted oncogene activation by site-specific recombination in transgenic mice. *Proc. Natl. Acad. Sci. USA* **89,** 6232–6236.
4. Soriano, P. (1999) Generalized lacZ expression with the Rosa26 Cre reporter strain. *Nat. Genet.* **21,** 70–71.
5. Bronson, S. K., Plaehn, E. G., Kluckman, K. D., Hagaman, J. R., Maeda, N., and Smithies, O. (1996) Single-copy transgenic mice with chosen-site integration. *Proc. Natl. Acad. Sci. USA* **93,** 9067–9072.
6. de Boer, J., Williams, A., Skavdis, G., et al. (2003) Transgenic mice with hematopoietic and lymphoid specific expression of Cre. *Eur. J. Immunol.* **33,** 314–325.
7. Magin, T. M., McWhir, J., and Melton, D. W. (1992) A new mouse embryonic stem cell line with good germ line contribution and gene targeting frequency. *Nucleic Acids Res.* **20,** 3795, 3796.
8. Hatada, S., Kuziel, W., Smithies, O., and Maeda, N. (1999) The influence of chromosomal location on the expression of two transgenes in mice. *J. Biol. Chem.* **274,** 948–955.
9. Kraus, M., Pao, L. I., Reichlin, A., et al. (2001) Interference with immunoglobulin (Ig)alpha immunoreceptor tyrosine-based activation motif (ITAM) phosphorylation modulates or blocks B cell development, depending on the availability of an Igbeta cytoplasmic tail. *J. Exp. Med.* **194,** 455–469.
10. Casola, S., Otipoby, K., Alimzhanov, M. et al. (2004) B cell receptor signal strength determines B cell fate. *Nat. Immunol.,* in press.

11. Schaft, J., Ashery-Padan, R., van der Hoeven, F., Gruss, P., and Stewart, A. F. (2001) Efficient FLP recombination in mouse ES cells and oocytes. *Genesis* **31,** 6–10.

12. Tucker, K. L., Wang, Y., Dausman, J., and Jaenisch, R. (1997) A transgenic mouse strain expressing four drug-selectable marker genes. *Nucleic Acids Res.* **25,** 3745, 3746.

13. Ramirez-Solis, R., Rivera-Perez, J., Wallace, J. D., Wims, M., Zheng, H., and Bradley, A. (1992) Genomic DNA microextraction: a method to screen numerous samples. *Anal. Biochem.* **201,** 331–335.

14. Chomczynski, P. (1992) One-hour downward alkaline capillary transfer for blotting of DNA and RNA. *Anal. Biochem.* **201,** 134–139.

15. Farley, F. W., Soriano, P., Steffen, L. S., and Dymecki, S. M. (2000) Widespread recombinase expression using FLPeR (flipper) mice. *Genesis* **28,** 106–110.

16. Rickert, R. C., Roes, J., and Rajewsky, K. (1997) B lymphocyte-specific, Cre-mediated mutagenesis in mice. *Nucleic Acids Res.* **25,** 1317, 1318.

17. Kuhn, R., Schwenk, F., Aguet, M., and Rajewsky, K. (1995) Inducible gene targeting in mice. *Science* **269,** 1427–1429.

18. Seibler, J., Zevnik, B., Kuter-Luks, B., et al. (2003) Rapid generation of inducible mouse mutants. *Nucleic Acids Res.* **31,** e12.

19. Kontgen, F., Suss, G., Stewart, C., Steinmetz, M., and Bluethmann, H. (1993) Targeted disruption of the MHC class II Aa gene in C57BL/6 mice. *Int. Immunol.* **5,** 957–964.

20. Robanus-Maandag, E., Dekker, M., van der Valk, M., et al. (1998) p107 is a suppressor of retinoblastoma development in pRb-deficient mice. *Genes Dev.* **12,** 1599–1609.

21. Noben-Trauth, N., Kohler, G., Burki, K., and Ledermann, B. (1996) Efficient targeting of the IL-4 gene in a Balb/c embryonic stem cell line. *Transgenic Res.* **5,** 487–491.

8

Analysis of the Germinal Center Reaction and In Vivo Long-Lived Plasma Cells

Kai-Michael Toellner, Mahmood Khan, and Daniel Man-Yuen Sze

Summary

This chapter describes the analysis of long- and short-lived plasma cells on tissue sections. Mice are immunized with 4-hydroxy-3-nitrophenyl acetyl (NP) coupled to a T-cell-dependent carrier. Antigen-specific germinal center cells and extrafollicular plasma cells are identified by immunohistology on tissue sections. Plasma cells labeled with 5-bromo-2′-deoxyuridine (BrdU) pulses given during different phases of B-cell response are identified on spleen sections. To identify mutated cells originating from cells in germinal centers, antigen-specific cells from spleen sections are isolated by microdissection. From these NP-specific recombined VDJ genes are amplified with family-specific primers by polymerase chain reaction (PCR) for deoxyribonucleic acid (DNA) sequencing. The methods described can be used to characterize origins and lifespan of plasma cells in vivo.

Key Words

Single-cell PCR; DNA sequencing; VDJ amplification; 4-hydroxy-3-nitrophenyl acetyl (NP)-chicken γ-globulin (CGG); immunoglobulin V gene mutation; carrier priming; BrdU labeling; antigen-specific B cells; immunohistology.

1. Introduction

Immunization with T-cell-dependent antigens induces the development of B cells along two pathways *(1–4)*. After interaction with primed T cells in the T zone, B cells can move to the border between T zone and red pulp and develop into early plasmablasts that differentiate into early plasma cells that produce low-affinity antibody (Ab). Plasmablasts and plasma cells first develop in extrafollicular foci where the splenic T zone contacts the red pulp. They can be identified by their expression of CD138 (syndecan-1). Plasmablasts proliferate and start producing cytoplasmic immunoglobulin. After a few days, they differentiate into nondividing plasma cells. Plasma cells syn-

From: *Methods in Molecular Biology, vol. 271: B Cell Protocols*
Edited by: H. Gu and K. Rajewsky © Humana Press Inc., Totowa, NJ

thesize and secrete large amounts of immunoglobulin. After 2–3 d most plasma cells are lost from the spleen and a minority of long-lived plasma cells survives *(5–7)*. The other pathway of B-cell development after cognate inter-action with T cells is the migration into follicles, where B cells form germinal centers. Here B cells proliferate and start the affinity maturation process with immunoglobulin variable-region gene hypermutation and selection of high-affinity B cells *(8)*. Germinal centers can be identified by their absence of IgD *(9)*, the presence of immune complex on follicular dendritic cells. They contain B cells, many of which are in cell cycle, and some T cells.

Hypermutation and affinity maturation can be followed after immunization with antigens that activate B cells with a restricted set of V genes. Coupled to a protein carrier that mediates T-cell help, 4-hydroxy-3-nitrophenyl acetyl (NP) induces a restricted response after primary immunization. The canonical response to NP coupled to T-cell-dependent antigens is V186.2-DFL16.1-Jh2 with λ light chain *(10–12)*. Secondary responses to haptens on T-cell-dependent carriers do activate a broader B-cell repertoire *(11,13,14)*, similar to T-cell-independent carriers *(15)*. Immunoglobulin variable gene mutations can be used to follow the relation and development of plasma cells and germinal center cells *(7,16,17)*. Immunoglobulin variable-region gene mutations in plasma cells that have developed from germinal center cells can be identified by isolating cells from lymphoid tissues, amplification of their immunoglobulin variable-region genes by polymerase chain reaction (PCR), and deoxyribonucleic acid (DNA) sequencing of the PCR products. The amplification of these genes from few or single cells with primers specific for this combination of genes is described in this chapter.

Immediate availability of T-cell help accelerates and synchronizes the response of naive B cells to hapten. Priming of mice with noncoupled carrier protein provides immediate T-cell help and also induces carrier-specific immunoglobulin that, on challenge, will clear free antigen within a short period. In this situation immunization with hapten carrier induces extrafollicular hapten-specific plasmablast proliferation from d 1 to d 4 after immunization, whereas proliferation of B cells in germinal centers is seen from d 3 and later *(3,7)*. The origin of long-lived plasma cells can be traced by labeling cells in S phase with radioactive thymidine *(5)*, or by pulse-chase labeling with the thymidine analog 5-bromo-2′-deoxyuridine (BrdU) *(7,18)*. BrdU given in the drinking water during early phases of the response labels the early expanding extrafollicular plasmablasts, which stop dividing at 96 h after immunization. Cells in germinal centers continue to proliferate after this time and rapidly dilute any BrdU taken up. This falls to levels undetectable by immunohistology within three cell divisions *(19)*. BrdU in plasma cells labeled during these dif-ferent phases can be stained on tissue sections.

2. Materials

2.1. Immunization

1. Chicken γ-globulin (CGG) (Jackson ImmunoResearch, West Grove, PA) is precipitated in alum using the method of Chase *(20)*. CGG is made up as a 5 mg/mL solution in sterile water and added to an equal volume of 9% $AlK(SO_4)_2$ (Sigma-Aldrich, Poole, UK) in water. The protein/alum mix is adjusted to pH 6.5 with 10 M NaOH and left for 30 min at room temperature to allow for maximum precipitation. The precipitate is washed twice in sterile phosphate-buffered saline (PBS) to adjust to a neutral pH. The pellet is resuspended to a final concentration of 250 µg/mL in sterile physiological saline with 5×10^9 chemically inactivated *Bordetella pertussis* bacteria (Lee Laboratories, Grayson, GA) as an adjuvant.
2. Prepare NP_{18}–CGG and NP-sheep immunoglobulin as follows: CGG or sheep immunoglobulin (Jackson ImmunoResearch) is dissolved at 10 mg/mL in 0.1 M $NaHCO_3$ on ice. Succinimide ester of NP (Biosearch Technologies, Novato, CA) is dissolved at 10 mg/mL in dimethylformamide (DMF) (Sigma-Aldrich). NP is added drop-wise, with stirring, to the protein solution at a ratio of 1 mg NP: 20 mg protein and incubated at room temperature for 2 h on a rotating mixer. NP–protein solution is extensively dialyzed against PBS, pH 7.4, using size 5 dialysis tubing (Medicell International, London, UK). This removes any nonconjugated NP from the mixture. This procedure gives an approximate substitution rate of 18 NP molecules per molecule protein *(21)*.
3. BrdU (Sigma-Aldrich) (*see* **Note 1**).

2.2. Immunohistology

1. Tissue-Tek OCT compound (Sakura, Zoeterwoude, The Netherlands).
2. Freezer spray (Seme, Melton Mowbray, UK).
3. Four-spot polytetrafluoroethylene (PTFE)-coated glass slides (Hendley-Essex, Loughton, UK).
4. CGG–biotin is prepared as described by Johnson *(22)*. CGG is dissolved at 1 mg/mL in 0.1 M $NaHCO_3$. Make up (+)-biotin N-hydroxy-succinimide ester (Sigma-Aldrich) at 1 mg/mL in dimethyl sulfoxide (DMSO) (BDH, Leicester, UK) and add to the CGG at a ratio of 100 µg biotin to 1 mg protein. This is incubated at room temperature for 4 h on a rotator. The CGG–biotin mix is dialyzed using size five dialysis tubing (Medicell International, London, UK) against two changes of PBS plus 0.1% sodium azide at 4°C to remove any remaining free biotin.
5. Wash trough: a glass trough holding a stainless steel slide rack. This is supported by two plastic tubes, so that a magnetic stirrer bar can be placed under the slide rack.
6. Humid chamber to incubate slides.
7. Wash buffer: mix 1 vol 0.2 M Tris (Trizma base, Sigma-Aldrich), 1.4 vol 0.1 N HCl, 1.6 vol 0.85% NaCl. The resulting wash buffer should have pH 7.6.
8. Normal mouse serum (Binding Site, Birmingham, UK).
9. 3,3′-Diaminobenzidine hydrochloride (DAB) tablets (Sigma-Aldrich), H_2O_2.

10. Alkaline phosphatase (AP) substrate Fast Blue: dissolve 8 mg levamisole hydrochloride (Sigma-Aldrich) in 10 mL Tris-HCl, pH 9.2 (adjust 1 vol 0.2 M Tris to pH 9.2 with HCl, add 1 vol 0.85% NaCl). Mix with 340 µL DMF (Sigma-Aldrich) containing 3.8 mg naphthol AS-MX phosphate (Sigma-Aldrich). Add 10 mg Fast Blue BB salt (Sigma-Aldrich) and shake until dissolved. Filter and use immediately.
11. Water-based mounting medium: ImmuMount (Thermo Shandon, Pittsburgh, PA)
12. 22-mm × 65-mm Glass cover slips (Surgipath, Richmond, IL).

2.3. BrdU Staining

AP substrate Fast Red: as AP substrate Fast Blue (**Subheading 2.2., step 10**), but Tris is adjusted to pH 8.2 and dye used is Fast Red TR (Sigma-Aldrich).

2.4. Micromanipulation

1. Needle puller (e.g., PC-10; Narishige, Tokyo, Japan).
2. Glass capillaries (e.g., G-1; Narishige).
3. Micromanipulator M (Leica, Wetzlar, Germany) (*see* **Note 2**).
4. Eight-well strips of 200 µL PCR tube with single lids (Alpha Laboratories, Eastleigh, UK).
5. Lysis buffer: 0.5 mg/mL proteinase K (Roche, Basel, Switzerland) in 0.5X PBS.
6. Heating block or water bath.

2.5. PCR

1. 96-Well PCR machine (e.g., MJ Research, Waltham, MA).
2. PfuTurbo DNA polymerase (Stratagene, La Jolla, CA), deoxynucleotide-triphosphate (dNTP) (Promega, Madison, WI), 2% gelatin type B from bovine skin (Sigma-Aldrich), mineral oil (PCR grade) (Sigma-Aldrich).
3. 1X TAE buffer *(23)*: 40 mM Tris-acetate, 1 mM ethylenediaminetetraacetic acid (EDTA), pH 8.0.
4. Horizontal gel apparatus (Geneflow, Fradley, UK).
5. Agarose (Sigma-Aldrich).
6. Loading buffer: bromophenol blue in TAE buffer with 20% sucrose.
7. 100-bp ladder DNA molecular weight (mol wt) standard (Invitrogen, Paisley, UK).
8. MinElute extraction kit (Qiagen, Crawley, UK).
9. ZERO Blunt TOPO PCR cloning kit for sequencing (Invitrogen).
10. QIAprep Miniprep kit (Qiagen).
11. Biohit Proline electronic 8-channel pipettor (Biohit, Helsinki, Finland).

3. Methods

3.1. Immunization and Tissue Preparation

1. Mice are primed by ip injection with 50 µg CGG, precipitated in alum with 10^9 chemically inactivated killed *B. pertussis* bacteria as an adjuvant.
2. After 5 wk, boost by intraperitoneal injection of 50 µg soluble CGG coupled with NP (NP$_{18}$–CGG) in saline.

Table 1
Antibody Combinations for Immunohistology

1st step	2nd step	3rd step	Use for identification of
CD3 IgD (sheep)	Rabbit antirat biotinylated Donkey antisheep peroxidase	Strept AB Complex/AP	T zone, follicles, germinal centers and red pulp
CD138 IgD (sheep)	Rabbit antirat biotinylated Donkey antisheep peroxidase	Strept AB Complex/AP	Plasmablasts and plasma cells
NP-sheep IgG3 (rat)	Donkey antisheep biotinylated Rabbit antirat peroxidase	Strept AB Complex/AP	Antigen-specific germinal center B cells
NP-sheep IgD (rat)	Donkey antisheep biotinylated Rabbit antirat peroxidase	Strept AB Complex/AP	Antigen-specific B cells, plasma cells

3. To label cells in proliferation at the time when the tissue is taken, mice are given an injection with 2 mg BrdU in 200 µL saline 2 h before sacrifice (*see* **Note 1**).
4. For pulse-chase experiments with BrdU, mice are given 2-mg injections of BrdU at the start of the pulse and BrdU at 1 mg/mL in the drinking water for the duration of the pulse (*see* **Note 1**).
5. Spleens are taken, placed on a piece of aluminum foil (*see* **Note 3**), and snap-frozen in liquid N_2 (*see* **Note 4**). After wrapping in the aluminum foil, spleens can be stored at –70°C in grip-sealed polythene bags until sections are cut.
6. The frozen tissues are mounted in a cryostat using O.C.T. compound (*see* **Note 5**). Four 5-µm-thick longitudinal sections are cut and mounted onto four-spot glass slides. The glass-mounted sections are air-dried under a fan for 1 h, then fixed in analytical-grade acetone at 4°C for 20 min, air-dried under a fan for 10 min, and stored in sealed polythene bags at –20°C until use.

3.2. Immunohistological Staining of Antigen-Specific B Cells and Plasma Cells

All staining is carried out on multispot slides carrying four acetone-fixed tissue sections with a combination of enzyme-conjugated or biotinylated secondary reagents. Optimum dilutions of all reagents are predetermined by serial dilution on tissue sections to give clear positive staining with minimum background. **Tables 1–3** give four combinations of Abs that can be used to identify different zones on spleen sections: T zone and follicles can be identified after staining for CD3 and IgD. Germinal centers are IgD⁻ areas within follicles containing proliferating B cells that label with a 2-h BrdU pulse and few

Table 2
Primary Antibody Reagents Used in Immunohistology

Specificity	Host Species, Clone	Source
CD3	Rat, KT3	Serotec, Oxford, UK
CD138	Rat, 281-2	BD Pharmingen, San Diego, CA
IgD	Sheep antiserum	The Binding Site, Birmingham, UK
NP-sheep IgG	–	*See* **Subheading 2.1.**
IgG3	Rat, LO-MG3-13	Serotec
IgD	Rat, 11-26c.2a	BD Pharmingen
BrdU	Mouse, Bu20a	DakoCytomation, Ely, UK

T cells. CD138 stains plasmablasts and plasma cells at the border between T zone and red pulp. NP-specific B cells and plasma cells are identified using NP-conjugated sheep Ig. This also stains immune complex containing NP-specific Ab in germinal centers. To differentiate staining of immune complex from staining of antigen-specific germinal center B cells, sections can be double-stained for NP-binding and IgG3. Immune complex contains a mixture of NP-specific Abs and circulating IgG3 appears black. As IgG3 is mainly induced by T-cell-independent antigens, most NP-specific T-cell dependent germinal center B cells do not switch to IgG3 and will appear blue.

1. Defrost microscope slides in polythene bags for 10 min under a fan.
2. Wash slides by transferring to a wash trough containing wash buffer. Stir for 5 min (*see* **Note 6**).
3. Transfer slides onto a horizontal tray in a humid chamber.
4. Add 100 µL primary Abs diluted in wash buffer to the individual sections on the slide (*see* **Note 7**). Incubate for 45 min in a humid chamber.
5. During the incubation period, preabsorb the secondary Ab with wash buffer containing 10% normal mouse serum in a total volume of 100 µL.
6. Wash sections twice with wash buffer.
7. Add 100 µL secondary Abs to the individual sections on the slide. Incubate for 30 min in a humid chamber. During the incubation prepare StreptABComplex/AP.
8. Wash twice with wash buffer.
9. Add 100 µL StreptABComplex/AP to each section. Incubate for 30 min in a humid chamber.
10. Wash twice with wash buffer.
11. Prepare DAB substrate: add 1 DAB tablet to 15 mL wash buffer, filter. Add 1 drop H_2O_2 to 10 mL DAB substrate.

Table 3
Secondary Antibody Conjugates Used in Immunohistology

Antibody Specificity	Host Species	Conjugate	Source
Rat Ig	Rabbit	Biotin or peroxidase	DakoCytomation
Sheep Ig	Donkey	Biotin or peroxidase	The Binding Site
Mouse Ig	Goat	Biotin	DakoCytomation
StreptABComplex/AP		Alkaline phosphatase	DakoCytomation

12. Add 2–3 drops DAB substrate to each section and develop for about 5 min (*see* **Note 8**)
13. Wash twice with wash buffer.
14. Prepare AP substrate.
15. Add 2–3 drops of FastBlue substrate to each section and develop for up to 20 min (*see* **Note 8**).
16. Slides not used for micromanipulation or BrdU staining can be mounted in a water-based mountant (ImmuMount) and covered with a cover slip.
17. Leave slides for micromanipulation in trough with wash buffer on ice (*see* **Note 9**).

3.3. Immunoenzymatic Staining of BrdU-Labeled Tissue Sections

Labeling for uptake of the thymidine analog BrdU is done after hot acid hydrolysis of the tissue sections. This denatures DNA, making BrdU accessible, and also hydrolyses all immunoglobulin on the tissue sections. Therefore secondary antimouse Ig Abs can be used at this step. BrdU is a mutagen, therefore BrdU labeled tissues should not be used for immunoglobulin gene mutation studies.

1. Wash sections for 5 min with wash buffer, followed by three washes in distilled water.
2. Incubate sections for 25 min in 1 M HCl at 60°C.
3. Rinse sections three times in distilled water in glass trough.
4. Wash 5 min in distilled water, followed by one wash with wash buffer (*see* **Note 10**).
5. Add 100 µL mouse anti-BrdU and incubate 1 h in a humid chamber.
6. Wash twice with wash buffer.
7. Add 100 µL biotinylated goat antimouse Ab and incubate for 45 min in a humid chamber. During the incubation prepare the StreptABComplex/AP.
8. Wash sections twice with wash buffer.

9. Add 100 µL StreptABComplex/AP and incubate for 30 min in a humid chamber.
10. Wash twice with wash buffer.
11. During the washing step prepare the alkaline phosphatase substrate: Add 8 mg levamisole to 10 mL Tris, pH 8.2; add 3.8 mg naphthol AS-MX phosphate in 340 µL DMF; mix; add 10 mg Fast Red TR salt; shake until dissolved; filter.
12. Add 2–3 drops to each section and develop for up to 20 min (*see* **Note 8**).
13. Slides are mounted and covered with a cover slip.

3.4. Micromanipulation of Germinal Centers and Plasma Cells

Sections double-stained for NP-binding cells and IgD (to identify follicular areas) are used for micromanipulation on a bright-field microscope. The location of follicles, germinal centers, and T zones can be confirmed using adjacent sections stained for CD3 and IgD.

1. Needles for micromanipulation are pulled using a needle puller.
2. Prepare 8-well strips of 200 µL PCR tubes containing 25 µL lysis buffer.
3. Remove tissue around single cell to be manipulated with a needle.
4. Scrape off single cell with a fresh needle.
5. The needle tip is broken off into an 8-well strip of PCR tubes containing lysis buffer.
6. Include negative controls (*see* **Note 11**).
7. Incubate overnight at 37°C or 2 h at 56°C.
8. Heat-inactivate proteinase K by heating to 95°C for 10 min.

3.5. Amplification of V-Region Genes by PCR

The lysate of the dissected cells is subjected to two rounds of PCR as established by Jacob and Kelsoe *(16,17)*. The first forward primer Ke13 is complementary to the genomic DNA 5′ to the transcriptional start site of V186.2, whereas the first reverse primer Ke14 is complementary to a region between J_H2 and J_H3. The 5′ internal primer Ke12 is complementary to the first 20 nucleotides of the V186.2 gene segment, and in addition, it contains recognition sequences for the restriction enzymes *Xba*I and *Eco*RI. The 3′ internal primer binds to the J_H2 gene segment and contains an additional *Bam*HI recognition sequence. Alternative sets of primers designed to optimize the recovery of a broader set of V_H genes and of all J_H segments in NP-specific cells are listed in **Table 4** (*see* **Note 12**).

1. Using a multichannel pipet, add 25 µL PCR buffer (9.5 µL water, 5 µL 10X Pfu buffer, 1 µL 0.5% gelatin, 1 µL 20 m*M* dNTP, 2.5 µL $MgCl_2$ (10 m*M*), 4 µL forward primer (5 µ*M*), 4 µL reverse primer (5 µ*M*), 0.5 µL PfuTurbo DNA polymerase) to the digest (*see* **Note 13**).
2. Include negative controls (*see* **Note 11**).

Table 4
Primers Used for PCR

Primer Name	Sequence	Annealing Site	References and Remarks
Ke13	CCTGACCCAGATGTC CCTTCTTCTCCAGCA GG	V186.2 leader	*(17)*
Ke14	GGGTCTAGAGGTGT CCCTAGTCCTTCATG ACC	Between J_H2 and J_H3	*(17)*
Ke12	TCTAGAATTCAGGTC CAACTGCAGCAGCC	Beginning of V186.2	*(17)*
Ke17	ACGGATCCTGTGAG AGTGGTGCCT	Within J_H2 region	*(17)*
fi3	CCTGACCCAGATGTC CCTTCTTCTCCAGC	V186.2 leader	3 Nucleotides shorter than Ke13
fi4	CTCACCTGAGGAGA CAGTGACCGTGGTC CCT	J_H1, 2, 3, and 4	Binds all 4 J_H segments
fi2	GTGTCCACTCCGAG GTCCAACTGCAGCA G	Beginning of V186.2 homolog sequence	Binds V186.2 and homologs, (2 bp shorter than Ke12)
fi7	AGGAGACAGTGACC GTGGTCCCTTGGCCC CA	J_H1, 2, 3, and 4	Binds all 4 J_H segments

3. Overlay the reaction with mineral oil.
4. Perform PCR with an initial denaturation step of 2 min at 94°C, followed by 40 cycles of 30 s at 94°C, 30 s at 58°C, 1 min at 72°C. Finish PCR with 7 min at 72°C.
5. Carefully open tubes to avoid cross-contamination. Using a multichannel pipet transfer 2 μL from first PCR to a new 96-well PCR plate.
6. Include negative controls (*see* **Note 11**).
7. Using a multichannel pipet add 48 μL PCR buffer (30 μL water, 5 μL 10X buffer, 1 μL 0.5% gelatin, 1 μL 20 mM dNTP, 2.5 μL MgCl$_2$ (10 mM), 4 μL inner forward primer (5 μM), 4 μL inner reverse primer (5 μM), 0.5 μL PfuTurbo DNA polymerase).
8. Add 2 drops of mineral oil.
9. Amplify using the same program as for the first PCR.
10. Add 10 μL loading buffer to the 50 μL PCR products.
11. The complete PCR product is analyzed on a 1.5% agarose gel in 1X TAE buffer.

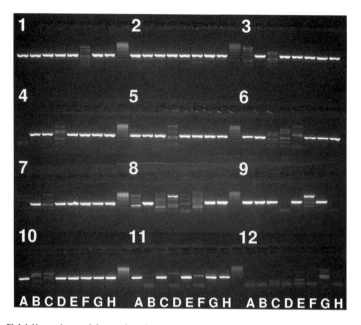

Fig. 1. Ethidium-bromide-stained agarose gel showing amplicons generated from rearranged NP-specific VDJ heavy chain genes. This experiment was done from non-activated NP-specific B cells of a mouse with a targeted insertion of a rearranged VDJ heavy chain gene sorted into wells of a 96-well PCR plate using a fluorescence-activated cell sorter (FACS) fitted with a cloning device *(26)*. Groups 1 to 11 show amplification from single sorted cells. Group 12 contains the product of eight wells not receiving a cell. This image illustrates the efficiency of the PCR, as a transgenic mouse will have the correct rearrangement recognized by the PCR primers in every sorted cell, and FACS cloning gives a near-100% yield of single cells per well. The frequency of reactions producing the correct 358-bp fragment was 75%. Some wells contain only nonspecific products (e.g., 1F, 3A). These and primer bands can be seen in the negative controls in group 12.

12. Bands showing good amplification of a specific product are cut out from the gel with a scalpel over an ultraviolet (UV) fluorescent lamp box and transferred to a clean microcentrifuge tube (**Figs. 1** and **2**).
13. The amplified DNA is purified using MinElute extraction kit (Qiagen).
14. If PCR was done from single cells, the purified and dried PCR products can be used for DNA sequencing (e.g., MWG-Biotech, Ebersberg, Germany) together with one of the primers of the second PCR (*see* **Note 14**).
15. If the PCR product is derived from many cells (e.g., whole germinal centers) the product has to be cloned before sequencing. As the product of Pfu DNA polymerase produces products with blunt ends, a blunt-end cloning vector like ZERO Blunt TOPO PCR cloning kit for sequencing (Invitrogen, Paisley, UK) can be used.

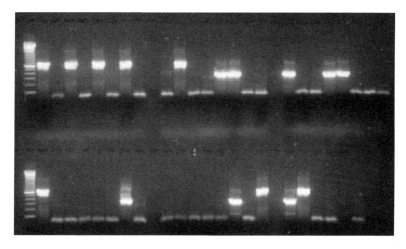

Fig. 2. PCR products from 46 single NP-specific plasma cells, dissected from a spleen section of a carrier-primed mouse 4 d after immunization with NP–CGG. Amplification was done with Ke13 and Ke14, followed by Ke12 and Ke17. Two types of PCR products are visible: bands at approx 370 bp showing VDJh2 heavy rearrangements and bands at approx 690 bp, showing VDJh1 rearrangements. Reactions with no specific product and negative controls (last wells, *bottom right*) show primer bands at <100 bp. Mol wt standard is a 100-bp ladder. Carrier priming activates a broader repertoire of B cells than primary immunization with NP–CGG. Jh1 rearrangements produce a longer product containing Jh1, the region between Jh1 and Jh2, and part of Jh2. The success rate of this PCR was 34%. On average PCR from micromanipulated single cells produces specific bands in 15% of reactions.

16. PCR products cloned in bacteria are cleaned for sequencing with QIAprep Miniprep kit (Qiagen) (*see* **Note 15**).

3.6. DNA Sequencing and Sequence Analysis

The analysis of DNA sequences is described in detail in the chapter by Berek.

1. The traces from the sequencing reactions are checked for mutations. Mutations can be confirmed by a second sequencing reaction in the opposite direction (*see* **Note 16**).
2. Sequences can be aligned and mutations identified by using a public Web-based tool like ClustalW (http://www.ebi.ac.uk/clustalw/).
3. Sequences can be compared to germline sequences by using IgBLAST (http://www.ncbi.nlm.nih.gov/igblast/) or IMGT/V-QUEST at the Immunogenetics database (http://imgt.cines.fr/) (*see* **Note 17**).

4. Notes

1. The thymidine analog BrdU can induce base transitions *(24)*. As a mutagen it should not be used for mutation studies. For the same reason care should be taken when handling BrdU.

2. Several alternatives to traditional glass needle micromanipulators are available: Leica and Palm (Bernried, Germany) produce microdissectors that cut very small tissue samples from stained and dried sections using a laser. These can be further processed for PCR.

3. Remove fatty tissue from the spleen. Fat causes problems when cutting frozen sections.

4. Place the spleen on a labeled piece of aluminum foil. Freeze by repeated dipping in liquid N_2.

5. Put a small strip of O.C.T. compound on the cold cutting block and place the spleen on top. Freeze immediately with freezer spray. Trim the spleen to produce a level surface containing white and red pulp. Using a scalpel remove any O.C.T. compound around the spleen that may have contact with the knife.

6. To wash the slide remove any excess fluid by dipping the sides of the slides onto paper tissue. Place in wash buffer and stir while washing. After the slides are removed from the wash buffer, excess wash buffer between the individual tissue sections is removed with a piece of folded filter paper.

7. Ab dilutions have to be carefully titrated before multiple stainings can be done. Do serial dilutions of the primary Ab, including a buffer blank on adjacent sections to find a dilution that produces a maximal specific staining with minimal background.

8. The development of the dye deposition should be controlled under a light microscope. FastBlue and FastRed form a precipitate in light. Therefore the developing sections that are currently not observed on the microscope should be kept under a dark cover.

9. Micromanipulation works best on freshly stained sections. If sections are not processed immediately they can be stored on ice for a few hours.

10. Sections can be stored overnight in wash buffer at 4°C at this stage.

11. Contamination can be a major problem with PCR from single molecules. Therefore several negative controls are essential: Use several tubes containing a clean needle at the lysis step, tubes with TE buffer instead of lysate added to the first PCR, tubes with no product added to the second PCR.

12. Ke12, 3, 4, and 7 amplify V186.2 and related V genes recombined to Jh2, the canonical Jh segment in a primary NP-specific response *(17)*. Fi2, 3, 4, and 7 are variations of these primers that bind a broader repertoire of V genes and all Jh segments. This can be useful when studying T-cell independent *(15)* or secondary responses to NP *(11,13,14)*.

13. When amplifying single molecules of DNA care has to be taken to avoid contamination from earlier PCRs. Preparation of tissue sections, micromanipulation and setup of the first PCR should be done in a different part of the building than the setup of the second PCR. The PCR machine can be located in this second room. The analysis and preparation of amplified PCR products has to be done in a third area. Every laboratory must have its own set of equipment and consumables. Different lab coats and gloves have to be worn in different laboratories. Single-use plastic materials should be bought sterile and not be autoclaved

on site. Use racked sterile filter tips. Bench surfaces can be cleaned with DNAZap (Ambion, Austin, TX).

14. Commercial sequencing services produce very high quality DNA sequences over 350 bp. Therefore single sequencing with one primer in one direction is usually sufficient to detect point mutations. Mutations can be confirmed by a second sequencing reaction in the opposite direction with the second PCR primer. As sequencing reactions only become readable 10–20 bp from the 3′ end of the sequencing primer, a part of the sequence will be lost if sequencing is only done in one direction.

15. Cloning PCR products before sequencing also has the advantage that primers in the cloning vector can be used for sequencing. This produces full-length sequences of the PCR product.

16. If PCR is done from single cells only, mutations introduced within the first few PCR cycles will have an impact on the DNA sequence read. Mutations introduced later will only be present in a minority of PCR products and not be seen during DNA sequencing. The measured error rate of the Pfu has been reported as 1.3×10^{-6} misincorporations/bp per cycle (*25*). The frequency of polymerase-induced mutations was determined by the sequence analysis of amplified VDJ DNA recovered from single cells of the hybridoma clone, B1-8 (V186.2, DFL16.1, J_H2). After a fixation and amplification protocol identical to that used for histological sections, no mutation was observed in about 6-kb sequence. Also, 86 amplified VDJ sequences (30 kb) of nonactivated cells from a mouse with a targeted insertion of a rearranged VDJ locus did not show any mutations (*26*).

17. The mouse immunoglobulin variable genes have not been completely sequenced. Mismatches of query sequences with germline gene sequences in current Ig gene databases do not necessarily indicate mutated sequences.

References

1. Jacob, J., Kassir, R., and Kelsoe, G. (1991) *in situ* studies of the primary immune response to (4-hydroxy-3- nitrophenyl)acetyl, I: the architecture and dynamics of responding cell populations. *J. Exp. Med.* **173,** 1165–1175.

2. Liu, Y. J., Zhang, J., Lane, P. J. L., Chan, E. Y. T., and MacLennan, I. C. M. (1991) Sites of specific B cell activation in primary and secondary responses to T cell-dependent and T cell-independent antigens. *Eur. J. Immunol.* **21,** 2951–2962.

3. Toellner, K.-M., Gulbranson-Judge, A., Taylor, D. R., Sze, D. M.-Y., and MacLennan, I. C. M. (1996) Immunoglobulin switch transcript production in vivo related to the site and time of antigen-specific B cell activation. *J. Exp. Med.* **183,** 2303–2312.

4. MacLennan, I. C. M., Gulbranson-Judge, A., Toellner, K.-M., et al. (1997) The changing preference of T and B cells for partners as T-dependent antibody responses develop. *Immunol. Rev.* **156,** 53–66.

5. Ho, F., Lortan, J. E., MacLennan, I. C., and Khan, M. (1986) Distinct short-lived and long-lived antibody-producing cell populations. *Eur. J. Immunol.* **16,** 1297–1301.

6. Smith, K. G., Hewitson, T. D., Nossal, G. J., and Tarlinton, D. M. (1996) The phenotype and fate of the antibody-forming cells of the splenic foci. *Eur. J. Immunol.* **26**, 444–448.

7. Sze, D. M.-Y., Toellner, K.-M., Garc'a de Vinuesa, C., Taylor, D. R., and MacLennan, I. C. M. (2000) Intrinsic constraint on plasmablast growth and extrinsic limits of plasma cell survival. *J. Exp. Med.* **192**, 813–821.

8. MacLennan, I. C. M. (1994) Germinal centers. *Annu. Rev. Immunol.* **12**, 117–139.

9. Stein, H., Bonk, A., Tolksdorf, G., Lennert, K., Rodt, H., and Gerdes, J. (1980) Immunohistologic analysis of the organization of normal lymphoid tissue and non-Hodgkin's lymphomas. *J. Histochem. Cytochem.* **28**, 746–760.

10. Jack, R. S., Imanishi-Kari, T., and Rajewsky, K. (1977) Idiotypic analysis of the response of C57BL/6 mice to the (4-hydroxy-3-nitrophenyl)acetyl group. *Eur. J. Immunol.* **7**, 559–565.

11. Reth, M., Hämmerling, G. J., and Rajewsky, K. (1978) Analysis of the repertoire of anti-NP antibodies in C57BL/6 mice by cell fusion, I: characterization of antibody families in the primary and hyperimmune response. *Eur. J. Immunol.* **8**, 393–400.

12. Bothwell, A. L., Paskind, M., Reth, M., Imanishi-Kari, T., Rajewsky, K., and Baltimore, D. (1981) Heavy chain variable region contribution to the NPβ family of antibodies: somatic mutation evident in a λ2a variable region. *Cell* **24**, 625–637.

13. McHeyzer-Williams, M. G., Nossal, G. J. V., and Lalor, P. A. (1991) Molecular characterization of single memory B cells. *Nature* **350**, 502–505.

14. Tao, W., Hardardottir, F., and Bothwell, A. L. (1993) Extensive somatic mutation in the Ig heavy chain V genes in a late primary anti-hapten immune response. *Mol. Immunol.* **30**, 593–602.

15. Maizels, N., Lau, J. C., Blier, P. R., and Bothwell, A. (1988) The T cell independent antigen, NP-Ficoll, primes for a high affinity IgM anti-NP response. *Mol. Immunol.* **25**, 1277–1282.

16. Jacob, J., Kelsoe, G., Rajewsky, K., and Weiss, U. (1991) Intraclonal generation of antibody mutants in germinal centres. *Nature* **354**, 389–392.

17. Jacob, J. and Kelsoe, G. (1992) *In situ* studies of the primary immune response to (4-hydroxy-3- nitrophenyl)acetyl, II: a common clonal origin for periarteriolar lymphoid sheath-associated foci and germinal centers. *J. Exp. Med.* **176**, 679–687.

18. Manz, R. A., Thiel, A., and Radbruch, A. (1997) Lifetime of plasma cells in the bone marrow. *Nature* **388**, 133–134.

19. Wynford-Thomas, D. and Williams, E. D. (1986) Use of bromodeoxyuridine for cell kinetic studies in intact animals. *Cell Tissue Kinet.* **19**, 179–182.

20. Chase, M. W. (1967) Preparation of immunogens. In Williams, C. A. and Chase, M. W., eds., *Methods in Immunology and Immunochemistry* vol. 1, Academic, New York, pp. 197–209.

21. Nossal, G. J. and Karvelas, M. (1990) Soluble antigen abrogates the appearance of anti-protein IgG1-forming cell precursors during primary immunization. *Proc. Natl. Acad. Sci. USA* **87**, 1615–1619.

22. Johnson, G. D. (1989) Immunofluorescence. In Catty, D., ed., *Antibodies, A Practical Approach,* vol. 2, IRL Press, Oxford, UK, pp. 179–200.
23. Sambrook, J., Fritsch, E. F., and Maniatis, T. (1989) *Molecular Cloning: A Laboratory Manual.* 2nd ed., Cold Spring Harbor Laboratory Press, Cold Spring Harbor, NY.
24. Morris, S. M. (1991) The genetic toxicology of 5-bromodeoxyuridine in mammalian cells. *Mutat. Res.* **258,** 161–188.
25. Cline, J., Braman, J. C., and Hogrefe, H. H. (1996) PCR fidelity of pfu DNA polymerase and other thermostable DNA polymerases. *Nucleic Acids Res.* **24,** 3546–3551.
26. Toellner, K.-M., Jenkinson, W. E., Taylor, D. R., Sze, D. M.-Y., Vinuesa, C. G., and MacLennan, I. C. M. (2002) Low-level hypermutation in T cell-independent germinal centers compared with high mutation rates associated with T cell-dependent germinal centers. *J. Exp. Med.* **195,** 383–389.

9

T-Cell-Dependent Immune Responses, Germinal Center Development, and the Analysis of V-Gene Sequences

Claudia Berek

Summary

To determine whether an immune system is able to support T-cell-dependent affinity maturation one needs antigens that induce well-characterized immune responses. This chapter describes the response of the Balb/c mouse to hapten 2-phenyl-4-ethoxymethylene-oxazolone (phOx) coupled to carrier protein chicken serum albumin (CSA). Immunization with this hapten carrier complex will induce the formation of germinal centers (GCs), the microenvironment where the process of affinity maturation takes place. Immunohistological techniques to label germinal centers and examine the development of the GC structure during the primary immune response are described. The process of affinity maturation can be studied by the analysis of the mutational pattern in the $V_\kappa Ox1$ L-chain in single GC B cells directly dissected from frozen tissue sections. The $V_\kappa Ox1$ L-chain sequences are amplified by polymerase chain reaction (PCR), cloned, and sequenced. Methods are described to analyze V-gene sequences. Potential difficulties in the interpretation of the mutational pattern are discussed.

Key Words

Immunohistology; V-gene analysis; somatic mutation; 2-phenyl-oxazolone; T-cell-dependent immune response; germinal center; R/S value; affinity maturation.

1. Introduction

In the splenic tissue of the normal healthy mouse B and T cells are compartmentalized (**Fig. 1**). T cells form a periarteriolar lymphocyte sheath to which the primary B-cell follicles are associated as large cuffs. These B cells are embedded in a network of follicular dendritic cells (FDCs), which present antigen in the form of antigen–antibody complexes to the B cell *(1)*. This chapter describes the basic immunohistological techniques used to visualize these structures in murine lymphoid tissues.

From: *Methods in Molecular Biology, vol. 271: B Cell Protocols*
Edited by: H. Gu and K. Rajewsky © Humana Press Inc., Totowa, NJ

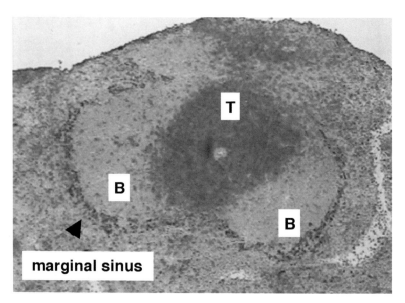

Fig. 1. Organization of lymphocytes in the murine splenic tissue. Labeling of T cells (primary Ab: anti-CD3-FITC (rat Ab); secondary Ab: anti-FITC–alkaline phosphatase conjugate; substrate: Neofuchsin) and metallophilic macrophages adjacent to the marginal zone (primary Ab: MOMA-1–biotin (rat Ab); secondary reagent: streptavidin-peroxidase; substrate: 3,3′diaminobenzidine). The B-cell *(B)* and T-cell *(T)* zones are indicated (courtesy of Ileana Voigt).

Immunization of animals with a T-cell-independent antigen will induce B-cell activation, which is followed by immediate differentiation into plasma cells. However, immunization with a T-cell-dependent antigen will induce a germinal center (GC) reaction *(2–4)*. Antigen-activated B cells enter the T-cell compartment of the splenic follicles through the marginal sinuses. Here, the B cells may come together with antigen-specific T cells and antigen-presenting dendritic cells that initiate GC development.

The GC is the microenvironment within which the affinity maturation of the humoral immune response takes place *(5,6)*. During B-cell proliferation, hypermutation is activated and the V-gene repertoire diversified by nucleotide exchanges. A single antigen-activated B cell may generate variants with different affinities. Antigen-dependent selection of these variants takes place on the FDC network and only those cells with high-affinity receptors are selected to differentiate into memory cells or antibody (Ab)-producing plasma cells.

To analyze the process of affinity maturation in a T-cell-dependent immune response, Balb/c mice (Ig allotype a) are immunized with the hapten

2-phenyl oxazolone (phOx) coupled to CSA (phOx–CSA) *(7)*. This immune response is dominated by B cells expressing a specific H-/L-chain combination, the $V_H Ox1$ and the $V_\kappa Ox1$ V-region genes *(8)*. However, the response may be quite different in mouse strains with a different Ig allotype. Thus, though C57BL/6 mice (Ig b allotype), will respond to phOx, the immune response is not restricted to the $V_H Ox1$ and the $V_\kappa Ox1$ combination *(9)*. To study T-cell-dependent affinity maturation in the C57BL/6 mice, immunization with the hapten 4-hydroxy-3-nitrophenyl acetyl (NP) coupled to the carrier chicken gamma globulin (CGG) is recommended *(10)* (for methods, *see* Chap. 8).

V-gene rearrangement is a feature unique to lymphocytes. During B-cell development in the bone marrow, the V-region genes of the H- and the L-chain of the Ig molecule are generated through the specific rearrangement of gene elements. Because the rearrangement process enhances diversity, the chance that two B cells will generate the same receptor is very small, and the rearranged gene can therefore be used as a clonotypic marker to follow the fate of individual antigen-activated B cells. Using microdissection, GC B cells can be directly isolated from frozen tissue sections, and by sequencing the rearranged genes, we can view the V-gene repertoire *(6,11)*. The mutational pattern in the $V_\kappa Ox1$ L-chain sequences can be used to study the process of affinity maturation in the immune response to phOx.

2. Material

2.1. T-Cell-Dependent Immune Response

1. 2-Phenyl-4-ethoxymethylene-oxazolone (Sigma).
2. CSA (Sigma).
3. BSA (Sigma).

2.2. Immunohistology

1. Tissue-Tek OCT compound (Miles).
2. Plastic molds (10 mm × 10 mm × 5 mm) to freeze tissue samples (Miles).
3. Liquid nitrogen.
4. Adhesive slides SuperFrost® Plus (Menzel-Gläser, Germany).
5. DAKO pen (DAKO).
6. Microtome/cryostat (Mikrom, Leica).
7. Acetone, ethanol, *N,N*-dimethyl formamide, 2-amino-2-methyl-1,3-propandiol (Merck).
8. Mayer's hematoxylin, (DAKO).
9. Kaiser's gelatin (Merck).
10. Peanut agglutinin (PNA)–biotin (Vector).
11. ExtrAvidin alkaline phosphatase conjugate (Sigma).

12. Levamisol (Sigma).
13. Neofuchsin (Merck).
14. Naphthol-AS-biphosphate (Sigma).
15. Light microscope and fluorescence microscope.
16. Biotin protein labeling kit (Roche).

2.3. Amplification of the V_κ L-Chains

1. AmpliTaq Gold™ (Perkin Elmer).
2. Trio block thermocycler (Biometra).
3. NuSieve GTG® (FCM Biozym Products).
4. 123-bp Deoxyribonucleic acid (DNA) ladder (Invitrogen).

3. Methods
3.1. T-Cell-Dependent Immune Response
3.1.1. Coupling of 2-Phenyl-4-Ethoxymethylene-Oxazolone to CSA

1. 15 mg 2-Phenyl-4-ethoxymethylene-oxazolone (**Caution:** Oxazolone may cause sensitization by skin contact; wear suitable protective clothing), 200 mg CSA, 200 mg NaHCO$_3$ and 500 µL sterile saline are mixed in a total volume of 4 mL, and stirred at 4° for approx 24 h *(12)*.
2. The precipitate is spun down by centrifugation at 17,000g for 30 min.
3. The supernatant containing phOx coupled to CSA is dialyzed against 4 L of 0.15 M NaCl, and 0.002% NaN$_3$ for 12 h in a cold room.
4. The dialysis buffer should be changed at least three times.
5. The molar extinction for phOx at optical density (OD)$_{352}$ is 32,000, the concentration of CSA is approx 10^{-3} M; as the coupling rate should be 10:1, one would expect an OD$_{352}$ of 320.
6. For primary immunization, the antigen is precipitated with alum: add equal volume of 10% potassium alum solution (25 g KAl(SO$_4$)$_2$ 12H$_2$O in 250 mL aqua dest) and adjust pH to 6.5 by adding 5 N NaOH dropwise as the buffering action of the potassium alum causes the pH to change suddenly.

The same protocol can be used to couple 2-phenyl-4-ethoxymethylene-oxazolone to BSA.

3.1.2. Immunization of Animals

For primary immunization, Balb/c mice are immunized intraperitoneally (ip) with 100 µg alum-precipitated phOx–CSA. A strong immune response is induced without the need for pertussis. For the secondary immunization animals are boosted 6 to 8 wk later with an intravenously (iv) injection of 100 µg soluble antigen *(13)*.

3.2. Immunohistology

3.2.1. Preparation of Tissue Sections

1. Spleens are removed, embedded in Tissue-Tek OCT compound and frozen over liquid nitrogen for 30 s.
2. Spleen sections of 6–8 μM are cut at a temperature between $-10°$ and $-20°C$ using a microtome/cryostat.
3. Sections are thaw-mounted onto adhesive slides.
4. They are dried on a heating plate for 5 min at 50°.
5. Finally, they are fixed in acetone for 10 min, air-dried, and stored at $-70°C$.

3.2.2. Labeling of Tissue Sections

It is advisable to start with single labeling of tissue sections and to titrate carefully each of the reagents to obtain specific staining of high quality. It is possible to do double, even triple, labeling with conventional immunohistological methods (an example for double staining is shown in **Fig. 1**). The combination of primary Abs from different species, labeling the Abs with different haptens (such as fluorescein isothiocyanate [FITC], digoxigenin, or NP), conjugating the primary or the secondary Abs with different enzymes, and using substrates that produce different precipitating colors allow the simultaneous labeling of different cell types or even of different structures on the same cell (*see* **Note 1**). The sensitivity of staining is further improved by using fluorescence colors and analyzing the sections with a fluorescence microscope.

3.2.2.1. GENERAL REMARKS

1. To prevent nonspecific binding, reagents are diluted in phosphate-buffered saline (PBS) containing 2% BSA (PBS/BSA).
2. To prevent drying of sections, a humid chamber is used for incubations.
3. To remove excess Ab after each incubation step, sections are washed three times by pipetting 150 μL PBS/BSA onto the section and incubating for 5 min.
4. All incubations are done at room temperature.

3.2.2.2. GENERAL PROCEDURE

1. Cover sections with blocking solution (PBS/BSA) for 15 min.
2. Sections are encircled by using a fatty pen (DAKO).
3. Incubate with 75 μL of appropriately diluted Ab for about 30 min.
4. Wash with PBS/BSA.
5. Incubate with 75 μL secondary Ab or reagent (enzyme-conjugated) for about 30 min.
6. Wash with PBS/BSA.
7. Incubate with 75 mL substrate solution in buffer as recommended. To avoid overstaining, the color reaction is followed under the light microscope.

8. The color reaction is stopped by washing with PBS/BSA.
9. Sections are counterstained with 100 µL hematoxylin (undiluted) for 30 s and then washed under running tap water.

3.2.3. Alkaline Phosphatase Labeling
Using the Substrate Neofuchsin

1. 10 µL 5% Neofuchsin (0.05 g Neofuchsin in 1 mL 2 N HCl, stored at 4°C) is mixed with 25 µL NaNO$_2$ solution (12 mg NaNO$_2$ in 250 µL aqua dest, freshly prepared) for 1 min, and added to a solution consisting of: 3.5 mL 0.05 M Tris-buffer, pH 9.7, 0.15 M NaCl, 1.25 mL 0.2 M 2-amino-2-methyl-1,3-propandiol, and 50 µL levamisol (40 mg/mL) (*see* **Note 2**).
2. Finally, 30 µL napthol solution (25 mg napthol AS-bisphosphate in 300 µL N,N-dimethyl formamide, freshly prepared) is added to the solution in **step 1**, using 100 µL 2N HCl; the pH is adjusted to pH 9. Mix well and filter before usage.

Alternatively, the Fuchsin substrate system (DAKO) may be used.

3.2.4. Preservation of Labeled Tissue Sections for Histology

1. Melt Kaiser's gelatin in a boiling water bath and let cool for 5 min.
2. Add 2 drops of Kaiser's gelantin to the section and cover with a cover slip.
3. Turn the slide and press onto a paper towel to remove excess gelatin.

3.2.5. Preservation of Labeled Tissue Sections for Microdissection

Microdissection works best on freshly labeled sections. If they are not processed immediately they should be dehydrated by washing with 30, 75, 90, and finally 100% ethanol. Each wash is for 5 min. The slides are then stored at –70°C.

For microdissection of GC cells, *see* Chaps. 8 and 16.

3.3. The GC Reaction

3.3.1. Development of the GC Structure

1. Ip immunization of Balb/c mice with a T-cell-dependent antigen phOx coupled to the carrier CSA leads to a strong activation of the immune system. Both an immune response specific for determinants on the carrier and specific for the hapten is induced.
2. Approximately 3–4 d after ip immunization, proliferation of the antigen-activated B cells is observed in the splenic tissue at the border of the T- and B-cell compartments *(14)*.
3. Within the next few days the network of the FDCs is filled with proliferating B cells. The primary follicle changes into a secondary follicle.
4. The classical GC with a dark and a light zone develops only in the second week after immunization. Proliferating B cells are mainly seen in the dark zone. A dense network of FDCs in which the GC B cells are embedded characterizes the

Table 1
Specific Reagents to Label Murine Lymphoid Tissue

Specificity	Source	Clone
Ki67	DAKO	Tec 3
FDCs	Becton Dickinson	M1
FDCs	Immunocontact	M2
CD21/CD35	Becton Dickinson	
PNA	Vector	
Macrophages	Bachem	MOMA-1

light zone. The naive B cells are localized at the outer regions of the FDC network. They form the so-called mantel zone.

3.3.2. Immunohistological Staining of GC

1. In the spleen, the marginal sinus surrounds the white pulp. Metallophilic macrophages in the sinuses can be labeled with the MOMA-1 Ab (**Table 1**).
2. GC B cells can be labeled with PNA *(15)* (*see* **Note 3**).
3. GC B cells are negative for IgD.
4. Proliferating B cells can be labeled with the Ki67 Ab specific for a nuclear antigen expressed only in proliferating cells *(16)*.
5. A number of different Abs are available to label the network of FDCs *(16)*:
   Ab M1: This labels FDCs, but only in secondary follicles *(14,17)*. In addition it will also label certain subtypes of macrophages, such as the tingible body macrophages, which are found in GC.
   Ab M2: This reagent specifically labels FDCs in primary and secondary follicles *(14,18)*.
   Abs CD21/CD35: These label the complement receptors CR1 and CR2, which are expressed on FDCs *(17)* and on B cells (*see* **Note 4**).
   Ab CD23: This labels the Fcε II receptor, which is expressed on FDC late in the GC reaction (about 14 d after immunization) and on naive B cells in the mantel zone *(14)*.

3.3.3. The Analysis of Affinity Maturation in T-Cell-Dependent Immune Responses

1. The process of affinity maturation in the immune response to phOx can be studied by cell dissection directly from frozen tissue sections *(11)*.
2. To define GC, sections are labeled with PNA–biotin, then incubated with streptavidin-phosphatase, and developed with the chromogenic substrate Neufuchsin.
3. To define hapten-specific GC, consecutive tissue sections are stained with phOx coupled to BSA *(11,14; see* **Note 5**).

Table 2
Primers for PCR Amplification of Murine V-Region Genes

Name	Sequence
$V_\kappa 1$	5'-GAT GTT TTG ATG ACC CAA ACT CCA-3'
$V_\kappa 4/5$	5'-CAA ATT GTT CTC ACC CAG TCT CCA-3'
$V_\kappa 10$	5'-GAT ATC CAG ATG ACA CAG ACT ACA-3'
$V_\kappa 24$	5'-GAT ATT GTG ATG ACG CAG GCT GCA-3'
J_κ	5'-TTC AAG CTT GGT CCC CGC ACC GAA CGT-3'
$V_\kappa Ox1$	5'-GGG CAG GGA ATT TGA GAT CAG AATA-3'
$J_\kappa 5$	5'-CGT TTC AGC TCC AGC TTG GTC CCA G-3

4. Sections are labeled with phOx–BSA–biotin (phOx–BSA is biotinylated using a biotin protein labeling kit), then incubated with streptavidin-phosphatase and the color reaction developed with the substrate Neofuchsin.
5. Methods to directly dissect GC cells from frozen tissue sections, amplification of V-gene sequences, cloning and sequencing are described in detail in Chap. 16. Additional information is given in Chap. 8.
6. Isolated cells (about 20 cells) are digested in 20 μL 1 mg/mL proteinase K at 50°C for 1 h, and then the enzyme is deactivated at 95°C for 10 min.
7. To analyze the canonical $V_\kappa Ox1$ sequence, the DNA is amplified by using primers specific for the $V_\kappa Ox1-J_\kappa 5$ rearrangement (*see* **Note 6**).
8. To amplify the V_κ-L-chain repertoire of single GC, DNA is amplified by using a primer mix specific for the different V_κ-germline genes together with a universal J_κ primer (**Table 2**) in a standard 50-μL PCR reaction. AmpliTaq Gold DNA polymerase (2 U) is recommended because it allows a hot start, and this helps ensure specific amplification. A thermocycler with a heated lid should be used so that it is not necessary to overlay the PCR reaction with oil; after 40 cycles a 20-μL aliquot is run on a 2% NuSieve GTG low-melting agarose gel. The region of the gel corresponding to roughly 270 bp (the size is determined by using the 123-bp DNA ladder as a marker) is excised. The isolated gel segment is melted in 250 μL of water in a heat block at 65°C, 0.5–1 μL of the DNA solution is then reamplified for 40 cycles, and the DNA fragments again separated on a NuSieve agarose gel.
9. V-gene fragments are cloned and sequenced (for details *see* Chap. 16).

3.4. V-Gene Analysis

1. The presumptive germline genes may be determined by using the *DnaPlot* program (www.dnaplot.de) from W. Müller, German Research Centre for Biotechnology, Braunschweig, Germany.
2. The program will indicate the V, D, and J segment used for rearrangement.
3. Differences in the presumptive germline gene are indicated (*see* **Note 7**).

3.5. The Analysis of Somatic Mutations

The pattern of somatic mutations is used to determine whether affinity maturation has taken place.

3.5.1. Key Mutation

In the immune response to phOx a His exchange at position 34 in the complementarity-determining region 1 (CDR1), either to Asn or Gln is characteristic for the high-affinity Abs (key mutation) *(7)*. In general, in the high-affinity Abs the His mutation is accompanied by a second mutation, a Tyr to Phe substitution at position 36 in the framework (FR) of the $V_\kappa Ox1$ L-chain.

3.5.2. R/S Values

A standard method is to determine the ratio of silent to replacement mutations. (R/S value) *(19)* (*see* **Note 8**). In the FR one usually finds that silent mutations accumulate. Selection against replacement mutations ensures that the over-all structure is kept intact. In contrast high R/S values are often found over the CDRs because here changes in the amino acid sequence may increase the affinity of the Ab (*see* **Note 9**).

4. Notes

1. The harshness of the peroxidase enzyme reaction can destroy epitopes. Therefore, when double staining, the labeling with alkaline phosphatase, including substrate incubation, has to precede peroxidase substrate reaction. When the phosphatase enzyme is used in both labelings, it is necessary to inactivate the first phosphatase-coupled reagent with 0.1 M HCl for 5 min prior to applying the second phosphatase-conjugated Ab.
2. Splenic tissue sections have endogenous alkaline phosphatase activity that should be blocked by adding levamisol to the substrate solution.
3. In addition to GC B cells, PNA stains reticular structures (**Fig. 2**).
4. Both B cells and FDCs are stained with an anti-CD21/CD35 Ab. However, the staining of FDCs is more prominent. Furthermore, FDCs and B cells can be distinguished by their characteristic pattern of staining, as the circular surface staining seen for B cells is clearly different from the network-like staining of FDCs.
5. This step is essential, as only approx 1 out of 10 GC is hapten specific; phOx coupled to BSA may also stain antigen–antibody complexes bound to FDCs. However, FDCs and B cells can be distinguished by their characteristic pattern of staining (*see* **Note 5**).
6. The $V_\kappa Ox1$ germline gene belongs to the $V_\kappa 4/5$ gene family, which consists of at least 30 different, but very similar, V-genes *(20)* To restrict PCR amplification to the $V_\kappa Ox1$ germline gene, a primer in the 5′ flanking region of the $V_\kappa Ox1$ gene (position 175) *(21)* can be used (**Table 2**, primer $V_\kappa Ox1$ *[11]*). PCR amplification

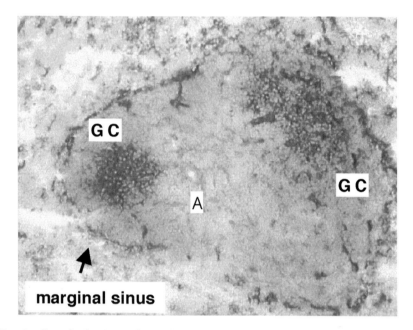

Fig. 2. Germinal center formation. Labeling of GC B cells (primary reagent: PNA–biotin; secondary reagent: ExtrAvidin alkaline phosphatase conjugate; substrate: Neofuchsin) in a murine splenic tissue section 13 d after immunization with phOx–CSA; in addition, unspecific staining is shown in the marginal sinuses. The white and red pulp can be seen, and GC and central artery (**A**) are indicated (courtesy of Ileana Voigt).

will result in a fragment of about 500 nucleotides. As a second step, the $V_\kappa Ox1$ rearrangement can be reamplified by using the $V_\kappa 4/5$ primer together with the $J_\kappa 5$ primer.

7. For the mouse the list of germline sequences is still not complete. The only systematic sequence analysis of the murine V-gene loci, is the one of the V_κ locus of the C57BL/6 mouse *(20)*. For the majority of sequences the corresponding germline genes are unknown. One has to be aware that there may be additional sequences that are more closely related than the one selected by the DnaPlot program.

8. For each of the codons an R/S value can be calculated by determining for each of the 3 nucleotides whether a base exchange will introduce a replacement or a silent mutation. With mutations occurring at random one would expect R/S values of 2.3. However, the analysis of germline gene sequences shows that the R/S values are quite different for the FR and the CDR regions. For example, triplets of rather different R/S values encode the amino acid Ser. For the TCN codons the R/S value is 2, whereas for the AGY (C or T) codons the R/S value is 8. Selection during evolution has ensured that the latter codons are mainly found in the CDRs of the V-region genes.

9. Care should be exercised in drawing conclusions from R/S values. A good example is the $V_\kappa Ox1$ germline gene. Here, the R/S value over the CDRIII is 8 *(7)*. This means that positive selection is only indicated in those cases where values higher than 8 are found. However, what one finds, is that practically all somatic mutations in the CDRIII of the $V_\kappa Ox1$ L-chain are silent mutations. X-ray analysis showed that the CDRIII is essential for binding to the hapten phOx. Therefore, in this particular case, positive selection is accompanied by a low R/S value in the CDRIII because most replacement mutations will lower the affinity for phOx.

Acknowledgments

I would like to thank Stephanie Camacho and Lucia Neure for establishing the protocols of the techniques described. This work was supported by the DFG Be 1171 /1–3. The Deutsche Rheuma ForschungsZentrum is supported by the Berlin Senate of Research and Education.

References

1. Tew, J. G., Wu, J., Fakher, M., Szakal, A. K., and Qin, D. (2001) Follicular dendritic cells: beyond the necessity of T-cell help. *Trends Immunol.* **22**, 361–367.
2. Nieuwenhuis, P. and Opstelten, D. (1984) Functional anatomy of germinal centers. *Am. J. Anat.* **170**, 421–435.
3. Berek, C. and Ziegner, M. (1993) The maturation of the immune response. *Immunol. Today* **14**, 400–404.
4. MacLennan, I. C. M. (1994) Germinal centers. *Ann. Rev. Immunol.* **12**, 117–139.
5. Berek, C., Berger, A., and Apel, M. (1991) Maturation of the immune response in germinal centers. *Cell* **67**, 1121–1129.
6. Jacob, J., Kelsoe, G., Rajewsky, K., and Weiss, U. (1991) Intraclonal generation of antibody mutants in germinal centres. *Nature* **354**, 389–392.
7. Berek, C. and Milstein, C. (1987) Mutation drift and repertoire shift in the maturation of the immune response. *Immunol. Rev.* **96**, 23–41.
8. Kaartinen, M., Griffiths, G. M., Markham, A. F., and Milstein, C. (1983) mRNA sequences define an unusually restricted IgG response to 2-phenyloxazolone and its early diversification. *Nature.* **304**, 320–324.
9. Kaartinen, M., Solin, M. L., and Makela, O. (1991) V genes of oxazolone antibodies in 10 strains of mice. *Eur. J. Immunol.* **21**, 2863–2869.
10. Allen, D., Cumano, A., Dildrop, R., et al. (1987) Timing, genetic requirements and functional consequences of somatic hypermutation during B-cell development. *Immunol. Rev.* **96**, 5–22.
11. Ziegner, M., Steinhauser, G., and Berek, C. (1994) Development of antibody diversity in single germinal centers: selective expansion of high-affinity variants. *Eur. J. Immunol.* **24**, 2393–2400.
12. Mäkelä, O., Kaartinen, M., Pelkonen, J. L., and Karjalainen, K. (1978) Inheritance of antibody specificity, V: anti-2-phenyloxazolone in the mouse. *J. Exp. Med.* **148**, 1644–1660.

13. Berek, C., Griffiths, G. M., and Milstein, C. (1985) Molecular events during maturation of the immune response to oxazolone. *Nature* **316,** 412–418.
14. Camacho, S. A., Kosco-Vilbois, M. H., and Berek, C. (1998) The dynamic structure of the germinal center. *Immunol. Today* **19,** 511–514.
15. Rose, M. L., Birbeck, M. S., Wallis, V. J., Forrester, J. A., and Davies, A. J. (1980) Peanut lectin binding properties of germinal centres of mouse lymphoid tissue. *Nature* **284,** 364–366.
16. Kosco-Vilbois, M. H., Zentgraf, H., Gerdes, J., and Bonnefoy, J.-Y. (1997) To 'B' or not to 'B' a germinal center? *Immunol. Today* **18,** 225–230.
17. Petrasch, S., Perez-Alvarez, C., Schmitz, J., Kosco, M., and Brittinger, G. (1990) Antigenic phenotyping of human follicular dendritic cells isolated from nonmalignant and malignant lymphatic tissue. *Eur. J. Immunol.* **20,** 1013–1018.
18. Taylor, P. R., Pickering, M. C., Kosco-Vilbois, M. H., et al. (2002) The follicular dendritic cell restricted epitope, FDC-M2, is complement C4; localization of immune complexes in mouse tissues. *Eur. J. Immunol.* **32,** 1888–1896.
19. Weigert, M. (1986) The influence of somatic mutation on the immune response. *Prog. Immunol.* **VI,** 138–142.
20. Kirschbaum, T., Jaenichen, R., and Zachau, H. G. (1996) The mouse immunoglobulin kappa locus contains about 140 variable gene segments. *Eur. J. Immunol.* **26,** 1613–1620.
21. Milstein, C., Even, J., Jarvis, J. M., González-Fernandez, A., and Gherardi, E. (1992) Nonrandom feature of the repertoire expressed by members of one Vk gene family and of the V-J recombination. *Eur. J. Immunol.* **22,** 1627–1634.

10

Introduction of Genes Into Primary Murine Splenic B Cells Using Retrovirus Vectors

Kuo-I Lin and Kathryn Calame

Summary

Primary murine splenic B cells can be cultured ex vivo and, following treatment with LPS, cytokines and/or CD40L proliferate, and undergo class switch recombination and terminal differentiation to become immunoglobulin-secreting plasmacytic cells. Methods are described here for introducing genes, encoding either normal or blocking forms of experimental proteins, into murine splenic B cells using retrovirus vectors. This makes it possible to study the effects of these proteins on late stages of B-cell development, including proliferation, class switch recombination, and plasmacytic differentiation.

Key Words

Retrovirus; splenic B cell; primary B cells; lipopolysaccharide (LPS); cytokine; immunoglobulin secretion; terminal differentiation.

1. Introduction

Treatment of primary murine splenic B cells cultured ex vivo with the polyclonal mitogen lipopolysaccharide (LPS) gives rise to a burst of proliferation followed by subsequent differentiation to immunoglobulin-secreting plasmacytic cells *(1,2)*. Such cultures can also be treated with cytokines and/or CD40L to induce isotype switching prior to plasmacytic differentiation. By using bicistronic retroviruses, wild-type, mutated, or blocking forms of various regulators can be introduced into these primary B cells to determine how particular proteins affect the processes of proliferation, isotype switching, and/or terminal differentiation to plasma cells. This is particularly useful if gene targeting causes embryonic lethality or an early block in B-cell development, making it difficult to study the role of the targeted gene in later stages of B-cell development.

From: *Methods in Molecular Biology, vol. 271: B Cell Protocols*
Edited by: H. Gu and K. Rajewsky © Humana Press Inc., Totowa, NJ

In this chapter we describe the methodology we have used to express genes in murine splenic B-cell cultures *(3,4)*. Two features of the system described here are worth noting. First, the retroviruses are designed to express two proteins from a bicistronic messenger ribonucleic acid (mRNA) having an internal ribosome entry site (IRES) for translational initiation of the second protein *(5)*. It is convenient to express a protein such as green fluorescent protein (GFP) that can be monitored by flow cytometry from the IRES. Thus, without sorting infected vs uninfected cells, one can analyze both populations by flow cytometry. Alternatively, if the experimental design requires, infected cells may be sorted by flow cytometry for further analysis of purified populations. Second, retroviral infection requires that the cells be cycling. This is easily accomplished by treating primary splenic B cells with the polyclonal mitogen LPS. However, because LPS causes proliferation and differentiation, for some experiments it is desirable to treat the cells in a way that causes proliferation but not differentiation. Treatment with LPS in conjunction with anti-IgM F(ab)$'_2$ achieves this situation *(2,6,7)*.

2. Materials

1. Packaging cell line (purchased from ATCC, or obtained from Dr. G. Nolan, Stanford).
2. Plasmids: including retroviral vector, expression vector for packaging proteins (obtained from Dr. O. Witte, UCLA), and expression vector for VSV-G protein (obtained from Dr. S. Goff, Columbia University).
3. Solutions for transfection: 100 mM chloroquine; 2 M CaCl$_2$; 2X HBS: 0.27 M NaCl, 1.5 mM Na$_2$HPO$_4$, and 50 mM HEPES, pH 7.0.
4. Cell culture medium.
5. Hank's balanced salt solution (HBSS).
6. Polybrene.
7. Mice.
8. Reagents for purification of splenic B cells: that is, anti-thy1.2 antibody.
9. Reagents: LPS (Sigma) and cytokines for the stimulation of proliferation of splenic B cells.
10. Flow cytometer.

3. Methods

The methods for transduction of exogenous genes via retrovirus vectors, described in **Subheading 3.1.**, include preparation of retroviruses, purification and stimulation of splenic B cells, and infection and selection of retrovirus-infected B cells. The method for preparation and infection of retrovirus essentially follows the protocol derived from the laboratory of Dr. G. Nolan; for more details please visit his Web site (http://www.stanford.edu/group/nolan/index.html).

3.1. Preparation of Retrovirus

3.1.1. Retroviral Vectors

The pGC–IRES retroviral plasmid, obtained from Drs. C. G. Fathman and G. Nolan, has been used for our experiments *(8)*. This plasmid was derived from the pLCAT backbone with the addition of 5′ and 3′ long terminal repeats (LTRs) and packaging signals from the Moloney murine leukemia virus (MoMLV)-based MFG vector. Two reporter proteins, GFP and yellow fluorescent protein (YFP) are expressed via an IRES derived from encephalomyocarditis virus (EMCV) to allow infected cells to be monitored or purified using flow cytometry. Meanwhile, an empty vector carrying only GFP or YFP after the IRES is used as a control *(8)*. Other murine leukemia virus-based vectors, such as murine stem cell virus (MSCV)–IRES–GFP can also be used.

3.1.2. Transfection

The ecotropic Phoenix packaging line, generated from 293T-cells, was chosen for transfection because of its high transfection efficiency *(9)*. However, because expression of the stably integrated genes for the retroviral *gag*, *pol*, and *env* genes often diminishes during culture of Phoenix cells, we introduce expression vectors for *gag* and *pol* by transient transfection. We also introduce by transient transfection an expression vector for the vesicular stomatitis virus G glycoprotein (VSV-G) *(10)*, to generate pseudotyped viruses and render them more stable to subsequent purification by ultracentrifugation.

Cells are cultured as described *(8)*. The night before transfection $4–5 \times 10^6$ cells are plated on 10-cm dishes. In our experience, using 15–20 µg of retrovirus vector, 15–20 µg of packaging deoxyribonucleic acid (DNA) (pSV-ψ⁻-E-MLV) *(11)* and 15–20 µg of VSV-G vector (pMD.G) *(12)* gives the best viral titers after concentration of virus supernatants. The procedure for Ca_2PO_4 transfection is described in detail elsewhere *(13)*. Briefly,

1. About 5 min before transfection, chloroquine is added to the cells to a final concentration of 25 µM.
2. A mixture containing DNA, $CaCl_2$ (to a final concentration of 31 mM) and water is mixed with an equal volume of 2X HBS. This solution is added drop by drop onto media, and the cells are incubated for 8 h at 37°C. The formula for preparing the reagents and sources from which they are obtained are detailed at a Web site (http://www.stanford.edu/group/nolan/index.html). After 8 h, the medium is changed, 10 mL fresh complete medium is added, and the cells are incubated at 32°C.

3.1.3. Concentration of Virus

Forty-eight hours posttransfection, the culture supernatant is collected for virus harvest and concentration. VSV-G, encoded by the pMD.G plasmid, interacts with a phospholipid component of cell membranes to mediate viral entry by membrane fusion *(14,15)* and therefore allows a wider host cell range for retroviral infection *(16)*. The other advantage of VSV-G pseudotyped retroviruses is that, owing to the structural stability of the pseudotyped virus particle, viral titers can be increased by concentration using ultracentrifugation without loss of infectivity *(9,17)*.

The viral supernatant is filtered through a 45-μm pore polyethersulfone (PES) filter to remove cellular debris and then subjected to ultracentrifugation in a Beckman ultracentrifuge using polyallomer tubes (9/16 × 3 1/2 inch) in a Beckman SW41Ti swinging bucket rotor at 107,000*g* (25,000 rpm) for 90 min at 4°C. After centrifugation, the supernatant is removed by aspiration and resuspended overnight at 4°C in an appropriate amount of HBSS. Viral particles in supernatant from a 10-cm plate culture are usually redissolved in 20–100 μL HBSS.

Retroviral stocks are unstable: Incubation at 37°C for 8 h decreases infectivity 10-fold and incubation at 4°C overnight decreases infectivity as much as 100-fold *(18)*. Viral stocks are also sensitive to repeated freeze–thaw cycles. Therefore, the resuspended retroviral stocks should immediately be aliquoted and frozen at –80°C; they should be used after thawing and not refrozen. This procedure usually causes less than 2X loss of infectivity.

Several cell lines, including 3T3, HeLa, and 293T-cells can be used for determining the concentration of infectious virus in the prepared stock. The stock is serially diluted and cells are infected with the diluted viruses in the presence of polybrene (5 μg/mL). The percentage of infected cells can be assayed 48 h after infection by determining the percentage of GFP- or YFP-positive cells using a fluorescent microscope or flow cytometry. The percentage of GFP- or YFP-positive cells falling in the linear range of infection efficiency is used to determine the titer of the stock where the numbers of infectious particles/ vol = cell number × fraction positive/vol of stock used for infection.

3.2. Purification and Stimulation of Splenic B Cells

Splenic B cells are isolated from 6–8-wk-old C57BL/6 mice. B cells can be isolated either by positive selection using anti-B220 beads or by negative selection using anti-thy1.2 beads. Briefly,

1. Spleens are dissected and immersed in 10 mL RPMI medium containing 10% fetal bovine serum (FBS), 50 μ*M* 2-mercaptoethanol (ME), and antibiotics (B-cell medium). Cell suspensions are obtained by pressing the tissue gently between two glass microscope slides.
2. Cells are spun down and red blood cells are lysed in red blood cell lysis buffer (Sigma) for 3–5 min at room temperature.
3. The remaining cells are pelleted; the color of resultant cell pellet should not be red. The cell pellet is resuspended in B-cell medium to a density of 10^6 cells/mL and plated in 15-cm culture dishes.

4. The cells are incubated at 37°C for at least 2 h to allow macrophages to adhere.
5. Cells remaining in suspension are collected, pelleted, resuspended in B-cell medium, and counted. B cells are purified using either positive or negative selection, following the protocols provided by manufacturers. Several companies, including Miltenyi Biotec and DYNAL, sell the beads for purification of mouse splenic B cells. The authors use anti-thy1.2 beads for negative selection. The amount of beads required is determined by the percentage of T cells in the splenocyte population and the numbers of B cells that we wish to obtain. In general, we use 10 µL of beads for 10^6 B cells. Either positive or negative selection will work for the purpose of purification; however, purification by positive selection may provide an activating signal and interfere with the experimental design *(19,20)*.
6. Determine the purity of B-cell population by fluorescence-activated cell sorting (FACS). Anti-B220 antibody is generally used at this step. The percentage of B220+ cells should be at least 90–95% after the purification.
7. Purified B cells are resuspended in 10^6/mL in B-cell medium containing mitogen or cytokine (*see* **step 8**), and 1 mL of cells are added to 12-well tissue-culture plates.
8. Cells are treated overnight at 37°C with mitogen or cytokine (*see* **Subheading 3.3.**) for inducing proliferation. Only replicating cells are infected with murine retroviruses. Splenic B cells stimulated with 10–25 µg/mL LPS undergo robust replication and are efficiently infected. In general, the efficiency of infection ranges from 20% to 50%. However, stimulation of splenic B cells with LPS also causes plasmacytic differentiation evidenced by increased expression of syndecan-1 on the surface and secretion of IgM. When the experimental design includes assessing the effect of expressed proteins on plasmacytic differentiation, it is necessary to find conditions that put the cells into cycle but do not cause differentiation. The combination of anti-IgM or anti-IgM F(ab)$'_2$ (5 µg/mL) (Southern Biotechnology) with LPS (10–25 µg/mL) allows B cells to proliferate but prevents their differentiation.

Other regulators, such as interleukin (IL)-2, IL-5, and CD40L that influence splenic B-cell proliferation and differentiation can also be used in combination with or without LPS. Certain cytokines, such as IL-4 (5 µg/mL) (R&D), when combined with LPS cause splenic B cells to undergoing class switching primarily to IgG_1 and can be used for studying the role of a gene in class switching.

3.3. Retroviral Infection and Monitoring Infected Cells

Splenic B cells stimulated with mitogens or cytokines overnight are infected with virus. Retrovirus stocks, removed from –80°C are thawed on ice and added immediately to splenic B cells in the presence of polybrene (5 µg/mL). The polycation polybrene facilitates the interaction of virus particles with the cell surface, thus aiding viral attachment and entry into cells *(21)*. The detailed procedure for infection is as follows (*see* **Note 1**):

1. Add viral stock directly to 1-mL cells cultured with LPS or LPS plus cytokine, at a multiplicity of infection (MOI) of 2–5 in the presence of polybrene (5 µg/mL). The MOI is determined by the titer of virus × vol of viral stock used)/ total number of cells for infection. We have found when using the protocol described that there is an upper limit of infection efficiency in each experiment, probably owing to variations in the cell cycle status in different cultures. The volume of virus stock that can be added is also limited. Usually, more than 100 µL concentrated virus stock will cause toxicity in a 1-mL culture.

2. In our experience spin infection is not usually necessary for infecting splenic B cells. Spin infection can improve the infection efficiency approximately twofold, which may be desirable when one plans to isolate GFP- or YFP-positive cells (*see* **step 7**). Spin infection involves centrifuging the cells and viruses at 800*g* for 90 min at room temperature in a Beckman J6 tabletop centrifuge.

3. Incubate the infected cells at 32°C overnight.

4. The next day, shift cultures to 37°C. It is not necessary to wash splenic B cells and replace with fresh B-cell medium at this step as toxicity is not usually observed.

5. Feed splenic B cells with B-cell medium containing the mitogen or cytokine 1 d after shifting the culture to 37°C.

6. Harvest B cells 48–72 h postinfection. The percentage of infected cells, indicated by expression of GFP or YFP, can be monitored by FACS analysis or fluorescent microscopy. **Figure 1** is a representative FACS figure showing the viral infection efficiency in splenic B cells after 3 d of infection. Other B-cell surface markers, such as B220, IgM, or syndecan-1 can be measured simultaneously to monitor ex vivo B-cell differentiation in both infected and uninfected cells.

7. Alternatively, GFP- or YFP-positive cells can be sorted by flow cytometry and then analyzed by other methods. In general, for a 1-mL splenic culture, at least 10^5 YFP- or GFP-positive cells can be recovered. Purified populations of infected cells can be analyzed in many ways. For example, we have sorted infected cells by FACS and studied expression of different transcription factors in the purified cells using semiquantitative reverse transcriptase polymerase chain reaction (RT-PCR) or quantitative RT-PCR. The ability of sorted cells to secrete immunoglobulin can also be analyzed by enzyme-linked immunosorbent spot-forming cell (ELISPOT) assay. Finally, sorted cells can be analyzed by immunohistochemistry.

3.4. Other Applications or Modifications of the Method

See **Notes 2–5**.

4. Notes

1. To achieve good infection efficiency, preparing a high titer of viral stock is the most critical step. The titer should be at least 2×10^7 to 1×10^8 infectious particles/mL. Some parameters are important in generating a good titer of viral stock, such as 2X HBS at precisely pH 7.0, using $CaCl_2$ from Mallinckrodt company and healthy

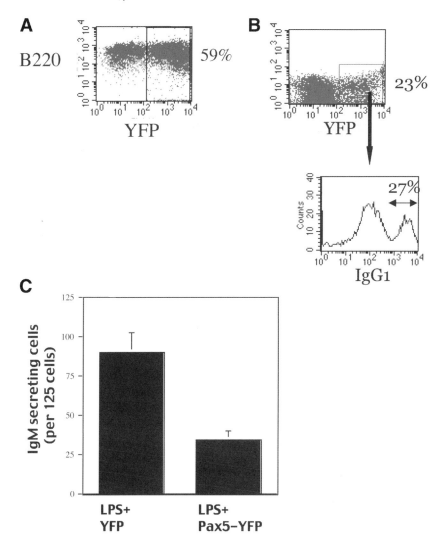

Fig. 1. Examples of retroviral infection in splenic B cells. (A) FACS shows 59% of B220+ cells stimulated with LPS express YFP 3 d after infection with the pGC–YFP retrovirus. (B) FACS shows 23% YFP+ cells in a culture treated with LPS and IL-4 following infections with pGC–YFP. In this experiment, 27% of the YFP+ cells underwent class switching to IgG1, demonstrated by expression of surface IgG1. (C) ELISPOT was used to compare the number of IgM-secreting cells in sorted YFP+ cells stimulated with LPS following infection with pGC–YFP or pGC–Pax5–YFP retrovirus vectors.

Phoenix cells. It is useful to do an initial serial titration of a virus stock in splenic cultures to establish the infection efficiency relative to the virus titer.

2. This protocol can be modified to infection of short-term bone marrow B-cell cultures stimulated with IL-7. However, pseudotype virus is toxic to bone marrow B-cell culture. In this circumstance, cotransfection with VSV-G construct and concentration of virus step should be omitted. The amount of polybrene should be tested for aiding optimal infection efficiency in different types of cell cultures. For infection of bone marrow B cells, we use 8 µg/mL. We have also found that spin infection, using the same conditions described in **Subheading 3.3.**, is required for infecting bone marrow B cells.

3. In addition to murine primary B-cell cultures, cultures of human tonsillar B cells have been described *(22–25)*. In a system described by Choi and colleagues, purified centrocytes (CD19$^+$ CD20$^+$ CD38^{high}) are cultured with CD40L, IL-2, and irradiated HK cells *(26)*. In the presence of IL-4, centrocytes differentiate to memory B cells (CD20$^+$ CD38^{low}), whereas IL-10 causes centrocytes in this culture to become plasma cells (CD20$^-$CD38^{high}) *(27,28)*.

4. There are potential disadvantages of using murine leukemia virus (MLV)-based retroviral vectors to infect mitogen/cytokine-stimulated mouse splenic B cells. The powerful toll-like receptor (TLR) signaling pathways activated by LPS may interfere with the experimental design. One way to avoid this would be to use attenuated lentivirus vectors in either the human tonsillar B-cell cultures or in murine splenic B cells without activation, because lentiviruses have the advantage of infecting nondividing cells *(29,30)*.

5. Recent success using lentivirus vectors to introduce interfering RNA to knock down the expression of a specific gene in primary T-cells suggests this can also be considered as a strategy to alter gene expression in primary splenic B cells *(31,32)*.

References

1. Sieckmann, D. G., Asofsky, R., Mosier, D. E., Zitron, I. M., and Paul, W. E. (1978) Activation of mouse lymphocytes by anti-immunoglobulin, I: parameters of the proliferative response. *J. Exp. Med.* **147**, 814–829.

2. Chen-Bettecken, U., Wecker, E., and Schimpl, A. (1985) IgM RNA switch from membrane to secretory form is prevented by adding antireceptor antibody to bacterial lipopolysaccharide-stimulated murine primary B-cell cultures. *Proc. Natl. Acad. Sci. USA* **82**, 7384–7388.

3. Piskurich, J. F., Lin, K. I., Lin, Y., Wang, Y., Ting, J. P., and Calame, K. (2000) BLIMP-I mediates extinction of major histocompatibility class II transactivator expression in plasma cells. *Nat. Immunol.* **1**, 526–532.

4. Shaffer, A. L., Lin, K. I., Kuo, T. C., et al. (2002) Blimp-1 orchestrates plasma cell differentiation by extinguishing the mature B cell gene expression program. *Immunity* **17**, 51–62.

5. Evstafieva, A. G., Ugarova, T. Y., Chernov, B. K., and Shatsky, I. N. (1991) A complex RNA sequence determines the internal initiation of encephalomyocarditis virus RNA translation. *Nucleic Acids Res.* **19**, 665–671.

6. Andersson, J., Bullock, W. W., and Melchers, F. (1974) Inhibition of mitogenic stimulation of mouse lymphocytes by anti-mouse immunoglobulin antibodies, I: mode of action. *Eur. J. Immunol.* **4,** 715–722.
7. Knodel, M., Kuss, A. W., Berberich, I., and Schimpl, A. (2001) Blimp-1 over-expression abrogates IL-4- and CD40-mediated suppression of terminal B cell differentiation but arrests isotype switching. *Eur. J. Immunol.* **31,** 1972–1980.
8. Costa, G. L., Benson, J. M., Seroogy, C. M., Achacoso, P., Fathman, C. G., and Nolan, G. P. (2000) Targeting rare populations of murine antigen-specific T lymphocytes by retroviral transduction for potential application in gene therapy for autoimmune disease. *J. Immunol.* **164,** 3581–3590.
9. Pear, W. S., Nolan, G. P., Scott, M. L., and Baltimore, D. (1993) Production of high-titer helper-free retroviruses by transient transfection. *Proc. Natl. Acad. Sci. USA* **90,** 8392–8396.
10. Emi, N., Friedmann, T., and Yee, J. K. (1991) Pseudotype formation of murine leukemia virus with the G protein of vesicular stomatitis virus. *J. Virol.* **65,** 1202–1207.
11. Muller, A. J., Young, J. C., Pendergast, A. M., Pondel, M., Landau, N. R., Littman, D. R., and Witte, O. N. (1991) BCR first exon sequences specifically activate the BCR/ABL tyrosine kinase oncogene of Philadelphia chromosome-positive human leukemias. *Mol. Cell Biol.* **11,** 1785–1792.
12. Cullen, B. R. (1986) Trans-activation of human immunodeficiency virus occurs via a bimodal mechanism. *Cell* **46,** 973–982.
13. Wigler, M., Pellicer, A., Silverstein, S., and Axel, R. (1978) Biochemical transfer of single-copy eucaryotic genes using total cellular DNA as donor. *Cell* **14,** 725–731.
14. Chen, S. T., Iida, A., Guo, L., Friedmann, T., and Yee, J. K. (1996) Generation of packaging cell lines for pseudotyped retroviral vectors of the G protein of vesicular stomatitis virus by using a modified tetracycline inducible system. *Proc. Natl. Acad. Sci. USA* **93,** 10,057–10,062.
15. Mastromarino, P., Conti, C., Goldoni, P., Hauttecoeur, B., and Orsi, N. (1987) Characterization of membrane components of the erythrocyte involved in vesicular stomatitis virus attachment and fusion at acidic pH. *J. Gen. Virol.* **68(Pt 9),** 2359–2369.
16. Yee, J. K., Friedmann, T., and Burns, J. C. (1994) Generation of high-titer pseudotyped retroviral vectors with very broad host range. *Methods Cell Biol.* **43(Pt A),** 99–112.
17. Ory, D. S., Neugeboren, B. A., and Mulligan, R. C. (1996) A stable human-derived packaging cell line for production of high titer retrovirus/vesicular stomatitis virus G pseudotypes. *Proc. Natl. Acad. Sci. USA* **93,** 11,400–11,406.
18. Burns, J. C., Friedmann, T., Driever, W., Burrascano, M., and Yee, J. K. (1993) Vesicular stomatitis virus G glycoprotein pseudotyped retroviral vectors: concentration to very high titer and efficient gene transfer into mammalian and nonmammalian cells. *Proc. Natl. Acad. Sci. USA* **90,** 8033–8037.
19. Asensi, V., Himeno, K., Kawamura, I., Sakumoto, M., and Nomoto, K. (1990) In vivo treatment with anti B-220 monoclonal antibody affects T and B cell differentiation. *Clin. Exp. Immunol.* **80,** 268–273.

20. George, A., Rath, S., Shroff, K. E., Wang, M., and Durdik, J. M. (1994) Ligation of CD45 on B cells can facilitate production of secondary Ig isotypes. *J. Immunol.* **152**, 1014–1021.

21. Conti, C., Mastromarino, P., and Orsi, N. (1991) Role of membrane phospholipids and glycolipids in cell-to-cell fusion by VSV. *Comp. Immunol. Microbiol. Infect. Dis.* **14**, 303–313.

22. Dahlenborg, K., Pound, J. D., Gordon, J., Borrebaeck, C. A., and Carlsson, R. (1997) Terminal differentiation of human germinal center B cells in vitro. *Cell Immunol.* **175**, 141–149.

23. Choi, Y. S. (1997) Differentiation and apoptosis of human germinal center B-lymphocytes. *Immunol. Res.* **16**, 161–174.

24. Arpin, C., Dechanet, J., Van Kooten, C., et al. (1995) Generation of memory B cells and plasma cells in vitro. *Science* **268**, 720–722.

25. Liu, Y. J. and Banchereau, J. (1997) Regulation of B-cell commitment to plasma cells or to memory B cells. *Semin. Immunol.* **9**, 235–240.

26. Kim, H. S., Zhang, X., Klyushnenkova, E., and Choi, Y. S. (1995) Stimulation of germinal center B lymphocyte proliferation by an FDC-like cell line, HK. *J. Immunol.* **155**, 1101–1109.

27. Choe, J., Kim, H. S., Zhang, X., Armitage, R. J., and Choi, Y. S. (1996) Cellular and molecular factors that regulate the differentiation and apoptosis of germinal center B cells: anti-Ig down-regulates Fas expression of CD40 ligand-stimulated germinal center B cells and inhibits Fas-mediated apoptosis. *J. Immunol.* **157**, 1006–1016.

28. Li, L., Zhang, X., Kovacic, S., et al. (2000) Identification of a human follicular dendritic cell molecule that stimulates germinal center B cell growth. *J. Exp. Med.* **191**, 1077–1084.

29. Zufferey, R., Nagy, D., Mandel, R. J., Naldini, L., and Trono, D. (1997) Multiply attenuated lentiviral vector achieves efficient gene delivery in vivo. *Nat. Biotechnol.* **15**, 871–875.

30. Naldini, L., Blomer, U., Gallay, P., et al. (1996) In vivo gene delivery and stable transduction of nondividing cells by a lentiviral vector. *Science* **272**, 263–267.

31. Rubinson, D. A., Dillon, C. P., Kwiatkowski, A. V., et al. (2003) A lentivirus-based system to functionally silence genes in primary mammalian cells, stem cells and transgenic mice by RNA interference. *Nat. Genet.* **33**, 401–406.

32. Qin, X. F., An, D. S., Chen, I. S., and Baltimore, D. (2003) Inhibiting HIV-1 infection in human T cells by lentiviral-mediated delivery of small interfering RNA against CCR5. *Proc. Natl. Acad. Sci. USA* **100**, 183–188.

11

Immunoglobulin Class Switching

In Vitro Induction and Analysis

Sven Kracker and Andreas Radbruch

Summary

During an immune response, B lymphocytes can switch expression of immunoglobulin (Ig) class (isotype) from IgM to IgG, IgE, or IgA. This Ig class switch is based on a deoxyribonucleic acid (DNA) recombination event that results in an exchange of the gene segments coding for the constant region of the Ig heavy chain, although the Ig heavy chain variable region is retained. This process changes the effector functions of the corresponding antibody (Ab). Much of our current understanding of the molecular mechanisms of class switch recombination is based on the analysis of B cells induced to switch class of Ig in vitro. In vitro, murine and human naive B cells can be activated with bacterial lipopolysaccharides, anti-CD40 or CD40L, to undergo class switch recombination. Cytokine signals can direct class switch recombination to distinct classes; for example, interleukin-4 will target murine IgG1 and IgE, and human IgG4 and IgE. Here we describe the technologies for the isolation of B lymphocytes, their activation to class switching, and the analysis of Ig class switching.

Key Words

Lipopolysaccharide; LPS; CFDA-SE; CFSE; IgG1; IgG3; IgE; interleukin-4; IL-4; proliferation; intracellular staining; surface staining; antibody class switch recombination; Ig isotypes; B cell; B lymphocyte.

1. Introduction

In 1964 Nossal and collaborators observed for the first time that progeny of individual B cells can switch the isotype of the immunoglobulin they produce in vitro *(1)*. Later, molecular analysis of transformed cells, representing B lineage plasma cells—that is, myelomas and hybridomas—expressing Ig classes other than IgM, revealed that in those cells the gene segments coding

From: *Methods in Molecular Biology, vol. 271: B Cell Protocols*
Edited by: H. Gu and K. Rajewsky © Humana Press Inc., Totowa, NJ

for the constant region of the IgM heavy chain (Cμ) had been replaced by the gene segments coding for the constant region of the Ig class expressed, through somatic recombination (class switch recombination) *(2–4)*. This recombination had occurred between highly repetitive DNA sequences located in front of the constant heavy (CH) genes involved—the switch regions *(5,6)*. In established myeloma, hybridoma, and B-cell lymphoma cell lines, class switch recombination does occur occasionally but may or may not be representative of the physiological process *(7)*. Recently, however, the B-cell lymphoma CH12F3 has been described to switch from IgM to IgA at frequencies of 50% on induction by CD40L, interleukin (IL)-4, and transforming growth factor (TGF)-β, and thus may represent a more valid model for physiological class switching *(8)*.

Murine naive IgM/IgD⁺ B cells can be activated in vitro with mitogens addressing pattern-recognition receptors, such as bacterial lipopolysaccharides (LPS), to switch to expression of IgG3 *(9,10)*. This in vitro system has been used extensively to analyze the frequencies of switching cells, switch dependency on activation and proliferation, sequential switching, and the control of switch recombination, in particular the question of whether and how the different classes of Ig are targeted by switch recombination—by chance or in a controlled fashion *(11)*. It became clear that cytokine signals induce transcription of distinct switch regions and the adjacent CH gene segments. This germline or "switch" transcription precedes switch recombination of the transcribed switch regions *(12,13)*. This recombination is targeted in individual B cells because both Ig heavy chain gene loci show the switch recombination to the same Ig classes *(14)*. With the use of murine B cells with mutated templates for switch transcription, it became clear that those switch transcripts do not only reflect a recombination-accessible state of chromatin; rather they are conditional for switch recombination and, to work, have to be processed *(15,16)*. Another molecular requirement for switch recombination to occur is expression of the gene for activation-induced deaminase (AID *[17–19]*). The precise action of this protein and its connection to the obligatory switch transcripts has not yet been clarified. It is clear, however, that switch transcription and AID expression result in double-stranded breaks of switch regions *(20)*. These double-stranded breaks are then repaired by systems involving DNA-dependent protein kinase (DNA-PK *[21–23]*), and perhaps the Nbs1/Mre11/Rad50 system *(20)* is involved in looping out and deleting the DNA between the two switch regions *(24)*.

Here, we describe methods for the analysis of Ig class switching in murine B cells isolated ex vivo and activated in vitro. These methods allow isolation and stimulation of murine naive splenic B lymphocytes, their monitoring for proliferation by labeling with succinimidyl ester (CFSE), their induction to

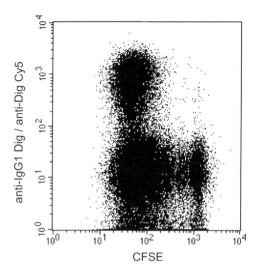

Fig. 1. Surface IgG1/CFSE staining of LPS plus IL-4 stimulated CD43-depleted B-cells after 4 d of culture.

undergo class switch recombination as such and specific targeting to distinct classes, and their cytometric evaluation by staining of intracellular, surface, or secreted Ig, as is also required for the isolation of B cells that had switched to a particular class of Ig, for further cellular and molecular analysis **(Fig. 1)**.

On the genomic level, the digestion–circularization polymerase chain reaction (DC-PCR) technique can be applied to estimate the amount of S region recombination that occurs in the course of class switching in primary B lymphocytes *(25)*. Secreted antibodies in culture supernatants can be measured with an enzyme-linked immunosorbent assay (ELISA *[26]*). The number of antibody (Ab) secreting cells can be quantified by plaque-forming cell assay and enzyme-linked immunospots (ELISPOTs; *[27,28]*). It should be noted, however, that the evaluation of Ig class switching on the level of Ig secretion, or secreting cells, can be grossly misleading because it requires not only the Ig class switch but also the switch from membrane bound Ig to secreted Ig, which is an independent molecular event. A better way to quantify the frequency of Ig class-switched B cells and plasma cells is their cytometric identification and enumeration according to staining of surface and intracellular Ig, as described here. An interesting option for the identification, quantification, and isolation of viable plasma cells according to antibodies they secrete is the cytometric secretion assay *(29,30)*. Cytometric evaluation offers the additional benefit of adding more parameters to the analysis, that is, the determination of proliferative history according to staining with carboxyfluorescein *(31)*.

2. Materials

2.1. Cell Culture

1. Murine spleen.
2. Complete medium: RPMI-1640 medium plus glutamine (PAA Laboratories, cat. no. E15-840) containing 100 U/mL penicillin and 100 µg/mL streptomycin (PAA Laboratories, cat. no. P11-010), 10% fetal calf serum (FCS; heat inactivated, Sigma, cat no. F7524), and 5×10^{-5} M 2-mercaptoethanol (ME) (Sigma, cat. no. M7552); store at 4°C.
3. 10 mg/mL *Escherichia coli* LPS serotype 055:B5 (Sigma, Taufkirchen, Germany) dissolved in phosphate-buffered saline (PBS)/bovine serum albumin (BSA). Store at –20°C.
4. Culture supernatant of murine myeloma cell line P3-X63 Ag.8.653 transfected with murine IL-4 complementary DNA (cDNA) or recombinant murine IL-4.
5. Cell strainers, 70 µm nylon (Becton Dickinson).
6. MidiMACS separation unit (Miltenyi Biotec; Bergisch Gladbach, Germany).
7. CD43 or CD5-microbeads; store protected from light at 4°C. Do not freeze (Miltenyi Biotec, Bergisch Gladbach, Germany).
8. MACS-LS$^+$/VS$^+$ column (Miltenyi Biotec, Bergisch Gladbach, Germany).
9. PBS, autoclaved and stored at 4°C.
10. PBS with 0.5% BSA (PBS/BSA), filter sterilized (0.2 µm); store at 4°C.
11. PBS/BSA plus 0.02% sodium azide (PBS/BSA/NaN$_3$), filter sterilized (0.2 µm); store at 4°C.

2.2. Cytometric Analysis

1. PBS/BSA/NaN$_3$ with 0.5% saponin (from Saponaria, Sigma, Taufkirchen, Germany) (saponin-buffer).
2. Fixative: 4% formaldehyde (Merck) in PBS.
3. Carboxyfluorescein diacetate, CFSE stock solution (25 mM) dissolved in tissue-culture grade dimethyl sulfoxide (DMSO). Store in suitably sized aliquots at –20°C.
4. Histopaque®-1083 (Sigma). Store at 4°C.
5. Antimouse FcγRIIb mAb (2.4G2) *(32)*.
6. Anti-IgG1-Dig *(16)*; store at –70°C.
7. Anti-IgG3-biotin clone: R40-82 (BD Bioscience; Heidelberg, Germany); store at 4°C.
8. Antistreptavidin-Cy5 (Amersham Pharmacia Biotech, cat. no. PA45001); store at –70°C.
9. Anti-Dig-Cy5; store at –70°C. Antidigoxigenin Fab fragments (Roche; Mannheim, Germany) were conjugated through Cy5™ *bis*-N-hydroxysuccinimide (NHS) ester (Amersham Bioscience, cat. no. PA15000, Freiburg, Germany) according to manufacturer's instructions.
10. 0.1 mg/mL propidium iodide (PI) stock solution; store at 4°C.
11. FACSCalibur™ and CELLquest™ research software (Becton Dickinson).

3. Methods

3.1. Isolation of Primary Naive Mature B Lymphocytes

Cell cultures of total splenic cells can be used to stimulate B cells to undergo Ab class switch recombination; however, the isolation of naive B cells prior to the stimulation of the B cells is recommended to ensure that other cells will not interfere with the stimulation. In addition, effector and memory B cells, which have already undergone switch recombination, should be removed.

Preparation of a lymphoid cell suspension from murine spleens (*see* **Note 1**).

1. Wet cell strainer with 10 mL cold (4°C) PBS in a 35-mm Petri dish.
2. Isolate murine spleen under sterile conditions.
3. Press spleen through cell strainer in cold PBS using the plunger from a 5-mL syringe (*see* **Note 2**).
4. Wash plunger and cell strainer with cold PBS and transfer the cell suspension from the Petri dish to a sterile 50-mL centrifuge tube.
5. Centrifuge cells 10 min at 350g at 4°C.
6. Remove supernatant carefully, leaving approx 500 µL of PBS.
7. Resuspend the cells by flicking the centrifuge tube, and add 40 mL cold PBS/BSA.
8. Optional: to remove fat and cell clumps pass cell suspension through a cell strainer.
9. Isolate B cells by depletion, using anti-CD43 or anti-CD5 microbeads from Miltenyi Biotec (*see* **Notes 3–5**).
10. Wash cells, remove supernatant completely, and resuspend cell pellet in 90 µL of PBS/BSA per 10^7 total cells (for fewer cells, use same vol).
11. Add 10 µL of MACS CD43 microbeads per 10^7 total cells, mix well, and incubate for 15 min at 6–12°C (for fewer cells use same volume). **Caution:** Do not put on ice.
12. Stop cell labeling by adding 45 mL of PBS/BSA.
13. Centrifuge cells for 10 min at 350g and 4°C.
14. Equilibrate the prechilled LS$^+$/VL$^+$ column with 5 mL cold PBS/BSA; let buffer pass column by gravity flow (*see* **Note 6**).
15. For depletion of cells apply a 23-gage × 0.6-mm needle to LS$^+$/VL$^+$ column, and check function of the column by adding 500 µL PBS/BSA.
16. Resuspend cells in 3 mL PBS/BSA, and apply the cells to the prepared MACS column. (Use a maximum cell number of 10^8 cells per 500 µL of buffer).
17. For analyzing the efficiency of MACS sort, remove an aliquot and store on ice, as a fraction from the original cell suspension.
18. Wash column two times with 3 mL PBS/BSA.
19. Remove needle carefully from the column and remove the column from the magnetic field. Place column on a suitable collection tube, pipet 5 mL of buffer onto the column, and firmly flush out the positive fraction using the plunger supplied with the column.

20. Check efficiency of sort by staining an aliquot of the presort, negative and positive fractions.

3.2. Monitoring of B-Cell Proliferation by Labeling With CFSE

This step can be done before or after B-cell enrichment.

To analyze the proliferation history of B cells that have undergone Ab class switch recombination, cells are labeled with CFSE. CFSE passively diffuses into cells and spontaneously and irreversibly binds to cellular proteins by reacting with lysine side chains and other available amines. When cells divide, CFSE labeling is distributed equally among the daughter cells. Each successive generation in a population of proliferating cells is marked by a halving of cellular fluorescence intensity, which can easily be discriminated by flow cytometry. Six to eight generations of asynchronously dividing B lymphocytes can be identified in this way *(33)*.

1. Wash cells with PBS to remove proteins.
2. Resuspend the cells gently in room temperature PBS at a cell concentration of 1×10^7 cells/mL.
3. Add CFSE at a final concentration of 5 μM to the cells, and incubate for 2–5 min at room temperature.
4. Stop labeling reaction by adding 40 mL of PBS/BSA or medium containing FCS.
5. Centrifuge cells for 10 min at 350g and 4°C; resuspend in warm medium.

3.3. Polyclonal Activation of B Lymphocytes Induces Them to Undergo Class-Switch Recombination

LPS is a polyclonal activator of B cells. In a low-density culture (i.e., 5×10^5 cells/mL or lower) murine splenic B cells can be stimulated with LPS alone to undergo class-switch recombination to IgG2b and IgG3. Activating B lymphocytes with LPS in the presence of IL-4 induces Ab class switch recombination to IgG1 and IgE.

3.3.1. Induction of Spleen B Cells to Undergo Class Switch to IgG2b and IgG3 by LPS

1. Count cells manually with a hemacytometer (*see* **Note 7**).
2. Add LPS at 40 µg/mL to complete medium, and filter sterilize the medium (*see* **Note 8**).
3. Culture B cells at a starting density of 2×10^5 cells/mL.
4. Incubate cells in a humidified 37°C, 5% CO_2 incubator.
5. Switching to IgG3 is detectable by surface staining after 60 h and reaches its maximum after 3–4 d.

3.3.2. Induction of Spleen B Cells to Undergo Class Switch to IgG1 and IgE

1. Add LPS at 40 µg/mL to complete medium and filter sterilize the medium (*see* **Note 9**).
2. Add 5–20 ng/mL of recombinant IL-4 to the medium containing LPS.
3. Culture B cells at starting density of 2×10^5 cells/mL.
4. Incubate cells in a humidified 37°C, 5% CO_2 incubator.
5. Switching to IgG1 is detectable by surface staining after 60 h and reaches its maximum after 3–4 d.

3.4. Evaluation of Immunoglobulin Class-Switch Recombination Through Intracellular and Surface Immunoglobulin Isotype Staining

3.4.1. Intracellular Staining of Immunoglobulin Isotypes

Owing to the large amount of immunoglobulins in the cytoplasma of plasmablasts, intracellular staining is more sensitive compared to surface staining. However, intracellular staining requires the cells to be fixed and permeabilized *(34)*. To remove dead cells a Ficoll gradient is recommended. The fixation makes all cells permeable to PI. Therefore PI cannot be used to exclude dead cells.

3.4.1.1. ISOLATION OF VIABLE CELLS

1. Prewarm Histopaque-1083 to 20°C.
2. Transfer cells in media into a 50-mL tube.
3. Carefully underlay the cells with the Histopaque-1083.
4. Centrifuge for 20 min at 600–700*g* and 20°C (accelerate and decelerate slowly).
5. Remove cells from the interphase and wash twice with PBS/BSA.

3.4.1.2. FIXATION PROTOCOL

1. Wash cells once with PBS to remove protein.
2. Resuspend cells in PBS at 2×10^6/mL.
3. Add formaldehyde/PBS to end conc. of 2%.
4. Incubate for 20 min at room temperature.
5. Wash the cells twice with PBS.
6. Resuspend cells in PBS/BSA/NaN$_3$ at $1–2 \times 10^6$ cells/mL, and store at 4°C in the dark until staining (*see* **Note 10**).

3.4.1.3. INTRACELLULAR STAINING

1. Use about 1.5×10^6 cells for each staining sample.
2. Spin down cells at 350*g* for 10 min.
3. Resuspend the cell pellet in 50–100 µL of permeabilization buffer containing the intracellular Ab (*see* **Note 11**).

4. Incubate at 4°C for 10 min in the dark.
5. Wash cells with 1 mL permeabilization buffer.
6. Wash cells with PBS/BSA/NaN$_3$.
7. Resuspend cells in 400 μL of PBS/BSA/NaN$_3$ and transfer to a fluorescence-activated cell sorting (FACS) tube.
8. FACS analysis.

3.4.2. Surface Staining

1. Use about 1.5×10^6 cells for each staining sample.
2. Spin down cells at 350g for 10 min.
3. Resuspend the cell pellet in 50 μL of PBS/BSA containing 50 μg/mL antimouse FcγRIIb Ab to block unspecific binding of the staining Ab.
4. Incubate for 15 min on ice.
5. Add 50 μL PBS/BSA containing staining anti-Ig Ab to cells (*see* **Notes 11** and **12**).
6. Incubate for 10 min at 4°C in the dark.
7. Wash cells with 1 mL PBS/BSA.
8. Resuspend cells in 400 μL of PBS/BSA and transfer to a FACS tube.
9. Immediately before FACS analysis add PI at an end conc. 1 μg/mL to cells to exclude dead cells.
10. FACS analysis.

4. Notes

1. Working quickly and keeping the cells at 4°C is important for cell survival.
2. Only press spleen through cell strainer; do not grind it.
3. Never incubate or put microbeads on ice.
4. Alternative techniques to isolate mature B cells: isolate B cells directly via anti-B220 or anti-CD19 microbeads from Miltenyi Biotec. T cells can also be depleted with antibodies against T-cell-specific surface antigens and complement. Small resting B cells can be selected through a discontinuous Percoll density gradient.
5. It is not necessary to remove red blood cells.
6. Cool magnet and LS$^+$/VS$^+$column to 4°C in the refrigerator before use.
7. Dilute a sample of the cell suspension in 3% acetic acid to lyse red blood cells, and count cells manually with a hemacytometer.
8. 10 mg/mL LPS stock solution should not be filter sterilized at this concentration. Therefore, the medium must be filter sterilized after LPS has been added at the final concentration. LPS concentration used in the assay depends on the mouse strain; a number of different mouse strains can be used; for example, C57BL.10, Balb/c, or CBA. These and many other strains are high responders to LPS. However C3H/HeJ are nonresponders to LPS and should not be used for LPS stimulation.
9. LPS stimulation can be replaced by CD40 stimulation using stimulatory Abs; for example 10 μg/mL anti-CD40 (IC10) or CD40 ligand.
10. If the fixed cells are used directly for intracellular staining, put the cells at 4°C in the dark for at least 15 min; otherwise still-reactive formaldehyde bound to the cells can couple antibodies to the cells, thus increasing staining background.

11. Titration of the staining Abs for optimal staining performance is recommended.
12. When staining for surface Ig the following should be considered: Ig of a second isotype could be derived from rare class switch variants that secrete this Ig, which could be taken up by nonswitched cells via Fc receptors. This problem can occur especially for IgE staining. Therefore, antibodies against the CH3 domain of IgE should be used for surface staining.

References

1. Nossal, G. J. V., Szenberg, A., Ada, G. L., and Austin, C. M. (1964) Single cell studies on 19S antibody production. *J. Exp. Med.* **119**, 485–502.
2. Honjo, T. and Kataoka, T. (1978) Organization of immunoglobulin heavy chain genes and allelic deletion model. *Proc. Natl. Acad. Sci. USA* **75**, 2140–2144.
3. Kataoka, T., Kawakami, T., Takahashi, N., and Honjo, T. (1980) Rearrangement of immunoglobulin γ1-chain gene and mechanism for heavy-chain class switch. *Proc. Natl. Acad. Sci. USA* **77**, 919–923.
4. Rabbitts, T. H., Forster, A., Dunnick, W., and Bentley, D. L. (1980) The role of gene deletion in the immunoglobulin heavy chain switch. *Nature* **283**, 351–356.
5. Kataoka, T., Miyata, T., and Honjo, T. (1981) Repetitive sequences in class switch recombination regions of immunoglobulin heavy chain genes. *Cell* **23**, 357–368.
6. Radbruch, A., Burger, C., Klein, S., and Müller, W. (1986) Control of immunoglobulin class switch recombination. *Immunol. Rev.* **89**, 69–84.
7. Sablitzky, F., Radbruch, A., and Rajewsky, K. (1982) Spontaneous immunoglobulin class switching in myeloma and hybridoma cell lines differs from physiological class switching. *Immunol. Rev.* **67**, 59–72.
8. Nakamura, M., Kondo, S., Sugai, M., Nazarea, M., Imamura, S., and Honjo, T. (1996) High frequency class switching of an IgM+ B lymphoma clone CH12F3 to IgA+ cells. *Int. Immunol.* **8**, 193–201.
9. Kearney, J. F., Cooper, M. D., and Lawton, A. R. (1976) B cell differentiation induced by lipopolysaccharide IV: development of immunoglobulin class restriction in precursors of IgG synthesizing cells. *J. Immunol.* **117**, 67–72.
10. Anderson, J., Coutinho, A., Lernhardt, W., and Melchers, F. (1977) Clonal growth and maturation to immunoglobulin secretion in vitro of every growth-inducible B lymphocyte. *Cell* **10**, 27–36.
11. Esser, C. and Radbruch, A. (1990) Immunoglobulin class switching: molecular and cellular analysis. *Annu. Rev. Immunol.* **8**, 717–735.
12. Stavnezer, J. and Sirlin, S. (1986) Specificity of immunoglobulin heavy chain switch correlates with activity of germline heavy chain genes prior to switching. *EMBO J.* **5**, 95–102.
13. Yancopoulos, G. D., DePinho, R. A., Zimmermann, K. A., Lutzker, S. G., Rosenberg, N., and Alt, F. W. (1986) Secondary genomic rearrangement events in pre-B cells: VHDJH replacement by a LINE-1 sequence and directed class switching. *EMBO J.* **5**, 3259–3266.

14. Radbruch, A., Muller, W., and Rajewsky, K. Class switch recombination is IgG1 specific on active and inactive IgH loci of IgG1-secreting B-cell blasts. *Proc. Natl. Acad. Sci. USA* **83**, 3954–3957.

15. Jung, S., Rajewsky, K., and Radbruch, A. (1993) Shutdown of class switch recombination by deletion of a switch region control element. *Science* **259**, 984–987.

16. Hein, K., Lorenz, M. G., Siebenkotten, G., Petry, K., Christine, R., and Radbruch, A. (1998) Processing of switch transcripts is required for targeting of antibody class switch recombination. *J. Exp. Med.* **188**, 2369–2374.

17. Muramatsu, M., Sankaranand, V. S., Anant, S., et al. (1999) Specific expression of activation-induced cytidine deaminase (AID), a novel member of the RNA-editing deaminase family in germinal center B cells. *J. Biol. Chem.* **274**, 18,470–18,476.

18. Muramatsu, M., Kinoshita, K., Fagarasan, S., Yamada, S., Shinkai, Y., and Honjo, T. (2000) Class switch recombination and hypermutation require activation-induced cytidine deaminase (AID), a potential RNA editing enzyme. *Cell* **102**, 553–563.

19. Petersen-Mahrt, S. K., Harris, R. S., and Neuberger, M. S. (2002) AID mutates *E. coli* suggesting a DNA deamination mechanism for antibody diversification. *Nature* **418**, 99–103.

20. Petersen, S., Casellas, R., Reina-San-Martin, B., et al. (2001) AID is required to initiate Nbs1/gamma-H2AX focus formation and mutations at sites of class switching. *Nature* **414**, 660–665.

21. Rolink, A., Melchers, F., and Andersson, J. (1996) The SCID but not the RAG-2 gene product is required for Sμ-Sε heavy chain class switching. *Immunity* **5**, 319–330.

22. Manis, J. P., Dudley, D., Kaylor, L., and Alt, F. W. (2002) IgH class switch recombination to IgG1 in DNA-PKcs-deficient B cells. *Immunity* **16**, 607–617.

23. Bosma, G. C., Kim, J., Urich, T., et al. (2002) DNA-dependent protein kinase activity is not required for immunoglobulin class switching. *J. Exp. Med.* **196**, 1483–1495.

24. von Schwedler, U., Jack, H. M., and Wabl, M. (1990) Circular DNA is a product of the immunoglobulin class switch rearrangement. *Nature* **345**, 452–456.

25. Chu, C. C., Max, E. E, and Paul, W. E. (1993) DNA rearrangement can account for in vitro switching to IgG1. *J. Exp. Med.* **178**, 1381–1390.

26. Kühn, R., Rajewsky, K., and Müller, W. (1991) Generation and analysis of interleukin-4 deficient mice. *Science* **254**, 707–710.

27. Sedgwick, J. D. and Holt, P.G. (1983) A solid-phase immunoenzymatic technique for the enumeration of specific antibody-secreting cells. *J. Immunol. Methods* **57**, 301–309.

28. Sedgwick, J. D. and Holt, P. G. (1986) The ELISA-plaque assay for the detection and enumeration of antibody-secreting cells. *J. Immunol. Methods* **87**, 37–44.

29. Manz, R., Assenmacher, M., Pfluger, E., Miltenyi, S., and Radbruch, A. (1995) Analysis and sorting of live cells according to secreted molecules, relocated to a cell-surface affinity matrix. *Proc. Natl. Acad. Sci. USA* **92**, 1921–1925.

30. Manz, R. A., Lohning, M., Cassese, G., Thiel, A., and Radbruch, A. (1998) Survival of long-lived plasma cells is independent of antigen. *Int. Immunol.* **10**, 101,703–101,711.

31. Hodgkin, P. D., Lee, J. H., and Lyons A. B. (1996) B cell differentiation and isotype switching is related to division cycle number. *J. Exp. Med.* **184,** 277–281.

32. Unkeless, J. C. (1979) Characterization of a monoclonal antibody directed against mouse macrophage and lymphocyte Fc receptors. *J. Exp. Med.* **150,** 580–596.

33. Lyons, A. B. and Parish, C. R. (1994) Determination of lymphocyte division by flow cytometry. *J. Immunol. Methods* **171,** 131–137.

34. Schmitz, J., Assenmacher, M., and Radbruch, A. (1993) Regulation of T helper cell cytokine expression: functional dichotomy of antigen-presenting cells. *Eur. J. Immunol.* **23,** 191–199.

12

In Vitro and In Vivo Assays of B-Lymphocyte Migration

Chantal Moratz and John H. Kehrl

Summary

The development and functional activities of B lymphocytes critically depend on their migratory capacity. Both in vitro and in vivo assays can be used to assess the migratory ability of B cells. Filter-based transwell assays measure both spontaneous and chemoattractant-induced cell migration in vitro. Flow cytometric analysis of labeled cell subsets present in the input and migratory cells from transwell assays allows the assessment of the migratory capacity of specific cell types in a complex mixture of cells. A similar approach can be used to assess the migratory capacity of B cells transfected with a green fluorescent fusion protein. Despite the success of the transwell assay, attempts to image directionally migrating B cells have been limited. The assessment of the in vivo migratory capacity of genetically modified or pharmacologically treated B cells usually involves their adoptive transfer into recipient mice. The localization of transferred cells in lymphoid organs can be determined by immunohistochemistry and/or flow cytometric analysis of harvested tissues or cells. Because the proper localization of B cells in lymphoid organs is a multistep process, abnormal positioning can occur even in the absence of an intrinsic defect in cell migration.

Key Words

Chemotaxis; chemokine; B lymphocyte; migration; adhesion; lymphoid organ; imaging; green fluorescent protein.

1. Introduction

B lymphocytes passage from the blood through B-cell-rich compartments in secondary lymphoid organs surveying for antigen. On antigen encounter, B cells change their migratory destination moving to locations that favor interactions with antigen-activated T cells. B cells enter B-cell follicles and on interaction with antigen-activated T cells can either directly differentiate into plasma cells or enter the germinal center *(1,2)*. Within the germinal center B cells acquire

From: *Methods in Molecular Biology, vol. 271: B Cell Protocols*
Edited by: H. Gu and K. Rajewsky © Humana Press Inc., Totowa, NJ

high-affinity antigen receptors and eventually leave as memory B cells or as plasma cell precursors *(3)*. The plasma cell precursors either remain in the lymphoid tissue of origin or leave to localize in the bone marrow or the gut *(4,5)*. Orchestrating many of these migratory patterns are chemoattractant molecules produced by other cell types, which serve as signposts for the migrating B lymphocytes *(6–11)*.

Although resting B lymphocytes have a spherical morphology, within seconds of exposure to a chemoattractant the cells become irregular with rapid protrusions and retractions of ruffles *(12)*. The microfilament system in individual cells rapidly reorganizes and filamentous actin (F-actin) accumulates adjacent to the cell membrane. During the next several minutes the cell adopts a polarized (head-to-tail) morphology in which the polymerized actin is concentrated at the leading edge of the cell, a region termed the *lamellipodium*. Behind this is the cell body, including the nucleus and organelles, which often tapers to a tail called the *uropod*. The cell moves in the direction of the leading edge.

Because of the complexities of B lymphocyte locomotion, multiple assays need to be employed to fully capture the responses to chemoattractants. Early biochemical events following the exposure to chemoattractants can be assayed, including measurements of calcium influx, the translocation to the leading edge of appropriate green fluorescent fusion proteins, and the formation of F-actin. Filter assays are popular for determining the portion of a cell population responsive to a chemoattractant; however, they provide little information on the speed and velocity of the responding cells (**Fig. 1**). Visual assays, including orientation assays in a chemoattractant gradient, can provide direct evidence of chemoattractant responses. Although neutrophils can be readily visualized migrating under agar, in similar assays B lymphocytes fail to enter the space between the microscope slide and agar. Although rarely used to visualize B-cell migration, three-dimensional matrices (made from collagen or fibrin) have often been used to study T-cell migration. These in vitro assays often need to be supplemented with in vivo assays that measure the capacity of B lymphocytes to migrate to specific sites and compartments. Here, several specific methods for the analysis of the mouse B-lymphocyte responses to chemokines are detailed.

2. Materials

2.1. Isolation of B Cells From Murine Lymphoid Tissue

1. 1X Phosphate-buffered saline (PBS), pH 7.4.
2. RPMI-1640 media (Gibco).
3. Bovine calf serum (BCS).
4. 40 μm Nylon mesh (Falcon).
5. 27- and 19-Gage needles.

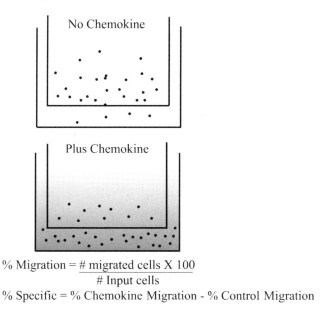

$$\% \text{ Migration} = \frac{\# \text{ migrated cells} \times 100}{\# \text{ Input cells}}$$

$$\% \text{ Specific} = \% \text{ Chemokine Migration} - \% \text{ Control Migration}$$

Fig. 1. Standard polycarbonate filter chamber assay. The upper schematic is of cells within the filter chamber, which is placed in a well. With media alone, a small percentage of cells spontaneously migrate through the filter into the lower chamber. The lower schematic is of the chamber assay set-up with chemokine added to the well and the filter chamber suspended in the well. Cells responding to chemokine will migrate through the filter and into the lower chamber. The filter chamber is removed at the termination of the assay, and the cells in the lower chamber are collected and enumerated.

6. 12-mL syringe.
7. AKC lysing buffer.
8. Bovine serum albumin (BSA) fraction V.
9. Flow cytometry cell sorter.
10. Magnetic negative selection systems: Dynal, Stem Cell Technologies, Miltenyi Biotech–MACS.
11. Fluorescence-activated cell sorting (FACS) buffer: PBS, pH 7.4, 1% BSA.

2.2. In Vitro Assessment of B-Lymphocyte Migrational Ability

1. 5-μm Transwell chamber, Costar (cat. no. 3421).
2. RPMI media.
3. BCS.
4. Chemokine.
5. FACS buffer: PBS, pH 7.4, 1% BSA.
6. Flow cytometer.

7. Staining buffer: HEPES-buffered sterile saline (HBSS), 10 mM HEPES buffer, pH 7.4, 1% BCS.
8. Fluorescein isothiocyanate (FITC)-phalloidin (Sigma or Molecular Probes).
9. Paraformaldehyde (PFA).
10. Saponin.
11. Fibronectin (Sigma, cat. no. F0895).
12. Zigmond Chamber Z02 (Neuro Probe).
13. Inverted microscope with heat and humidity chamber.
14. ImagePro software.

2.3. In Vivo Assessment of Cell-Migration Patterns

1. 5-and-6-Carboxyfluorescein diacetate succinimidyl ester (CFSE; Molecular Probes).
2. Lipophilic tracers (Molecular Probes).
3. PBS.
4. HBSS.
5. Fetal calf serum (FCS).
6. Flow cytometer.
7. PFA.
8. Sucrose.
9. Tissue-Tek® O.C.™ (Sakura/Fisher).
10. Section blocks (Fisher).

3. Methods
3.1. Isolation of B Cells From Murine Lymphoid Tissue
3.1.1. Lymphocyte Isolation

1. Mice are sacrificed and lymphoid tissue harvested. Peripheral lymph nodes, including subiliac, proper axillary, accessory axillary and cervical, mesenteric lymph nodes, Peyer's patches, and/or spleen are harvested.
2. The tissues are disrupted with forceps or needles in RPMI/10% BCS and the cells dissociated by gentle teasing followed by filtering through a 40 μm nylon mesh (Falcon, BD) to remove connective tissue cells.
3. Peritoneal cavity cells are harvested by injecting 7–10 mL of ice-cold RPMI/10% BCS into the peritoneal cavity using a 27-gage needle and removing the cell wash with a 19-gage needle.
4. Bone marrow cells are harvested by dissecting the femur and tibia, removing excess tissue, cutting the ends off the bones, and then flushing the marrow out with RPMI/10% BCS through both ends of the bones.
5. The bone marrow and spleen cells are depleted of red blood cells with AKC buffer and subsequent washing.
6. The cell suspensions are centrifuged at 300g for 5 min at 4°C, the supernatants removed, and the pellet resuspended in RPMI/10% BCS.
7. A count of the suspensions is made to determine cell number per milliliter.

3.1.2. Isolation of B Cells or Specific B-Cell Subsets by FACS Cell Sorting

1. Cells are pelleted and resuspended in FACS buffer: PBS, pH 7.4/1% BSA fraction V.
2. Cells are stained for specific B-cell markers to identify a specific subset of cells. For example, CD23, CD21, and B220 (Pharmingen, BD) can be used to identify newly transitional, follicular, and marginal zone B cells in the spleen.
3. After a brief incubation at 4°C, the cells are washed three times with FACS buffer and resuspended at 1×10^7/mL in FACS buffer to remove the excess monoclonal antibody (MAb).
4. The cells are then sorted by flow cytometry. Obtaining significant numbers of cells with this method may be problematic; however, it is the best method to collect a specific B-cell subset (*13*).
5. After sorting, the cells are incubated at 37°C for 1 h prior to any assay.

3.1.3. Isolation of B Cells by Negative Selection Using Magnetic Separation

1. B cells can be purified from bulk population by removing non-B cells. There are multiple methods to achieve this purification; this subheading describes one such procedure (refer to methods published by Dynal, Inc., Miltenyi Biotec–MACS, Stem Cell Technologies).
2. Cells are pelleted and resuspended in FACS buffer. Cells are stained with biotinylated MAb to non-B-cell markers such as CD4, CD8, GR-1, Mac-1, Ter119, CD11c, and DX5 in FACS buffer. These MAbs will bind T cells, macrophages, erythrocytes, NK cells, granulocytes, and dendritic cells.
3. Cells are incubated at 4°C for 15 min with 2X PBS and resuspended in FACS buffer. Dynabeads M-280 streptavidin beads (Dynal) are washed with PBS twice and with FACS buffer twice. The bead-to-cell ratio is determined by manufacturer's protocols (*14*).
4. The washed magnetic beads are added to the cell suspensions and incubated at 4°C, slowly rotating for 15 min. The suspensions are then attached to the magnet (Dynal) and allowed to separate.
5. The nonadherent suspension is collected and reapplied to the magnetic source, and again the nonadherent cells are collected. This population is washed, counted, resuspended in RPMI/10% BCS, and incubated at 37°C for 1 h prior to any assay.

3.2. In Vitro Assessment of B-Lymphocyte Migrational Ability

3.2.1. Standard Polycarbonate Filter Assay

1. Harvest the B cells, make single-cell preparation of the cells, and count.
2. Depending on the cell type (primary vs cell lines), the cells should be at 1×10^7/mL to 1×10^6/mL. For mouse B cells a 5-μm Costar transwell chamber (no. 3421) is a good choice (*see* **Note 1**).
3. Set up chemotaxis chambers:
   a. In bottom of the chamber add 600 μL of RPMI/10% FCS media with chemoattractant (i.e., 50–100 ng/mL of CXCL12 or 1000 ng/mL of CXCL13). A

control chamber should be set up without chemokine to measure spontaneous migration.

b. Place insert into the chamber and add 100 µL of cells to the insert.

4. Incubate chamber for 2–3 h at 37°C in a standard CO_2 incubator.

5. Following the incubation period remove the insert and collect cells in the lower chamber for analysis.

6. Spin the cells down; at this point they can be stained to subtype the population or resuspended in 300 µL FACS buffer to access migration.

7. On the flow cytometer collect data for 1 min for each sample at the high flow rate (important for calculations).

8. For controls use a 100-µL aliquot from the input cell sample, and run on flow the same as for the chemotaxis samples.

9. The percentage of migrating cells can be calculated by dividing the number of cells collected in the lower chamber by the number of input cells × 100. The percentage of cells that migrated in response to chemokine is the difference between the percentages obtained from the chemokine-loaded and control chambers. By determining the percentage of a specific cell subset (i.e., marginal zone, follicular B cells, transitional B cells, or plasma cells) in the input and migrated cell populations, the migratory capacity of that subset can be assessed (*15–19; see* **Note 2**).

3.2.2. Assay of Chemokine-Induced F-Actin Polymerization by Staining With FITC–Phalloidin

1. Cells previously stained for phenotypic marker are washed and warmed to 37°C for 10 min. The cells at 1.25×10^6 cells per milliliter suspended in HBSS/10 mM HEPES/1% BCS with Ca^{++} and Mg^{++}.

2. Chemokines are added at appropriate concentrations and the cells vortexed.

3. At varying time-points, a solution of 0.8 U/mL FITC–phalloidin (Molecular Probes), saponin 0.1%, and 23% PFA in PBS is diluted 1:5 into the sample of chemokine-stimulated cells.

4. The fixed and permeabilized cells are incubated on ice for 10 min, washed twice with PBS containing 1% BSA and 0.1% saponin, and resuspended in PBS.

5. Data is collected on a FACSCalibur (BD Biosciences); phalloidin staining is quantified on a linear scale *(20,21)*.

3.2.3. Use of a Zigmond Chamber to Monitor Chemoattractant-Induced B-Lymphocyte Polarization

1. Coat the cover slip with fibronectin (Sigma, cat. no. F0895, 1:200 dilution of 1% fibronectin in PBS) for 1 h at room temperature.

2. Rinse the cover slip with PBS.

3. Add cells to cover slip in a humidified chamber (Neuro Probe, no. Z02) at 37°C in 5% CO_2 for 30 min (cells were previously suspended in RPMI 2% BCS at a concentration of 1×10^6 cells per milliliter and prewarmed to 37°C; *see* **Note 3**).

4. Rinse cell layer on cover slip with migration media and drain excess liquid off slip.

No CXCL12   Exposed to CXCL12 for 15 Minutes

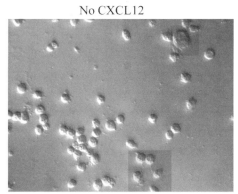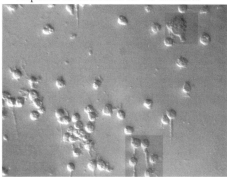

Fig. 2. Chemokine induced morphology change. Still frame images from time-lapse experiment with the Zigmond chamber examining the response of variant line, a sub-clone of the HS-Sultan human B-cell line. The still frame on the *left* is of cells on the bridge of the chamber when both grooves of chamber contain media alone. The still frame on the right is of the same cells on the bridge of the chamber after CXCL12 is introduced into one groove of the chamber, creating a chemokine gradient. The darkened squares in the two frames highlight cells with substantial morphologic changes in response to chemokine.

5. Invert the cover slip onto the chamber so that the cells line over the bridge, then tighten clamps (*see* **Note 4**). Add media to open ends of grooves (*see* **Notes 5** and **6**).

6. Observe the cells on the bridge using an inverted microscope. For long-term imaging the Zigmond chamber needs to be maintained at 37°C, ideally in a 5% CO^2 and humidified atmosphere.

7. To observe chemokine-induced cell polarization, wick away the liquid from one side of the bridge while simultaneously replacing it with media containing chemokine. It will take approx 30 min to generate a stable gradient across the bridge, although morphologic changes in the B cells can be observed as early as several minutes following the addition of chemokine.

8. The morphological changes can be documented by time-lapse photography, or alternatively, the chamber can be placed in the incubator for 30 min to allow the B cells to polarize and the cells to be photographed through an inverted microscope (**Fig. 2**).

3.3. In Vivo Assessment of Cell-Migration Patterns

3.3.1. Cell-Labeling Choices

Two types of dyes can be used to trace cells in vivo. They are intracellular dyes, such as CFSE, or lipophilic tracers, such as the lipophilic carbocyanines

DiI, DiO, DiD, and DiR. CFSE is stably incorporated into the cell and can be used to monitor short- and long-term (up to several months) migration and cell division. Lipophilic dyes are incorporated into the cell membrane. These tracers can be used for relatively long-term cell tracing in vivo and in vitro. If possible, the donor origin of the transferred lymphocytes should be monitored using mismatched allelic markers such as Ly5a/b or Ly9.1

3.3.2. Basic Labeling Procedure With CFSE

1. For high cell numbers, resuspend the cells at 50×10^6/mL in either PBS (without serum) or HBSS (without serum). For low cell numbers resuspend the lymphocytes in PBS containing 5% FCS at 0.5–10×10^6 cells per milliliter. The FCS is needed at the low cell concentration to buffer the toxic effect of the concentrated dye.
2. Dilute the stock CFSE dye (Molecular Probes) to 50 μM in PBS. Add 25 μL of dye solution per milliliter of cells—final concentration is 1.25 μM—and mix rapidly. Incubate 5 min at room temperature and wash with 10 vol of PBS 5% FCS; centrifuge cells 5 min at 300g at 20°C. Remove supernatant and repeat wash three times.

3.3.3. Basic Labeling Procedure for Lipophilic Tracers

1. Resuspend cells at 5–10×10^6 cells per milliliter in buffered salt solution (HBSS).
2. Add dye (Molecular Probes) to cell solution so final dye concentration is 0.5–50 μM, and incubate at room temperature for 5 min.
3. Wash out the dye by centrifuging the cells for 5 min at 300g at 20°C. Remove supernatant and add 10 vol of HBSS with 5% FCS. Repeat the wash three times.

3.3.4. In Vivo Transfer of Labeled Lymphocytes

Resuspend labeled lymphocytes in tissue-culture medium lacking serum and inject intravenously into lateral tail vein of a mouse. Use cell concentration of 1–40×10^6 cells in 0.2 mL vol (**22**).

3.3.5. Analysis of Tissue Cells by Flow Cytometry

1. Lymphoid organ cells of recipient mice are typically harvested 30–90 min after transfer, counted, and subjected to flow cytometric analysis. If the analysis of the transferred cells is to be delayed, CSFE labeling is a better choice. If the analysis is to occur more than 24 h after transfer, a higher concentration of CSFE should be used (5 μM) in the initial labeling. Despite their loss of fluorescence with time, CSFE-labeled cells can be detected for several months following transfer.
2. Cells from lymphoid tissues of recipient mice are harvested and prepared as described in **step 1**.
3. Cells are first incubated with rat anti-CD16/32 (BD PharMingen) to block Fc receptors, or with other appropriate blocking reagents on ice for 15 min, washed, and then incubated with MAb in FACS buffer.

4. After an additional incubation on ice the cells are washed three times with FACS buffer.
5. Flow cytometry analysis can be carried out on a FACSCalibur (BD Systems) with CELLquest software *(23,24)* (*see* **Notes 7** and **8**).

3.3.6. Analysis of Tissues by Fluorescence Microscopy of Tissue Sections

1. Tissues from recipient mice are harvested, rinsed once in cold PBS, and fixed in 4% PFA in PBS for 4 h at room temperature (*see* **Note 9**).
2. Wash the tissues twice in PBS and incubate in 20% sucrose/PBS in cold room, rocking for 1 h. Afterward wash tissues five times in cold PBS and remove excess PBS.
3. Snap freeze sections in Tissue Tek O.C.T. compound and cut (8 μM) cryostat sections.
4. Thawed sections can be fixed in 4% PFA, washed in PBS, and quenched in 50 mM NK$_4$Cl in PBS.
5. After the sections are washed and blocked with appropriate buffer the sections can be stained with labeled MAbs.

4. Notes

1. A simple modification of the assay is to overlay the polycarbonate filter with fibronectin, collagen, or a monolayer of endothelial cells.
2. This assay depends on the ability of the migrating B cells to adhere to the 2D polycarbonate filter and then to migrate along the surface toward a gradient diffusing from the nearest hole. It provides no information about the velocity of the migrating cells, but it allows an estimate of the proportion of the population that can respond to the chemoattractant.
3. Do not let cover slip or cells dry.
4. Make sure chamber is clean and dry before use.
5. Avoid air bubbles when inverting cover slip onto chamber and when replacing the media with media plus chemokine.
6. Because the gap between the cover slip and the bridge is less than one cell diameter, care must be taken in placing the cover slip on the chamber as the cells can be damaged.
7. Care needs to be taken not to gate out highly fluorescent CSFE-labeled cells during the acquisition of the flow cytometry data.
8. The number of events that need to be collected depends on the organ of origin of the cells being analyzed, the percentage of CSFE-labeled cells, and the time since transfer. Typically the highest proportion of CSFE-labeled cells appears in the spleen and peripheral blood.
9. Organic fixatives cannot be used (alcohol, acetone, xylene); they will remove the dye from the tissue. Only water-based fixatives can be used, such as PFA. Frozen sections are preferred for analysis owing to the nature of the dyes; paraffin embedding is not acceptable *(25)*.

References

1. MacLennan, I. C. M., Gulbranson-Judge, A., Toellner, K.-M., et al. (1997) The changing preference of T and B cells for partners as T-dependent antibody responses develop. *Immunol. Rev.* **156**, 53–66.
2. Melchers, F., Rolink, A. G., and Schaniel, C. (1999) The role of chemokines in regulating cell migration during humoral immune responses. *Cell* **99**, 351–354.
3. McHeyzer-Williams, M. G. and Ahmed, R. (1999) B cell memory and the long-lived plasma cell. *Curr. Opin. Immunol.* **11**, 172–179.
4. Wehrli, N., Legler, D. F., Finke, D., et al. (2001) Changing responsiveness to chemokines allows medullary plasmablasts to leave lymph nodes. *Eur. J. Immunol.* **31**, 609–616.
5. Hargreaves, D. C., Hyman, P. L., Lu, T. T., et al. (2001) A coordinated change in chemokine responsiveness guides plasma cell movements. *J. Exp. Med.* **194**, 45–56.
6. Kunkel, E. J. and Butcher, E. C. (2002) Chemokines and the tissue-specific migration of lymphocytes. *Immunity* **16**, 1–4.
7. Campbell, J. J., Qin, S., Bacon, K. B., Mackay, C. R., and Butcher, E. C. (1996) Biology of chemokine and classical chemoattractant receptors: differential requirements for adhesion-triggering versus chemotactic responses in lymphoid cells. *J. Cell. Biol.* **134**, 255–266.
8. Wright, D. E., Bowman, E. P., Wagers, A. J., Butcher, E. C., and Weissman, I. L. (2002) Hematopoietic stem cells are uniquely selective in their migratory response to chemokines. *J. Exp. Med.* **195**, 1145–1154.
9. Legler, D. F., Loetscher, M., Roos, R. S., Clark-Lewis, I., Baggiolini, M. and Moser, B. (1998) B cell-attracting chemokine, 1: a human CXC chemokine expressed in lymphoid tissues, selectively attracts B lymphocytes via BLR1/CXCR5. *J. Exp. Med.* **187**, 655–660.
10. Gunn, M. D., Ngo, V. N., Ansel, K. M., Ekland, E. H., Cyster, J. G., and Williams, L. T. (1998) A B-cell-homing chemokine made in lymphoid follicles activates Burkitt's lymphoma receptor-1. *Nature* **391**, 799–803.
11. Warnock, R. A., Campbell, J. J., Dorf, M. E., Matsuzawa, A., McEvoy, L. M., and Butcher, E. C. (2000) The role of chemokines in the microenvironmental control of T versus B cell arrest in Peyer's patch high endothelial venules. *J. Exp. Med.* **191**, 77–88.
12. Sanchez-Madrid, F. and del Pozo, M. A. (1999) Leukocyte polarization in cell migration and immune interactions. *EMBO J.* **18**, 501–511.
13. Sharrow, S. O. (2002) Immunofluorsence and cell sorting, Chapter 5, in *Current Protocols in Immunology* (Coligan, J. E., Krusibeek, A. M., Margulies, D. H., Shevach, E. M., and Strober, W., eds.), Wiley, New York, pp. 5.1.1.–5.1.8.
14. Dynal Biotech Inc. (1998) Biomagnetic Applications in Cellular Immunology, in *Dynal Methods Publication*, Chapter 3.9, pp. 35. www.dynal-biotech.com.
15. Moratz, C., Kang, V. H., Druey, K. M., et al. (2000) Regulator of G protein signaling 1 (RGS1) markedly impairs Gi alpha signaling responses of B lymphocytes. *J. Immunol.* **164**, 1829–1838.

16. Hauser, A. E., Debes, G. F., Arce, S., et al. (2002) Chemotactic responsiveness toward ligands for CXCR3 and CXCR4 is regulated on plasma blasts during the time course of a memory immune response. *J. Immunol.* **169,** 1277–1282.
17. Reif, K., Ekland, E. H., Ohl, L., et al. (2002) Balanced responsiveness to chemoattractants from adjacent zones determines B-cell position. *Nature* **416,** 94–99.
18. Bowman, E. P., Campbell, J. J., Soler, D., et al. (2000) Developmental switches in chemokine response profiles during B cell differentiation and maturation. *J. Exp. Med.* **191,** 1303–1318.
19. Forster, R., Schubel, A., Breitfeld, D., et al. (1999) CCR7 coordinates the primary immune response by establishing functional microenvironments in secondary lymphoid organs. *Cell* **99,** 23–33.
20. Rabin, R. L., Alston, M. A., Sircus, J. C., et al. (2003) CXCR3 is an inflammatory chemokine receptor that bridges central and effector functions. *J. Immunol.,* submitted.
21. Kim, C. H., Qu, C. K., Hangoc, G., et al. (1999) Abnormal chemokine-induced responses of immature and mature hematopoietic cells from motheaten mice implicate the protein tyrosine phosphatase SHP-1 in chemokine responses. *J. Exp. Med.* **190,** 681–690.
22. Parish, C. R., and Warren, H. S. (2002) In vivo assays for lymphocyte function, Chapter 4, in *Current Protocols in Immunology,* (Coligan, J. E., Krusibeek, A. M., Margulies, D. H., Shevach, E. M., and Strober, W., eds). Wiley, New York, pp. 4.9.1.–4.9.10.
23. Okada, T., Ngo, V. N., Ekland, E. H., et al. (2002) Chemokine requirements for B cell entry to lymph nodes and Peyer's patches. *J. Exp. Med.* **196,** 65–75.
24. Forster, R., Mattis, A. E., Kremmer, E., Wolf, E., Brem, G. and Lipp, M. (1996) A putative chemokine receptor, BLR1, directs B cell migration to defined lymphoid organs and specific anatomic compartments of the spleen. *Cell* **87,** 1037–1047.
25. Horan, P. K., Melnicoff, M. J., Jensen, B. D., and Slezak, S. E. (1990) Fluorescent cell labeling for in vivo and in vitro cell tracking. In *Methods in Cell Biology,* vol. 33 (Darzynkiewicz, Z. and Crissman, H. A., eds.), Academic, New York, pp. 469–490.

13

Analysis of Antigen-Specific B-Cell Memory Directly Ex Vivo

Louise J. McHeyzer-Williams and Michael G. McHeyzer-Williams

Summary

Helper T-cell-regulated B-cell memory develops in response to initial antigen priming as a cellular product of the germinal center (GC) reaction. On antigen recall, memory response precursors expand rapidly with exaggerated differentiation into plasma cells to produce the high-titer, high-affinity antibody(Ab) that typifies the memory B-cell response in vivo. We have devised a high-resolution flow cytometric strategy to quantify the emergence and maintenance of antigen-specific memory B cells directly ex vivo (**Subheading 3.1.**). Extended cell surface phenotype establishes a level of cellular diversity not previously appreciated for the memory B-cell compartment (**Subhedeading 3.2.**). Using an "exclusion transfer" strategy, we ascertain the capacity of two distinct memory B-cell populations to transfer antigen-specific memory into naive adoptive hosts (**Subheading 3.3.**). Finally, we sequence expressed messenger ribonucleic acid (mRNA) from single cells within the population to estimate the level of somatic hypermutation as the best molecular indicator of B-cell memory (**Subheading 3.4.**). In this chapter, we describe the methods used in each of these four sections that serve to provide high-resolution quantification of antigen-specific B-cell memory responses directly ex vivo.

Key Words

B-cell memory; helper T-cell-regulated B-cell memory; humoral immunity; antigen-specific B-cell memory; immunity; adaptive immunity; Th-cell memory; somatic hypermutation; affinity maturation; germinal centers.

1. Introduction

B-cell memory develops in the germinal center (GC) reactions of the primary immune response to thymus-dependent (TD) antigens *(1–4)*. Populations of affinity-matured memory B cells persist for long periods of time in multiple cellular forms *(5–9)*. These post-GC memory B cells can be broadly categorized into terminally differentiated long-lived plasma cells that will not respond to

From: *Methods in Molecular Biology, vol. 271: B Cell Protocols*
Edited by: H. Gu and K. Rajewsky © Humana Press Inc., Totowa, NJ

antigen recall and the precursors for the memory response *(4,10,11)*. We have recently demonstrated that antigen-specific memory response precursors exist in at least two phenotypically distinct cellular compartments *(8,12)*. One major distinction between these two compartments is the differential glycosylation of B220, the B-cell glycoform of the transmembrane phosphatase CD45R that can be distinguished by the binding of the monoclonal antibody (mAb), RA3-6B2 (**Subheading 3.1.**). Among extensive further phenotypic distinctions (**Subheading 3.2.**), the two memory B-cell populations also behave differently on adoptive transfer into naive hosts (**Subheading 3.3.**). Both populations contain memory B cells that express somatically mutated messenger ribonucleic acid (mRNA) at the single cell level (**Subheading 3.4.**). The B220$^+$ memory population displays greater proliferative potential, whereas the B220$^-$ memory population displays greater differentiative capacity in producing more plasma cells and serum Ab over the same period of time. A linear progression of differentiation from prememory cells (GC B cells) to memory B cells (B220$^+$) to preplasma memory cells (B220$^-$) to plasma cells (CD138$^+$) provides the simplest framework to help integrate the totality of our data in this antigen-specific model (**Fig. 1**).

In this chapter, we will focus on the memory response to the TD antigen 4-hydroxy-3-nitrophenyl acetyl (NP)–keyhole limpet hemocyanin (KLH) *(13–20)*. This hapten–protein conjugate, using monophosphoryl lipid A as an adjuvant, produces robust hapten-specific B-cell responses of the IgG1, IgG2a, and IgG2b subclasses *(8,12)*. Intraperitoneal immunization promotes splenic responses in which antigen-specific clonal expansion peaks by d 7 after initial priming, and GC reactions persist for approx 3 wk, producing memory B cells that can remain for the life of the animal *(12,18,20)*. On antigen recall, memory response precursors vigorously expand and promote plasma cell development as the major memory Th-cell-regulated response to antigen *(1,4,8,10,11,21)*. There is also evidence for a GC reaction during antigen recall; however, the origins of the cells that drive these secondary GCs remain poorly resolved.

2. Materials

1. 6–10-wk-old C57BL/6 mice (Jackson Laboratories, Bar Harbor, ME).
2. NP–KLH in Ribi adjuvant (Ribi Immunochem Research; Hamilton, MT).
3. 0.14 *M* NH$_4$Cl red cell lysis buffer.
4. FACS-Vantage SE Flow Cytometer: 488- and 647-nm lines/six-color detection.
5. Fluorophore conjugated antibodies (Abs): many as listed in **Subheading 3.1.3.**
6. NP-allophycocyanin (APC; *see* **Note 7** for production).
7. 6–10-wk-old C57BL/6 Rag-1-deficient mice.
8. Complementary deoxyribonucleic acid (cDNA) reaction mix. 4 U/mL Murine leukemia virus-reverse transcriptase (MLV-RT; Gibc-BRL, Gaithersburg, MD) with recommended 1X RT buffer, 0.5 n*M* spermidine (Sigma Chemical),

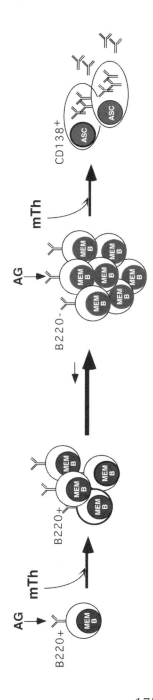

Memory → Pre-Plasma → Plasma

Fig. 1. Linear progression of cellular differentiation in the memory B-cell response to antigen. B220+ memory B cells produce B220− preplasma memory B cells as an intermediate to terminally differentiated CD138+ plasma cells. Each of the memory B cells is thought to require antigen and memory Th cells to regulate cell fate in vivo.

100 μg/mL bovine serum albumin (BSA; Boehringer-Mannheim, Indianapolis, IN), 10 ng/mL oligo (dT) (Becton Dickinson), 200 μM each deoxynucleotide-triphosphate (dNTP; Boehringer-Mannheim), 1 mM dithiothreitol; DTT) (Promega, Madison, WI), 220 U/mL RNasin (Promega), 100 μg/mL *Escherichia coli* transfer RNA (tRNA; Boehringer-Mannheim) and 1% Triton X-100.

9. First-round polymerase chain reaction (PCR) reaction mixture: 2 U/mL *Taq* polymerase with the recommended 1X reaction buffer (Promega), 0.1 mM of each dNTP (Boehringer-Mannheim), 1 mM MgCl$_2$, 0.8 μM primer LAMext3(sense) (5′-TACTC TCTCTCCTGGCTCTCAGCTC-3′) and 0.8 μM primer LAMext3 (anti) (5′-GTTGTTGCTCTGTT TGGAAGGCTGG-3′).

10. Second-round PCR reaction mixture: 2 U/mL *Taq* polymerase with the recommended 1X reaction buffer (Promega), 0.1 mM of each dNTP (Boehringer-Mannheim), 1 mM MgCl$_2$, 0.8 μM primer LAM.sens(int) (5′-CCATTTCCCAGG CTGTTGTG-3′) and 0.8 μM primer LAM.anti(int) (5′-CTCCATACCCTGAGTG ACAG-3′).

11. Direct sequencing mixture: 4 μL of PCR product, 4 μL Dye Terminator ready reaction mix (PerkinElmer, Foster City, CA), 1.6 pmol primer LAM.seq (5′-GGCTGTTGTGACTCA GGAAT-3′).

3. Methods

The methods described here outline six-color flow cytometric analysis used to quantify and isolate antigen-specific B cells, extended phenotypic analysis of target population, exclusion transfer of single-memory B-cell subsets into adoptive hosts to measure memory activity, and single-cell repertoire analysis to ascertain the presence of somatic hypermutation within memory B cells of interest.

3.1. High-Resolution Flow Cytometry for Antigen-Specific Memory B Cells

3.1.1. Immunization Regimen

We use 6–10-wk-old C57BL/6 mice (Jackson Laboratories, Bar Harbor, ME) immunized with 400 μg NP–KLH intraperitoneally (ip) in Ribi adjuvant (Ribi Immunochem Research; Hamilton, MT; *see* **Note 1**). At least 6 wk after the initial injection, the animals are rechallenged with an identical immunization (*see* **Note 2**). Animals are sacrificed across the acute phase of the memory response (wk 1), or rested for another 6 wk as an excellent source of quiescent memory B cells after sufficient time for the recall response to subside.

3.1.2. Cell Preparation

Spleens are harvested, pushed through metal mesh into red cell lysis buffer (0.14 M NH$_4$Cl) and vigorously resuspended to form single-cell suspensions (*see* **Note 3**). After 5 min at room temperature, cells are pelleted by centrifugation and resuspended in phosphate-buffered saline (PBS) + 5% fetal calf serum (FCS) (fluorescence-activated cell-sorting [FACS] wash) at 4×10^8 cells/mL at a 2X cell concentration ready for Ab labeling (*see* **Note 4**).

3.1.3. Labeling Cells for Flow Cytometry

We outline the highest resolution strategy that we currently use to provide the clearest access in one sample to the three major memory B-cell responders (*see* **Note 5**). This approach requires a FACS-Vantage SE cytometer fitted with two lasers using 488-nm and 647-nm excitation lines. The optical bench must be equipped for six-color/eight-parameter detection (*see* **Note 6**).

Cell suspensions were labeled for flow cytometry at 2.0×10^8 cells per milliliter on ice for 45 min, with predetermined optimal concentrations of fluorophore-labeled mAbs (purchased from Pharmingen, unless otherwise noted) or antigen. The following Abs were used for staining cells: Cy5PE-H129.19 (anti-CD4), Cy5PE-53-6.7 (anti-CD8) (Caltag Laboratories, Burlingame, CA), fluoroscein (fluorescein isothiocyanate [FITC])-11.26 (anti-IgD), phycoerythrin (PE)-281.2 (anti-CD138/syndecan), Cy7PE-6B2 (anti-CD45R/B220), allophycocyanin (APC)-NP (4-hydroxy-3-nitrophenyl acetyl; *see* **Note 7**), and biotin-HM79b (anti-CD79b, Igβ). After staining, cells were washed twice in PBS with 5% FCS and resuspended in streptavidin–Cy7APC for 15 min. Cells were then washed twice and resuspended in 2 μg/mL propidium iodide (PI) (for dead cell exclusion) in PBS with 5% FCS and remained on ice until analysis on the cytometer.

3.1.4. Calibration and Compensation Setup

The cytometer is calibrated each day using beads prelabeled to fluoresce in each of the six fluorescence channels used. Voltages on the photomultiplier tubes (PMTs) for each channel are kept constant from day to day, and the stream, laser, and optical elements are adjusted to achieve a standard signal of predetermined height and spread for each channel (set up as standardized small targets on an acquisition template).

Each of the reagents used in the experiment is set up as a single color control with only PI included for setting "compensations" (adjusting for spectral overlap). In some cases, the target population is too small to use the particular reagent/sample combination for compensating (e.g., no NP-binding in naive animals). In these cases, we use other reagents that have the similar labeling properties to the reagent that will be used, but have a sizable target population (e.g., APC-B220 in this case instead of APC-NP, as it is known to require the same level of compensation). In the label combination described, the omnicomp cross-beam compensation adjustment is required for Cy5PE and Cy7PE into the APC and Cy7APC channels.

3.1.5. Data Acquisition and Analysis

Files of 500,000 to 1×10^6 events already excluding PI and very small events are routinely collected for analysis. Files may vary in size depending on the frequency of the target population of interest. The example presented in **Fig. 2** outlines the way to analyze that data; it is useful to describe the process sequentially, as doing so outlines the essence of the strategy itself as it pertains to isolating antigen-specific memory B cells.

The first and most important feature of the high-resolution strategy is the notion and importance of the "exclusion channel" (*see* **Note 6**). In the example presented in **Fig. 2**, we have included Abs to CD4 and CD8 directly conjugated to Cy5PE into this channel. This channel not only excludes T cells but also excludes cells that have the propensity to nonspecifically bind Abs and proteins in general. In previous studies, we have included F4/80 as a macrophage marker here with success (*see* **Note 8**). The next step is the inclusion of plasma cells (syndecan expression) and B cells (CD79b expression) (**Fig. 2**, *upper row, second panel*). Then we look for cells with these attributes that have lost sIgD (and IgM; in some experiments both reagents are included in this channel) and can bind NP (**Fig. 2**, *upper row, last two panels*, represents quiescent memory 42 d after priming and 5 d after recall, respectively). The graph in the second row summarizes the total numbers of NP$^+$ B cells in the spleens with this phenotype prior to antigen challenge, recall, and the emergent memory response.

Finally, to categorize the three major NP$^+$ memory B-cell subsets (PI$^-$CD4$^-$CD8$^-$CD79b$^+$IgD$^-$NP$^+$) we can display their B220 vs syndecan (CD138) profiles (**Fig. 2**, *lower row, second panel*, d 5 memory example). There are NP-specific CD138$^+$B220^{int} plasma cells (when sorted into enzyme-linked immunosorbent spot-forming cell [ELISPOT] assays, >60% spontaneously secrete NP-specific IgG). There is evidence for two subsets of CD138-NP-specific B cells that both express CD79b but at different levels of B220. In the memory response, very few of the B220$^+$ memory B cells express GL7, a marker associated with the GC reaction, and would more likely represent expanding memory B cells. The largest memory B-cell population at d 5 is the B220-NP-specific cells of the preplasma memory cell phenotype. This distribution is consistent with the expected functional balance of memory responders associated with antigen-specific recall responses.

3.2. Extended Cell-Surface Phenotype

It is not sufficient to identify subpopulations of memory B cells by any one set of criteria. This subsection summarizes some extended phenotypic analyses

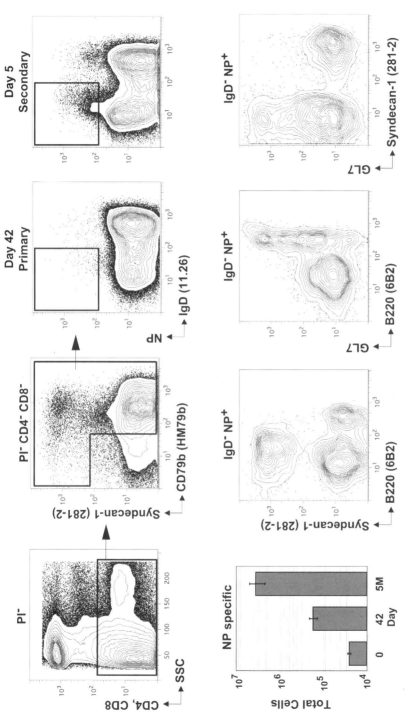

Fig. 2. High-resolution identification of antigen-specific memory B cells directly ex vivo. *Upper panels* display 5% probability contour plots with outliers for the parameters exhibited. *Insert panels* describe the progressive "gates" applied at analysis to focus on the subpopulations of interest. Animals have been immunized with NP–KLH in Ribi adjuvant, rested for 6 wk, and reimmunized with a priming dose of antigen in adjuvant. Profiles present one representative animal for the time-points indicated.

179

Table 1
Extended Phenotype of NP-Specific B-Cell Subpopulations

	T cell	IgD+ B220+ naive	NP-specific B cells B220+	B220−	CD138+
B220	−	+++	++++	−	+
CD138	−	−	−	−	+++++
CD79b	−	++	++	++	+
CD11b	−	−	−	+++	+
CD19	−	+	++	−	+
CD21	−	+	+	+	−
CD22	−	++	++	++	+
CD24	−	+	+++	−	+
CD43	+	−	−	+++	+++
CD45R	+	+	++	+	+
PNA	+	+	++	++	+
GL7	−	−	++ (& −)	−	−
BLA-1	−	++	++	+	+

MFI: −, 0–20; +, 21–100; ++, 101–300; +++, 301–600; ++++, 601–900

that may be useful for verifying the characteristics of the antigen-specific populations. **Table 1** summarizes some of the more extended phenotypic characteristics of the antigen-specific B-cell subsets. These data provide a stylized summary of phenotype from d 7 and d 14 of the primary response to antigen *(12)*. They represent some of the distinctive features of the three subsets seen at this early stage of the primary response and are largely recapitulated after antigen recall. Similar modes of analyses were undertaken in this five-color strategy, excluding CD79b in each of the assays, as well as using multiple staining combinations. Separate staining combinations focused either on the CD138+ NP-specific population at d 7 or the B220+ and B220− subpopulations at d 14 when CD138+ numbers were low.

The basic procedure follows what has been presented in **Subheading 3.1.** with permutations in reagent usage depending on the phenotype examined. The ideal approach would use seven-color resolution that would allow the use of the initial strategy to identify all three subsets simultaneously and then add any new parameter in a seventh color to unequivocally assign the expression level of the new molecule across the three antigen-binding subsets. The new digital instruments from Becton Dickinson (FACS DiVa) will allow optical bench extension with digital processing of the signals expandable to 10–12 colors where the fluorophores and optical bench configuration are available.

3.3. Exclusion Adoptive Transfers of Single-Memory B-Cell Subsets

Verification of memory B-cell activity is an important and necessary facet of these studies. Owing to the low target cell frequencies and our poorly defined understanding of the regulatory requirements of memory B-cell restimulation in vivo, special technical issues arise for adoptive transfer experiments. We have devised a method referred to as *exclusion transfer* that is described in **Subheadings 3.3.1–3.3.3.**

3.3.1. Cell Sorting and Reanalysis

Setting up the cell sorter for bulk population sorting is a detailed procedure that is beyond the scope of this chapter to describe. We will assume that the instrument is set up accurately for sorting and the populations identified carefully using the appropriate acquisition software (Cell Quest in our situation). The focus of our studies was the two populations of nonsecreting memory B cells (evaluated directly by antigen-specific ELISPOT analysis). Phenotypically, these cells were NP-specific (CD4⁻CD8⁻F4/80⁻IgD⁻NP⁺) and CD138⁻ B220⁺ or CD138⁻B220⁻ for the experiments described. The initial sort entails exclusion of all NP-specific B cells and plasma cells from the spleen cell preparation (*see* **Note 9**). Next, we sort memory B cells from each NP-specific subpopulation. Ideally, sorting is undertaken with sample and collection tubes kept at 4°C using a recirculating apparatus. After each population is collected, a small sample of cells is diluted in PI and reanalyzed to check the sort purity. Further, once cell mixtures are made for injection (see below), a small sample is reanalyzed immediately prior to iv injection into adoptive hosts. Thus, although cell counts are often underestimates, cell purities are often at the idealized level (95–99% purity). If there is >5% contamination with the inappropriate memory B-cell population the experiment is aborted.

3.3.2. Adoptive Transfer and Immunization

Using sets of sorted cells, mixtures are made for iv transfer into Rag-1-deficient animals. We found that $5–8 \times 10^6$ background cells were sufficient for the transfer experiments, based on minimal background and capacity to support memory B-cell responses in vivo. Further, $4–6 \times 10^4$ NP-specific cells from each memory B-cell subset was sufficient to mount a recall response 5 d later in the recipient. We also focused on d 4 and d 42 postrecall as the timepoints for collecting memory B cells with the appropriate phenotype. Cell mixtures were pelleted and resuspended in 100 µL PBS for iv injection into the tail vein. Shortly after iv injection (approx 15 min), each animal was primed with 50 µg NP–KLH ip without adjuvant to initiate an immune response in the adoptive host.

Table 2
Adoptive Transfer D 42 Postrecall

	Before	After Transfer	
	% Total Spleen NP$^+$ IgD$^-$	Cell Recovery % of transfer	Relative Ab titer NP-specific IgG
NP$^-$ Spleen	0.02	N/A	<1.0
B220$^+$ Mem	0.10	670	20
	0.44	39	500

3.3.3. Reanalysis Postexclusion Transfer

Animals were sacrificed 5 d later with spleens taken for reanalysis of the NP-specific response. The ideal strategy would follow the identical format as used for isolation in the first place. Emphasis is placed on total cell numbers as an index of cell expansion in vivo. Serum is collected for assessment of antigen-specific Ab production. Phenotypic distribution is used to evaluate cell subset balance and was used in the initial studies to help establish a parent–progeny relationship across memory B-cell subsets. The results presented in **Table 2** highlight the comparative analysis of Ab production and cell expansion between B220$^+$ and B220$^-$ memory B-cell subsets, which allows us to postulate the existence of the model presented in **Fig. 1**.

3.4. Repertoire Studies With Single-Cell Resolution

The behavior of isolated memory B-cell subsets in vivo is an important device in understanding the distinguishing attributes of different memory B cells. However, these assays are by their design population studies that provide no resolution to the frequency of the target cells with particular attributes. It is important to complement assays of population behavior with those that estimate population purity with single-cell resolution. This last section describes one such assay under the **Subheadings 3.4.1.–3.4.5.** The sample data set presented in **Table 3** outlines the mutation frequency across the different subsets of memory B cells and lists the efficiencies for reverse transcriptase polymerase chain reaction (RT-PCR) from different populations of memory responders.

Table 3
Expressed λ Light Chain is Somatically Mutated

| Subset | % cells | Single cell repertoire studies | | |
		# per mut seq	% replace	PCR efficiency
B220⁺ memory	85	3.1	92	30%
B220⁻ memory	81	4.3	82	15%
ABS: CD138⁺	88	3.3	92	60%

3.4.1. Automated Single-Cell Deposition

Single-cell repertoire studies require the automated cell deposition unit (ACDU) attached to a stream-in-air flow cytometer. These units can be used as the collection device for cells identified for sorting. The basic setup procedures for sorting using this unit are universal and beyond the scope of this chapter to describe. However, when sorting into 10 μL 72-well microculture plates (Terasaki and Robbins), some specific precautions must be taken. The instrument is set up optimally to sort populations as usual. Then, the sort mode is switched to "counter," and a plate is positioned to receive the sorted cells. For the setup, a large population of cells from a control sample is selected, and 100–500 cells per well selected to be sorted using CloneCyt software. As this sort proceeds, the operator must watch the stream position as in enters each well of an empty plate. The "home" position and fine angle of the stream should be adjusted until the stream is entering the center of each well. At this point, the stage and advance should be set for single-cell deposition, which can be selected for in the software (*see* **Note 10**).

3.4.2. cDNA Reaction

In these repertoire studies, single cells are sorted directly into 5 μL cDNA reaction mixtures. Using a 72-well plate, cells are only sorted into the center 60 wells with the remaining 12 wells acting as the most appropriate negative controls for these experiments. Further, samples should be processed in rows, with the first and last sample in each row acting as a negative control for processing errors in this design. Each cell is sorted into an oligo-(dT)-primed 5 μL cDNA reaction mixture as just described. Set up in low profile 72-well microtiter trays (Robbins Scientific, Sunnyvale, CA) immediately incubated at 37°C for 90 min, then stored at –80°C until further use.

3.4.3. Nested PCR Amplification

First round of PCR (PCR-a): 2 µL of cDNA from each single-cell cDNA reaction is used in a 25 µL reaction to amplify the λ light chain. Each reaction set begins with 5 min × 95°C, then 40 cycles of 95°C × 15 s, 50°C × 45 s, 72°C × 90 s, and ending with 72°C × 5 min (*see* **Note 11**). *Second round of PCR (PCR-b):* 1 µL of the first round PCR product is used for an additional 25 µL amplification reaction using primers nested medially to the primers used in PCR-a Each reaction set begins with 95°C for 5 min then 40 cycles of 95°C for 15 s, 50°C for 45 s and 72°C for 90 s and ending with 72°C for 5 min. At least two negative cDNA samples are processed per 10 single-cell samples across rows as described in the preceding paragraph. The negatives interspersed with positives are critical to control for contamination during sample preparation (*see* **Note 12**).

3.4.4. Direct Sequence Analysis From Single Cells

A 5-µL-amount of PCR-b product is run on a 1.5% agarose gel to screen for positives (single bands of the right size). The PCR product is then separated from primers using a CL-6B Sepharose (Pharmacia, Piscataway, NJ) column. The PCR product is then directly sequenced using a linear amplification protocol for 25 cycles of 96°C for 10 s, 50°C for 5 s, and 60°C for 4 min on a 9600 GeneAmp PCR system (PerkinElmer). Samples are separated on a 6.5% acrylamide gel after ethanol precipitation of sequencing reaction products and run on an ABI 373 sequencing system and processed using ABI Prism sequence 2.1.2 for collection and analysis (PerkinElmer).

4. Notes

1. The Ribi adjuvant is a monophosphoryl lipid A (MPL) synthetic derivative of lipopolysaccharide (LPS) in squalene (a metabolizable oil). Hence this is a TLR4 agonist that acts in a nondepot manner to promote an immune response best assayed in the spleen. Use a high conjugation ratio for the hapten–protein that results in approx 15–20 NP molecules per KLH. This is not possible to calculate exactly as the KLH is known to form higher order structures that make it difficult to estimate molecular weight.

2. Although more classical recall responses use an iv injection of low-dose antigen (5–50 µg), we chose a second primary to more closely compare the cellular behavior of memory B cells to naive B cells (using the same external stimuli). This regimen results in the recruitment of memory B cells as determined by rapid reemergence of the cellular response (peaks at d 3 compared to d 7 in primary). Further, the plasma cells produced in this memory response all express mutated B-cell antigen receptor (BCR) light chains, and most carry the W33L mutation

in the $V_H 186.2$ V_H gene segment known to increase NP-binding affinity *(16,19, 22,23)*. The RT-PCR efficiency for plasma cells at this time-point averages 60%; hence, we sample a substantial number of the plasma cells that emerge in the recall response (more details in **Subheading 3.4.**).

3. A 1/50 dilution of cells is used to estimate cell count using a hemocytometer. These counts aim to estimate total white blood cell count at the time of sacrifice, excluding red blood cells by size and morphology, but not excluding dead cells. PI is used to exclude dead and dying cells at the time of flow cytometric analysis. Target cell frequencies are calculated based on live cells and back calculated to the total cell numbers in the animals using the cell count. The assumption here and inherent in these calculations is that cell death ex vivo owing to the cell preparation procedures is random.

4. This is an atypically high cell concentration for labeling. All reagents have been titrated for optimal working concentrations using these high cell concentrations and 45 min staining conditions as described.

5. Antigen binding alone is insufficient criteria to isolate antigen-binding B cells. Many reagents (antigens and Abs alike) are nonspecifically "sticky" to macrophage and dying cells, or they bind through FcR or other surface molecules for which there are no blocking Abs. Those using two- or three-color strategies to isolate antigen-specific B cells often strike this difficulty and focus analysis on spurious cell populations. The six-color strategies can be modified to four- and five-color systems but with decreased resolution (i.e., increased contamination within each subset) for each population studied. When modified it is best to focus analysis on one memory B-cell subset at a time.

6. The combination of reagents used can be modified depending on the Ab–fluorophore combination available. Greatest difficulties arise when using duochrome reagents (Cy5PE, Cy7PE, and Cy7APC). Commercially available Cy7 duochromes have highly variable fluorescence properties from batch to batch as there appears to be little regard to standardization with respect to conjugation ratios used. Cy5PE is particularly difficult to work with as it is primarily collected after excitation with the 488-nm laser (excites the PE and this leads to energy transfer and excitation of the Cy5 moiety). However, when this reagent passes through the 647-nm line, the Cy5 is directly excited and emission is collected in the same channel as the APC signal. This problem can be overcome using an omnicomp correction system that allows for cross-beam compensation. Because the system of cross-beam compensation is never ideal (full adjustment is usually incomplete), we always use the Cy5PE channel as our exclusion channel and "dump" all positive signals we collect into this channel. The signal from PI can also be collected into this channel and excluded along with all other signals unimportant to the focus of the strategy.

7. The making of NP–APC requires low-conjugation ratios and empirical testing. The high conjugation ratios lead to inordinately high background staining of cells that do not amplify mRNA for Ig. We use the succinimide ester of NP (Biosearch

Technologies) or the iodinated-form with higher binding, NIP (Genosys) and PhycoPro APC (Prozyme). Different sources for phycobiliproteins are aggregated to different levels that can also elevate background. The APC is dialyzed against 3% NaHCO$_3$ overnight, then purity and concentration are calculated using A280, A655, and A620. The 655:620 ratio should be >1.4; the 655:280 ratio should be >4; conc. = [A655 × dilution] / 5.9 = mg/mL. Dissolve 10 mg 4-hydroxy-3-nitrophenylacetyl-hydroxysuccinimide ester (NP Osu) in 1 mL anhydrous *N,N*-dimethylformamide (while vortexing), then add 20 μg, 40 μg, and 80 μg NP Osu per mg APC (while vortexing); cover tubes and place on rocking platform for 2 h at room temperature; dialyze against 3% NaHCO$_3$ and then against PBS, and store at 4°C. Determine A280, A620, and A655 for each conjugation reaction; the 655:280 ratio is an indicator of NP haptenation. Ratio decreases with increasing haptenation. These reagents must be tested empirically for optimal concentration using d-7- or d-14-immunized animals and full phenotypic display as described. In our recent experience, all three conjugates work well at dilutions of 1:100 to 1:400 final, but the higher ratios require low dilutions to decrease noise (events that pile up on the diagonal between NP and IgD labeling as described).

8. Caution must be taken in what is to be added to the exclusion channel. In our earlier studies, we used CD11b in this channel and inadvertently excluded all the B220$^-$ memory B cells in the process.

9. As we do not know what accessory cells and memory Th cells are needed for the splenic memory B-cell response, it is important to provide these elements to the Rag-1-deficient animals to support recall responses. However, it is key to test these populations as a source of noise in these assays. Hence, the appropriate and most critical negative control is the transfer of background cells alone with antigen as the measure of background activity.

10. Each instrument should be checked for accuracy prior to repertoire studies using deposition of single cells into microculture medium. Once sorted, the cells are centrifuged in the plates and simply visualized by microscopy. The standard 10-μL wells have angled sides that cannot be adequately visualized and occasionally cells will remain up the sides even after centrifugation. In general, even when RT-PCR efficiency is 80–90% (as it is for a TCRβ chain in some experiments not described here), the visual check only reveals 70–80% single cells in the plates. The sort-gate selection can also be broadly checked in this manner to estimate the probability of sorting doublets (which is negligible in our experience).

11. The primers are listed for the light chain amplification. For these studies, we have found that each new set of primers even from the same source needs to be optimized for Mg^{2+} concentration and optimal *Taq* concentration. Although it is important to test a titration of Mg^{2+}, it is also important to titrate each concentration across multiple 10-fold dilutions so that the decision on concentration is based on sensitivity of each reaction—not simply a presence/absence call.

12. The processing between PCR-a and PCR-b is the most susceptible to cross-contamination. These reactions are all undertaken in a laminar flow hood to help reduce droplet contamination between samples. Glove changes at the end of each

row of samples are needed to reduce contamination. Where practicable, racks and equipment are dedicated to PCR and never used for the processing of DNA samples from the same gene for sequence analysis (large source of potential contamination); 0.1 M HCl can be used on occasions to clean down equipment used for PCR. The potential for contamination following two sets of 40 amplification cycles cannot be emphasized enough as a source of difficulty in this protocol.

References

1. MacLennan, I. C. M. and Gray, D. (1986) Antigen-driven selection of virgin and memory B cells. *Immunol. Rev.* **91,** 61–85.
2. Rajewsky, K. (1996) Clonal selection and learning in the antibody system [Review]. *Nature* **381,** 751–758.
3. Kelsoe, G. (1996) The germinal center: a crucible for lymphocyte selection. *Semin. Immunol.* **8,** 179–184.
4. McHeyzer-Williams, M. G. and Ahmed, R. (1999) B cell memory and the long-lived plasma cell. *Curr. Opin. Immunol.* **11,** 172–179.
5. Liu, Y. J., Oldfield, S., and MacLennan, I. C. (1988) Memory B cells in T cell-dependent antibody responses colonize the splenic marginal zones. *Eur. J. Immunol.* **18,** 355–362.
6. Cascalho, M., Ma, A., Lee, S., Masat, L., and Wabl, M. (1996) A quasi-monoclonal mouse. *Science* **272,** 1649–1652.
7. Lam, K. P., Kuhn, R., and Rajewsky, K. (1997) In vivo ablation of surface immunoglobulin on mature B cells by inducible gene targeting results in rapid cell death [see comments]. *Cell* **90,** 1073–1083.
8. McHeyzer-Williams, L. J., Cool, M., and McHeyzer-Williams, M. G. (2000) Antigen-specific B cell memory: expression and replenishment of a novel B220-memory B cell compartment. *J. Exp. Med.* **191,** 1149–1166.
9. O'Connor, B. P., Cascalho, M., and Noelle, R. J. (2002) Short-lived and long-lived bone marrow plasma cells are derived from a novel precursor population. *J. Exp. Med.* **195,** 737–745.
10. McHeyzer-Williams, M. G. (2003) B cells as effectors. *Curr. Opin. Immunol.* **15,** 354–361.
11. McHeyzer-Williams, M., McHeyzer-Williams, L., Panus, J., et al. (2003) Helper T-cell-regulated B-cell immunity. *Microbes Infect.* **5,** 205–212.
12. Driver, D. J., McHeyzer-Williams, L. J., Cool, M., Stetson, D. B., and McHeyzer-Williams, M. G. (2001) Development and maintenance of a B220-memory B cell compartment *J. Immunol.* **167,** 1393–1405.
13. Jack, R. S., Imanishi-Kari, T., and Rajewsky, K. (1977) Idiotypic analysis of C57BL/6 mice to the (4-hydroxy-3-nitrophenyl)acetyl group. *Eur. J. Immunol.* **7,** 559–565.
14. Makela, O. and Karjalainen, K. (1977) Inherited immunoglobulin idiotypes of the mouse. *Transplant Rev.* **34,** 119–138.
15. Reth, M., Hammerling, G. J., and Rajewsky, K. (1978) Analysis of the repertoire of anti-NP antibodies in C57BL/6 mice by cell fusion, I: characterization of anti-

body families in the primary and hyperimmune response. *Eur. J. Immunol.* **6,** 393–400.

16. McHeyzer-Williams, M. G., Nossal, G. J. V., and Lalor, P. A. (1991) Molecular characterization of single memory B cells. *Nature* **350,** 502–505.

17. Lalor, P. A., Nossal, G. J. V., Sanderson, R. D., and McHeyzer-Williams, M. G. (1992) Functional and molecular characterization of single, (4-hydroxy-3-nitrophenyl)acetyl (NP)-specific, IgG1+ B cells from antibody-secreting and memory B cell pathways in the C57BL/6 immune response to NP. *Eur. J. Immunol.* **22,** 3001–3011.

18. McHeyzer-Williams, M. G., McLean, M. J., Lalor, P. A., and Nossal, G. J. V. (1993) Antigen-driven B cell differentiation in vivo. *J. Exp. Med.* **178,** 295–307.

19. Jacob, J., Kelsoe, G., Rajewsky, K., and Weiss, U. (1991) Intraclonal generation of antibody mutants in germinal centres. *Nature* **354,** 389–392.

20. Jacob, J., Kassir, R., and Kelsoe, G. (1991) *in situ* studies of the primary immune response to (4-hydroxy-3-nitrophenyl)acetyl, I: the architecture and dynamics of responding cell populations. *J. Exp. Med.* **173,** 1165–1175.

21. MacLennan, I. C. M., Liu, Y. J., Oldfield, S., Zhang, J., and Lane, P. J. (1990) The evolution of B-cell clones. *Curr. Top. Microbiol. Immunol.* **159,** 37–63.

22. Bothwell, A. L. M., Paskind, M., Reth, M., Imanishi-Kari, T., and Rajewsky, K. (1981) Heavy chain variable region contribution to the NPb family of antibodies: somatic mutation evident in a γ2a variable region. *Cell* **24,** 625–637.

23. Cumano, A. and Rajewsky, K. (1986) Clonal recruitment and somatic mutation in the generation of immunological memory to the hapten NP. *EMBO J.* **5,** 2459–2468.

14

B-Cell Signal Transduction

*Tyrosine Phosphorylation, Kinase Activity,
and Calcium Mobilization*

Tilman Brummer, Winfried Elis, Michael Reth, and Michael Huber

Summary

Signal transduction by the B-cell antigen receptor (BCR) regulates development, survival, and clonal expansion of B cells. The BCR complex comprises the membrane-bound immunoglobulin molecule and the Ig-α/Ig-β heterodimer, and was shown to form oligomeric structures. Antigen-mediated engagement of the BCR results in the tyrosine phosphorylation of multiple signaling proteins leading to calcium mobilization and the activation of downstream serine/threonine kinases as well as transcription factors. In pervanadate (PV)-treated B cells, comparable pathways are activated on expression of the BCR, indicating that the BCR can signal in an antigen-independent fashion as well. In this chapter, we describe the analysis of antigen-dependent and -independent tyrosine phosphorylation events as well as a method to study calcium mobilization from differentially stimulated B cells. Furthermore, we emphasize the use of phospho-specific antibodies (Abs) and low-molecular-weight enzyme inhibitors in the process of mapping BCR-activated signaling pathways as well as determining activation states of signaling proteins.

Key Words

Antigen; B lymphocyte; B-cell antigen receptor; calcium mobilization; glutathione-*S*-transferase (GST) fusion proteins; low-molecular-weight inhibitors; mitogen-activated protein kinase; pervanadate; protein kinase B; protein tyrosine kinase; protein tyrosine phosphatase; tyrosine phosphorylation; Western blotting.

1. Introduction

Development, survival, and clonal expansion of B cells are dependent on signal transduction by the B-cell antigen receptor (BCR) *(1,2)*. The BCR comprises the membrane-bound immunoglobulin (mIg), which carries the two antigen-binding sites, and the disulfide-bridged Ig-α/Ig-β heterodimer,

From: *Methods in Molecular Biology, vol. 271: B Cell Protocols*
Edited by: H. Gu and K. Rajewsky © Humana Press Inc., Totowa, NJ

which is noncovalently bound to mIg in a 1:1 stoichiometry *(3)* and mediates the signal transduction of the BCR via immunoreceptor-tyrosine-based activation motifs (ITAM) *(4)*. During B-cell development, the BCR seems to be able to generate signals in an antigen-independent and -dependent fashion. In the absence of antigen, the BCR signals positive selection of immature B cells and the maintenance of mature B cells. On binding to antigen, the BCR emits a negative-selection signal for immature B cells and an activation signal for mature B cells *(5)*. An antigen-independent, BCR-expression-dependent signaling mode was observed in a study of protein-tyrosine kinase (PTK) substrate phosphorylation in J558L cells treated with the phosphatase inhibitor pervanadate (PV) *(6,7)*. The existence of an antigen-independent signaling mode was corroborated by several studies using transgenic and knockout models *(8–12)*.

Proximal BCR signaling elements are, among others, the PTK Syk and Lyn, the adapter protein SLP-65 (aka BLNK or BASH), and a protein-tyrosine phosphatase (PTP), most probably SHP-1 *(13,14)*. Recent studies using an approach of reverse genetics have demonstrated that Syk phosphorylates both tyrosine residues of the ITAM, whereas Lyn is only capable of phosphorylating the N-terminal tyrosine *(15)*. As a positive allosteric enzyme, Syk is strongly activated by binding to the phosphorylated ITAM tyrosines via its two N-terminal SH2-domains and establishes a positive feedback loop at the BCR *(15)*. This BCR/ITAM-dependent Syk activation and signal amplification is counterbalanced by PTP, the activity of which is regulated by H_2O_2 and the RedOx equilibrium via a double-negative feedback loop *(15)*.

The importance of Syk in BCR signaling is further demonstrated by Syk-deficient DT40 B cells, which exhibit a complete block in BCR-mediated activation of the mitogen-activated protein kinases (MAPK) extracellular signal-related kinase (ERK) and c-Jun N-terminal kinase (JNK) as well as in Ca^{2+} mobilization *(16,17)*. One important effector of Syk is the adapter protein SLP-65, which is rapidly phosphorylated by Syk on BCR engagement and functions as a key player in BCR-mediated activation of phospholipase C-γ2 (PLC-γ2) by Syk and the PTK Btk *(18)*. Activated PLC-γ2 generates the second messengers diacylglycerol (DAG) and inositol-1,4,5-trisphosphate (IP_3). Binding of IP_3 to its cognate receptor releases Ca^{2+} from the endoplasmic reticulum (ER) into the cytoplasm. This initial Ca^{2+} mobilization then promotes Ca^{2+} influx from the extracellular space leading to sustained Ca^{2+} oscillations in the cytoplasm *(19)*. DAG, on the other side, activates multiple protein kinase C (PKC) isoforms and guanine nucleotide exchange factors (GEFs) for the small GTPases Ras and Rap *(20)*. Ras, in its activated form (Ras-GTP) mediates activation of the ERK pathway through the B-Raf and Raf-1 kinases and also contributes, together with the transmembrane protein

CD19 or the docking protein BCAP, to phosphatidylinositol 3-kinase (PI3K) activation *(21–23)*. PI3K, which is an essential signaling element in B cells, generates phosphatidylinositol-3,4,5-trisphosphate (PIP_3). This phospholipid serves as ligand for many proteins with a pleckstrin homology (PH) domain like Btk or PLC-γ2, which are recruited to the plasma membrane in a PIP_3-dependent manner *(20)*. Likewise, a rise in PIP_3 levels results in PH-domain-dependent recruitment of protein kinase B (PKB; also known as Akt) to the plasma membrane, where it is phosphorylated and activated by the phospho-inositide-dependent kinase 1 *(24)*. Lastly, rising PIP_3 levels as well as tyrosine-phosphorylated CD19 and SLP-65 recruit Vav, a GEF for Rac-GTPases, to the membrane. Rac-GTP links the BCR to the ERK, JNK, and p38 pathways, which are implicated in the induction of immediate early genes *(25,26)*.

Another important BCR-triggered event is the activation of the transcription factors of the c-Rel/NF-κB family, which regulate the expression of survival factors like Bcl-X_L and c-IAPs *(27)*. Constitutive activation of this pathway has been implicated in several B-cell malignances *(28)*. In unstimulated B cells, NF-κB is restrained in the cytoplasm by the inhibitor proteins of NF-κB (IκB). BCR-mediated NF-κB activation involves several elements, such as Btk, PLCγ-2, and PKCs leading to the rapid phosphorylation of IκB by the IκB kinase (IKK) complex. This phosphorylation triggers the degradation of IκB by the ubiquitin/proteasome pathway and results in the release and nuclear translocation of NF-κB *(29–31)*.

In this chapter, we describe the analysis of antigen-dependent and -independent tyrosine phosphorylation events as well as a method to study calcium mobilization from differentially stimulated B cells. Furthermore, we emphasize the use of phospho-specific antibodies (Abs) and low-molecular-weight enzyme inhibitors in the process of mapping BCR-activated signaling pathways as well as determining activation states of signaling proteins.

2. Materials
2.1. Reagents and Buffers

Beside some standard reagents, the following chemicals (analytical grade) and materials may be required:

1. 4 Fast-Flow Protein Sepharose beads (Amersham Pharmacia).
2. Anti-Ig Abs for cell stimulation, for example, mouse antichicken IgM M4 (Southern Biotechnologies) or goat antimouse κ (Southern Biotechnologies).
3. Enhanced ChemiLuminescent (ECL) detection system (Amersham Pharmacia).
4. ECL Hyperfilm (Amersham Pharmacia).
5. Glutathione-Sepharose beads (Amersham Pharmacia).
6. Hydrogen peroxide (H_2O_2; Fluka).

Table 1
Separating Gel

	8%	10%	12%	15%
ddH$_2$O	15.4 mL	13.4 mL	11.3 mL	8 mL
Separating gel buffer	8 mL	8 mL	8 mL	8 mL
(1.5 *M* Tris-HCl, 0.1% SDS; pH 8.8)				
30% Premade acrylamide/	8.6 mL	10.6 mL	12.7 mL	15.0 mL
bis-acrylamide solution 37.5/1,				
e.g., Rotiphorese Gel 30				
(30% acrylamide, 0.8%				
bis-acrylamide solutions; Roth)				
TEMED	30 µL	30 µL	30 µL	30 µL
Ammonium persulfate	200 µL	200 µL	200 µL	200 µL

7. Indo-1 (Molecular Probes).
8. Sodium orthovanadate (Na$_3$VO$_4$; Sigma).
9. Luria-Bertani (LB) agar plates: 2g agar/100 mL LB medium, store at 4°C.
10. Nonidet P-40 (NP40; also known as Igepal; Calbiochem).
11. Pluronic acid (Molecular Probes).
12. Polyvinylidene difluoride (PVDF) membrane (Immobilon P®, Millipore).
13. Protease inhibitors: aprotinin, leupeptin, AEBSF (Roche).
14. Standard lysis buffer: 50 m*M* Tris-HCl, pH 7.4, 1% NP40, 137 m*M* NaCl, 1% glycerol, 1 m*M* sodium orthovanadate, 0.5 m*M* ethylenediaminetetraacetic acid (EDTA), pH 8.0, supplemented with protease inhibitors (0.1 mg/mL aprotinin, 0.5 mg/mL leupeptin, 1 m*M* AEBSF).
15. RIPA buffer: 50 m*M* HEPES, pH 7.4, 1% Triton X-100, 0.5% sodium deoxycholate, 0.1% sodium dodecyl sulfate (SDS), 50 m*M* NaF, 10 m*M* sodium orthovanadate, 5 m*M* EDTA and supplemented with protease inhibitors (0.1 mg/mL aprotinin, 0.5 mg/mL leupeptin, 1 m*M* AEBSF).
16. NP40/RIPA buffer: 50 m*M* Tris-HCl, pH 7.4, 1% NP40, 0.5% sodium-deoxycholate, 0.1% SDS, 137.5 m*M* NaCl, 1% glycerol, 1 m*M* sodium-orthovanadate, 0.5 m*M* EDTA, pH 8.0, supplemented with protease inhibitors (0.1 mg/mL aprotinin, 0.5 mg/mL leupeptin, 1 m*M* AEBSF).
17. 2X Reducing sample buffer: 20% glycerol, 6% SDS, 6% β-mercaptoethanol, 0.6% bromophenol blue.
18. Lysis buffer (*see* **Subheading 3.3.2.**): 50 m*M* Tris-HCl, pH 7.5, 150 m*M* NaCl, 1% Triton X-100, 10 µg/mL aprotinin, 2 µg/mL leupeptin, 100 µ*M* phenyl-methylsulfonyl fluoride (PMSF), 1 mg/mL lysozyme, 5 m*M* dithiothreitol (DTT).
19. Wash buffer (*see* **Subheading 3.3.2.**): 50 m*M* Tris-HCl, pH 7.5, 150 m*M* NaCl, 0.1% Triton X-100, 10 µg/mL aprotinin, 2 µg/mL leupeptin, 100 µ*M* PMSF.
20. Separating gel (*see* **Table 1**).

Table 2
Stacking Gel

ddH$_2$O	8.9 mL
Stacking gel buffer (480 mM Tris-HCl, pH 6.8)	3.8 mL
30% Premade acrylamide/*bis*-acrylamide solution 37.5/1, e.g., Rotiphorese Gel 30 (30% acrylamide, 0.8% *bis*-acrylamide solutions; Roth)	2.5 mL
TEMED	15 µL
Ammonium persulfate	150 µL

21. Stacking gel (*see* **Table 2**).
22. 10X Sodium dodecyl sulfate-polyacrylamide gel electrophoresis (SDS-PAGE) running buffer: Dissolve 288 g glycine, 60.7 g Tris-base, 20 g SDS in 2 L ddH$_2$O.
23. 10X transfer buffer (without methanol): dissolve 225 g glycine and 60 g Tris-base in 2 L ddH$_2$O. Combine 200 mL of 10X transfer buffer with 200 mL methanol and fill up to 2 L to yield 1X transfer buffer.

2.2. Cell Lines

The following cell lines represent well-established model systems for the study of antigen-dependent and -independent BCR signal transduction:

2.2.1. DT40

The chicken B-cell line DT40 is an avian leukosis virus (ALV)-induced bursal lymphoma and corresponds to the bursal stem cell stage of avian B lymphopoiesis *(32,33)*. As outlined in Chapter 18, DT40 cells represent a powerful tool to dissect intracellular signaling pathways by biochemical methods on a genetically well-defined basis. In addition, we have generated a transgenic DT40 line, DT40MCM, which expresses the tightly controlled, 4-hydroxy-tamoxifen (4-HT) inducible Cre-recombinase MerCreMer *(34)*. DT40MCM cells allow the rapid and efficient recombination of *lox*P-flanked (floxed) gene segments and can be employed for the inducible deletion of floxed gene segments within the DT40 genome as well as for the inducible expression of transgenes (**ref. 23** and unpublished data).

2.2.2. WEHI-231

WEHI-231 is a murine B-cell lymphoma from a (Balb/c × NZB)F1 intercross *(35)*. These cells express exclusively a BCR of the IgM isotype (µ/κ) and are considered a model for immature B cells. Strong BCR engagement results in apoptosis *(36)*.

2.2.3. K46

K46 cells used in this study are a class switch variant (μ/κ to $\gamma2a/\kappa$) of the originally described K46 cell line isolated from a B-cell lymphoma of a Balb/c mouse.

2.2.4. J558L

J558L was isolated from a plasmacytoma of a Balb/c mouse and represents the plasma cell stage of B-cell development. Because expression of the immunoglobulin heavy chain (α-isotype) and the BCR signaling subunit Ig-α has ceased, J558L cells lack an endogenous BCR and secrete only $\lambda1$ light chains. Therefore, J558L cells allow the reconstruction of BCR of any desired isotype by transfection with expression vectors for Ig-α and a heavy chain *(6,7,37,38)*.

2.2.5. Ramos

Ramos B cells are derived from an Epstein-Barr virus genome called negative Burkitt lymphoma isolated from a 3-yr-old boy *(39)*. As Ramos B cells express only IgM on their surface, they are considered representative of immature B cells.

2.3. Cell Culture

B-cell lines are usually cultivated in RPMI-1640 supplemented with 10% fetal calf serum (FCS), 2 mM L-glutamine, 100 U/mL penicillin, 100 µg/mL streptomycin, 10 mM HEPES, pH 7.4, and 50 µM β-mercaptoethanol. All cell lines are kept at 37°C in a humidified incubator with a 5% CO_2 atmosphere. For the culture of DT40 cells, the RPMI-1640-based medium is supplemented with 1% chicken serum (CS). Both FCS and CS are heat-inactivated at 60°C for 30 min to denature components of the complement system.

3. Methods

3.1. Stimulation of B Cells

3.1.1. Stimulation of B Cells Through the BCR
With Anti-Ig Antibodies

Unless transgenic mouse models or cell lines with a defined idiotype (e.g., J558L transfectants) are used, most B-cell lines and polyclonal B-cell populations isolated from lymphatic organs must be stimulated with anti-Ig Abs to mimic antigen encounter. The following protocol can be applied to B-cell lines (**Subheading 2.2.**) as well as to freshly isolated B cells from spleen and bone marrow. The optimal amount of anti-Ig Abs for a successful stimulation varies between different cell lines and should be determined experimentally for the individual antibody. However, we found that 2–10 µg of the stimulating Ab

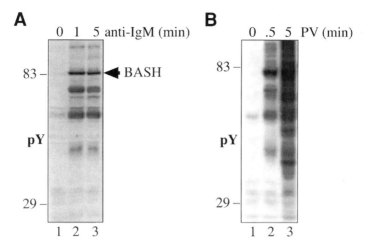

Fig. 1. Analysis of tyrosine phosphorylation events in differentially stimulated B cells. (A) DT40 B cells were stimulated with 5 µg anti-IgM Ab M4 for the indicated times and processed to total cellular lysates. Tyrosine-phosphorylated proteins such as BASH were detected with anti-phospho-tyrosine Ab 4G10. (B) DT40 B cells were stimulated with PV (37.5 µM sodium orthovanadate; 7.5% H_2O_2) for the indicated times and processed to total cellular lysates. Tyrosine-phosphorylated proteins were detected with anti-phospho-tyrosine Ab 4G10. Note that anti-IgM stimulation and a 0.5-min stimulation with PV results in a similar pattern of tyrosine-phosphorylated proteins.

is usually sufficient to trigger multiple signaling pathways in murine splenic B cells and the B cell-lines DT40, K46, or Ramos. Chicken DT40 B cells are stimulated with the monoclonal anti-IgM Ab M4 (Southern Biotechnologies) **(Fig. 1)**. Likewise, Ramos B cells can be stimulated with either goat antihuman IgM F(ab′)₂ fragments or goat antihuman IgM Abs (Southern Biotechnologies). K46 cells can be stimulated with goat antimouse IgG Abs. Alternatively, K46, WEHI-231 cells, and the majority of primary murine B cells can be stimulated with goat anti-κ Abs. The following protocol can be used for the biochemical analysis of BCR signaling:

1. Harvest B cells and adjust the cell suspension in ice-cold RPMI-1640 medium (without any supplements) to a density of 10^7 cells/mL. Keep the cells on ice until stimulation.
2. Prewarm the suspension to 37°C with gentle agitation for 1 min prior to the addition of anti-Ig Abs, and incubate for the indicated times.
3. Following incubation, the cells are centrifuged in a tabletop centrifuge for 10 s and resuspended in 200 µL ice-cold lysis buffer. The lysis is performed on ice for 30 min, and debris is subsequently removed by centrifugation (11,600g; 4°C; 10 min).

4. The supernatants are mixed with an equal volume of 2X sample buffer and proteins are denatured by boiling for 5 min and subsequently chilled on ice.
5. Thereafter, 7.5×10^5 to 10^6 cell equivalents are separated by SDS-PAGE, transferred to PVDF membranes, and analyzed by Western Blotting (**Subheading 3.4.**).

3.1.2. Stimulation of B Cells With the Phosphatase-Inhibitor PV

Stimulation of B cells with the phosphatase-inhibitor PV (*see* **Note 1**) inhibits protein-tyrosine phosphatases (PTPs) and thereby shifts the intracellular equilibrium between kinases and phosphatases toward the kinases (for review, *see* **ref. 40**). The following protocol is based on the publication by Wienands et al. (1996) and can in general be applied to all B-cell lines and primary B cells as well as to nonlymphoid cell lines (*[6]*; **Fig. 1**). However, we have noticed strong differences in the sensitivity of a cell line toward PV, and therefore we highly recommend performing dose-response and time-course experiments for the determination of an optimal PV dose.

1. Mix 1 mL of a stock solution of 10 mM sodium orthovanadate with 330 μL of 30% H_2O_2 and incubate for 5 min at room temperature. Successful formation of the PV complex is indicated by the yellow color of the solution, which becomes immediately visible on addition of H_2O_2. The obtained PV-solution equals to 6 mM PV plus excess H_2O_2.
2. As a starting concentration, this solution is further diluted 1:5 with water, and 5 μL of this solution is added to 10^7 cells (suspended in 1 mL RPMI-1640 medium without supplements) to yield a final concentration of 25 μM PV.
3. Subsequently, the cells are centrifuged in a tabletop centrifuge for 10 s and resuspended in 200 μL ice-cold lysis buffer. The lysis is performed on ice for 30 min and debris is subsequently removed by centrifugation (11,600g; 4°C; 10 min).
4. The supernatants are mixed with an equal volume of 2X sample buffer, and proteins are denatured by boiling for 5 min and subsequently chilled on ice.
5. Thereafter, 7.5×10^5 to 10^6 cell equivalents are separated by SDS-PAGE, transferred to PVDF membranes, and analyzed by Western blotting (**Subheading 3.4.**).

In the aforementioned protocols (**Subheadings 3.1.1.** and **3.1.2.**), cell pellets are resuspended in lysis buffers containing detergents such as Triton X-100, NP40, or digitonin (for review on detergents, *see* **ref. 41**). The choice of the detergent is dependent on the purpose of the experiment and the characteristics of the protein of interest. For example, some specific protein–protein interactions are only detectable by coimmunoprecipitation under mild lysis conditions using low detergent concentrations (e.g., 0.5% NP40) or particular detergents such as digitonin. In contrast, the use of mild detergents is not advisable if the activity of protein kinases should be assayed in immune complexes derived from these lysates because of the high risk of copurified "contaminating" kinases, and therefore the use of RIPA buffer is recommended. Mild detergents may not solubilize membrane proteins and cytoskeleton-associated or nuclear proteins. Indeed, the solubilization of membrane proteins, which are located in specialized lipid microdomains, or lipid rafts, requires the use of special detergents and nuclear proteins such as the immediate early gene products Egr-1 or c-Fos, are best solubilized with a combination of the detergents NP40, sodium deoxycholate, and sodium dodecylsulfate (RIPA buffer; *(23)*).

3.2. Immunoprecipitation

Affinity purification using Abs directed against the protein of interest (immunoprecipitation) can be used for enzymatic in vitro assays or for the analysis of posttranslational modification—for example, tyrosine phosphorylation—of the protein of interest. To this end, total cellular lysates (TCLs), which are prepared as described in **Subheading 3.1.**, are incubated with an appropriate amount of Abs recognizing the protein of interest. The required amount of Abs is dependent on the source and should be determined empirically. However, an amount of 0.5–5 µg of a monoclonal Ab or affinity-purified antiserum is usually sufficient to isolate enough protein for subsequent Western blot analyses or enzymatic assays. The following protocol for the purification of protein kinases for immune-complex kinase assays can be applied to both B lymphoma lines and primary B cells. Alternatively, the purified proteins can also be analyzed for their phosphorylation status using phospho-specific Abs.

1. $1–2 \times 10^7$ B cells are lysed in the lysis buffer of choice as described in **Subheading 3.1.2.**
2. TCLs are precleared with protein G Sepharose beads rotating at 4°C for 1 h.
3. Incubation of precleared TCLs with 0.5–5 µg Abs is done on ice for 1 h. If using Abs already bound to Sepharose, then incubate on a wheel at 4°C overnight (proceed directly with **step 5**).
4. Add the same amount of protein G Sepharose beads as used for the preclearing step (**step 2**), and incubate on a wheel at 4°C overnight.
5. The beads are washed at least three times with 1 mL of lysis buffer.
6. Subsequently, the beads are resuspended in 50 µL lysis buffer, mixed with 50 µL 2X sample buffer, and boiled for 5 min (for SDS-PAGE and Western blotting, *see* **Subheading 3.4.**) or are equilibrated with an appropriate buffer for kinase assays (*see* **Note 2**).

3.3. Protein Purification Via GST Fusion Proteins

Many signaling proteins comprise a modular structure enabling them via their different domains to participate in various protein complexes *(42)*. Among the protein–protein interaction domains studied best are the SH2 and SH3 domains, which depending on the sequence context bind to phosphorylated tyrosine residues and proline-rich regions, respectively *(43)*. To gain informa-

PV

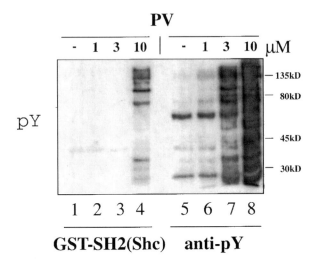

Fig. 2. Studying subsets of tyrosine-phosphorylated proteins by means of GST fusion protein pull-down experiments. K46 B cells were either left unstimulated or stimulated with different concentrations of PV (1, 3, and 10 µM) for 3 min. TCLs were subjected to precipitation with GST–SH2(Shc) fusion protein (*lanes 1–4*) or immuno-precipitation with anti-phospho-tyrosine (pY) Abs (*lanes 5–8*). Precipitates were separated by SDS-PAGE, and precipitated proteins were analyzed by anti-pY Western blotting. The anti-pY immunoprecipitation is more sensitive in terms of total recovery of tyrosine-phosphorylated proteins. The GST–SH2(Shc) pull-down, however, is more selective and enables one to study the subset of tyrosine-phosphorylated proteins, which bind—directly or indirectly—to the SH2-domain of the adaptor protein Shc.

tion about specific binding partners of a given domain, the respective comple-mentary deoxyribonucleic acid (cDNA) can be fused to the cDNA of the enzyme glutathione-*S*-transferase (GST). The resulting GST-signaling domain chimera can be expressed in bacteria and purified from bacterial lysates. This GST fusion protein is then used to precipitate binding partners from lysates of differentially stimulated cells. To this end, TCLs (prepared as described in **Subheading 3.1.**) are incubated with an appropriate amount of GST fusion protein containing the signaling domain of interest. The required amount of fusion protein is between 2 and 10 µg for 1–2 × 10^7 cells. This amount usually enables the isolation of a sufficient amount of protein for subsequent Western blot analysis. Because a given SH2 or SH3 domain only binds to a limited number of (tyrosine-phosphorylated) proteins, pull-down experiments using GST fusion proteins can be used to visualize only a specific subset of all tyro-sine-phosphorylated proteins (**Fig. 2**).

3.3.1. Transformation of Competent Escherichia coli

1. Thaw 50 µL of competent bacteria (e.g., DH10B or DH5α) on ice. Mix them with 100 ng of the expression plasmid coding for the respective GST fusion protein (e.g., pGEX-2T-Shc-SH2 *[44]*) and incubate on ice for 30 min.
2. Then heat the mixture to 42°C for 2 min (heat shock) and, after the addition of 1 mL LB medium, incubate at 37°C for 30 min.
3. Subsequently, plate the bacteria onto selective LB agar plates containing 100 µg/mL ampicillin (LB/amp agar plates), and then incubate at 37°C overnight.

3.3.2. Preparation of GST Fusion Proteins

This protocol is designed for 250-mL batches and should yield approx 500 µg of GST fusion protein. When scaling up, multiple batches are better than doing one large batch.

Day 1 (afternoon)

1. Inoculate a 200-mL culture of LB/amp (50 µg/mL) with one colony (ideal) or from a stab of frozen culture.
2. Incubate overnight in a shaker at 37°C.

Day 2

1. Add the 20 mL overnight culture to 230 mL of LB/amp, grow to an optical density $(OD)_{600}$ of 0.8–1.0 (approx 2–3 h).
2. Add isopropyl-β-D-thiogalactopyranoside (IPTG) to a final concentration of 100 µM, and incubate at 37°C for 3–4 h or at room temperature overnight.
3. Pellet the bacteria (2500g, 15 min, 4°C), pour off the supernatant, and freeze the pellet at –20°C (pellet should be pale yellow, not brown or black).

Day 3

1. Thaw pellet on ice and resuspend in 10 mL of cold lysis buffer.
2. Lyse for 30 min on ice with occasional mixing.
3. Sonicate on ice (20 bursts of 10 s each), maintaining the culture temperature below 10°C.
4. Centrifuge (11,600g, 30 min, 4°C).
5. Optional: Freeze lysate containing GST fusion protein at –80°C for future use. (Particular GST fusion proteins, once they are purified, are rather unstable, and it is advantageous to process only the amount of bacterial lysate needed for any particular experiment).
6. Wash glutathione Sepharose beads (glut beads) three times in wash buffer.
7. Coincubate lysate with 500 µL of washed glut beads for 3 h at 4°C on a nutator or rotator (use 15-mL conical Falcon tubes).

8. Save supernatant for reextraction and SDS-PAGE analysis.
9. Wash glut beads three times with 10 mL of wash buffer and determine yield by comparing GST fusion protein purified by 5 μL of glut beads to a bovine serum albumin (BSA) standard via SDS-PAGE and subsequent Coomassie staining of the gel.

3.3.3. Precipitation Using GST Fusion Proteins

1. 1–2×10^7 B cells are lysed as described above (**Subheading 3.1.**).
2. TCLs are precleared with 10 μL glut beads rotating at 4°C for 1 h.
3. Incubate precleared TCLs with 2–10 μg of GST fusion protein bound to glut beads on a wheel at 4°C overnight.
4. The beads are washed at least three times with 1 mL of lysis buffer.
5. Subsequently, resuspend the beads in 50 μL lysis buffer mixed with 50 μL 2X sample buffer, and boil for 5 min for SDS-PAGE and Western blotting (**Subheading 3.4.**).

3.4. SDS-PAGE and Western Blot Analysis

Denatured protein samples are separated according to the principle of the discontinuous (DISC) SDS-PAGE protocol by Laemmli *(45)*. Depending on the size of the protein of interest, separating gels with 8, 10, or 12% polyacrylamide can be employed.

1. Prepare a gel according to the tables in **Subheading 2.1., steps 20–21**. After pouring, the separating gel is polymerized for at least 2 h before it is overlaid by a 5% stacking gel.
2. Perform electrophoresis according to the recommendations for your electrophoresis device.
3. After electrophoretic separation, the gels are equilibrated for 2 min in transfer buffer, and the separated proteins are transferred to PVDF membranes (Immobilon P; Millipore) by wet electrotransfer using a tank blot device (TransBlot system; Bio-Rad).
4. After transfer, the membranes are blocked with 30–50 g/L skim milk powder dissolved in phosphate-buffered saline (PBS) containing 0.1% Tween-20 (PBST) for 30 min.
5. For protein detection, membranes are incubated with primary Abs diluted in blocking solution for at least 1 h at room temperature or overnight (4°C).
6. Wash three times for 15 min with PBST, and then incubate with horseradish peroxidase (HRP)-conjugated secondary Abs at room temperature for 1 h.
7. Wash four times with PBST.
8. Detect immunoreactive bands by chemiluminescence using the ECL/Hyperfilm ECL system (Amersham).

3.5. Phosphospecific Abs As an Alternative
to In Vitro Kinase Assays

The invention of phospho-specific Abs has revolutionized the field of signal transduction during the last 5 yr. As many phosphorylation events correlate strongly with enzymatic activity, the monitoring of the phosphorylation state of a protein at a particular amino acid residue can often be used as a surrogate marker for enzymatic activity. Using phospho-specific Abs, it is therefore possible to assay the activity of a particular enzyme in total cellular lysates, which makes time- and material-consuming in vitro enzymatic assays unnecessary (*see* **Note 2**) and therefore also reduces the risk of in vitro artifacts. The comparison of the signal generated by the phospho-specific Ab relative to the signal generated by the Ab recognizing the protein, irrespective of its phosphorylation state, indicates the extent of activation of the protein of interest.

Figures 3 and **4** provide several examples of how the activation of intracellular signaling pathways can be monitored using phospho-specific Abs. We have listed in **Table 3** several phospho-specific Abs that have been successfully used in our laboratory, as well as Abs recognizing the nonphosphorylated protein. **Table 4** lists conjugated secondary Abs for Western blotting.

3.6. Low-Molecular-Weight Enzyme Inhibitors As a Tool
for the Functional Analysis of Intracellular Signaling Pathways

A growing number of low-molecular-weight enzyme inhibitors that target specific key signaling elements are now widely used as tools for the functional analysis of a given signal transduction pathway in a biological response toward an extracellular stimulus such as BCR engagement. For example, the importance of the RAF/MEK/ERK pathway in pre-B cells and in the effector functions of mature B cells has been demonstrated by using the highly MEK-specific inhibitors PD98059 and U0126 *(46–48)*. As outlined above, BCR engagement results in the simultaneous activation of several signaling pathways, which often converge on downstream events such as the assembly of active transcription factor complexes. By using a panel of small molecule inhibitors, it is therefore possible to dissect the individual contribution of signaling pathways to such complicated events *(49)*. For an overview of the variety, mechanism of inhibition, and specificity of these inhibitors, the reader is referred to the work by Davies et al. *(50)*. An in vivo proof of the unspecificity of a widely used, so-called PKC-δ-specific inhibitor (Rottlerin) is demonstrated in **ref. *51***.

The following protocol has been successfully applied for the complete inhibition of the Raf/MEK/ERK and PI3K/PKB pathways in DT40 cells by the inhibitors U0126 and LY294002, respectively.

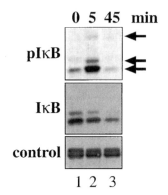

Fig. 3. Using phospho-specific Abs to study the posttranslational modifications of signaling proteins as exemplified by the BCR-stimulated phosphorylation and degradation of IκBα. DT40 cells were stimulated with 5 μg anti-IgM Ab M4 for the indicated time-points, and total cellular lysates were subject to Western blot analysis (10% SDS-PAGE). The phosphorylation of IκBα, which occurs at Ser32 and peaks 5 min after the onset of BCR engagement, was detected by using a phospho-specific Ab for phosphorylated Ser32 (*top panel*). This phosphorylation event triggers the polyubiquitinylation of IκBα, which in turn results in its rapid degradation by the 26S proteasome *(29)*. This degradation is indicated by the decrease of total IκBα (*middle panel*) relative to the control protein (*bottom panel*), which is unspecifically recognized by the IκBα Abs. Note that the anti-phospho-Ser32 Abs recognize IκBα proteins with decreased electrophoretic mobilities (*arrows*), which represent most likely phosphorylated and polyubiquitinylated IκBα species.

1. Adjust DT40 cells to a density of 10^7 cells/mL by centrifugation and resuspension.
2. Add the MEK inhibitor U0126 (Promega) or the PI3K inhibitor LY249002 (Calbiochem) to the cell suspension to achieve a final concentration of 50 μ*M* or 10 μ*M*, respectively. As a vehicle control, treat an aliquot of cells with the same volume of dimethyl sulfoxide (DMSO).
3. Incubate the cells for 20 to 30 min at 37°C.
4. Stimulate the cells with 5 μg anti-IgM M4 Ab.
5. Following incubation, the cells are centrifuged in a tabletop centrifuge for 10 s and resuspended in 200 μL ice-cold lysis buffer. The lysis is performed on ice for 30 min, and debris is subsequently removed by centrifugation (11,600*g*; 4°C; 10 min).
6. The supernatants are mixed with an equal volume of 2X sample buffer, and proteins are denatured by boiling for 5 min and subsequently chilled on ice.
7. Thereafter, 7.5×10^5 to 10^6 cell equivalents are separated by SDS-PAGE, transferred to PVDF-membranes, and analyzed by Western blotting.
8. Confirm inhibition of the Raf/MEK/ERK pathway and PI3K/PKB pathway by Western blot analysis for the activated, phosphorylated forms of ERK and PKB, respectively (*see* **Fig. 4**).

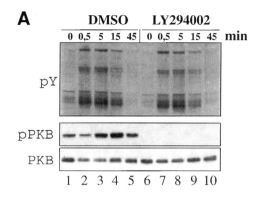

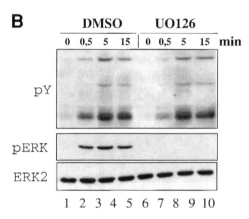

Fig. 4. The use of low-molecular-weight signaling enzyme inhibitors. (**A**) DT40 B cells were pretreated with either DMSO (vehicle control) or 10 μ*M* of the PI3K inhibitor LY294002 for 30 min and then stimulated with 5 μg anti-IgM Ab M4 for the indicated times. Tyrosine-phosphorylated proteins such as BASH were detected with anti-phospho-tyrosine Ab 4G10. Activated PKB was detected with an Ab raised against PKB phosphorylated at Ser473 (pPKB), whereas the total amount of PKB was detected with an anti-PKB Ab. Note that LY294002-treated DT40 cells display a normal pattern of PTK-substrate phosphorylation, whereas the PI3K-dependent phosphorylation of PKB at Ser473 is completely abolished. (**B**) DT40 B cells were pretreated with either DMSO (vehicle control) or 50 μ*M* of the MEK inhibitor U0126 for 20 min and then stimulated with 5 μg anti-IgM Ab M4 for the indicated times. Tyrosine-phosphory- lated proteins were detected with anti-phospho-tyrosine Ab 4G10. Activated ERK was detected with an Ab raised against the dual phosphorylated Thr-Glu-Tyr (TEY) motif (pERK), whereas the total amount of ERK2, the predominantly expressed ERK isoform in B lymphocytes, was detected with an anti-ERK2 Ab. Note that U0126-treated DT40 cells display a normal pattern of PTK-substrate phosphorylation, whereas the MEK- dependent phosphorylation of ERK is completely abolished.

Table 3
List of Antibodies (applied working concentration, by supplier or determined experimentally by titration)

Antibody	Origin/Isotype	Immunogen/Recognized epitope	Species specificity	Source	Applied concentration
Anti-phospho-Btk (Tyr 223)	Polyclonal rabbit IgG-fraction	Phosphorylated Tyr223 (human Btk)	Human, mouse	Cell Signaling Technology/ New England Biolabs	1:1000
Anti-Btk (M-138): sc-1696	Polyclonal rabbit IgG-fraction	Amino acids 1–391 of Btk (mouse)	Human, mouse, rat	Santa Cruz Biotechnology	1:200
Antiactivated phospho-p38	Polyclonal rabbit IgG fraction	Dual phosphorylated p38 (pThr180/pTyr 182)	Human, mouse	Cell Signaling Technology/ New England Biolabs	1:1000
Anti-p38	Polyclonal rabbit IgG fraction	Amino acids 341–360 (human)	Human, mouse	Cell Signaling Technology/ New England Biolabs	1:1000
Anti-phospho-PKCδ	Polyclonal rabbit IgG-fraction	PKCδ phosphorylated at Thr505	Human, mouse	Cell Signaling Technology/ New England Biolabs	1:1000
Anti-PKCδ (C-17)	Polyclonal rabbit IgG-fraction	C-terminus of rat PKCδ	Human, mouse, rat	Santa Cruz Biotechnology	1:1000
Anti-phospho-GSK3 α/β	Polyclonal rabbit IgG fraction	Dual phosphorylated GSK3 α/β	Human, mouse, rat	Cell Signaling Technology/ New England Biolabs	1:1000
Anti-phospho-PLC-γ1 (sc 1294-R) Tyr 783	Polyclonal rabbit IgG fraction	Human PLCγ1 phosphorylated at Tyr783	Human, mouse	Santa Cruz Biotechnology	1:400
Anti-PLC-γ1 (530)	Polyclonal rabbit IgG fraction	Amino acids 530–850 (rat PLCγ1)	Human, mouse, rat	Santa Cruz Biotechnology	1:400
Antiactivated MAP-Kinase 12D4	Murine, monoclonal IgG1κ	Dual phosphorylated p42MAPK and p44MAPK pTEpY-Motif (position 183–185)	Human, mouse, rat, chicken	NanoTools	1:1000
Anti-ERK2	Murine, monoclonal, IgG$_{2b}$	Rat ERK2/unknown	Human, mouse, rat, chicken	Transduction Laboratories	1:1000
Anti-phospho-IκB-α (Ser32)	Polyclonal rabbit IgG-fraction	IκBα phosphorylated at serine 32	Human, mouse, chicken	Cell Signaling Technology/ New England Biolabs	1:1000
Anti-IκB-α	Polyclonal rabbit IgG fraction	Unknown	Human, mouse, chicken	Cell Signaling Technology/ New England Biolabs	1:1000

Antibody	Type	Raised against	Reactivity	Source	Dilution
Anti-phospho-MEK	Polyclonal IgG fraction	Raised against phosphopeptide corresponding to residues around Ser 217/221 of human MEK 1/2	Human, mouse, rat, chicken	Cell Signaling Technology/ New England Biolabs	1:1000
Anti-MEK-1 (C-18) cat. no. sc-219	Polyclonal rabbit IgG fraction	AS 376–393 (rat)	Human, mouse, rat, chicken	Santa Cruz Biotechnology	1:2000
Anti-phospho-PKB/Akt (S473)	Rabbit IgG	Raised against a synthetic phosphopeptide corresponding to residues around S473	Human, mouse, chicken	Cell Signaling Technology/ New England Biolabs	1:1000
Anti-phospho-PKB/Akt (T308)	Rabbit IgG	Raised against a synthetic phosphopeptide corresponding to residues around T308	Human, mouse, chicken	Cell Signaling Technology/ New England Biolabs	1:1000
Anti-PKB/Akt	Rabbit IgG	Raised against a synthetic peptide corresponding to residues 466 to 479 of murine PKB	Human, mouse, chicken	Cell Signaling Technology/ New England Biolabs	1:1000
Anti-phospho-Raf-1 (Ser259)	Polyclonal rabbit IgG-fraction	Raf-1 phosphorylated at serine 259	Human, mouse, chicken	Cell Signaling Technology	1:1000
Anti-phospho-Raf-1 (Ser338)	Rat monoclonal IgG$_1$	Raf-1 phosphorylated at Ser338, irrespective of the phosphorylation state of Tyr341; B-Raf phosphorylated at Ser445; raised against synthetic peptide GQRDpSSYpYWEIEAS of Raf-1	Raf-1 and B-Raf (human, mouse, rat, chicken)	Upstate Biotechnology	1:1000
Anti-Raf-B (H-145)	Polyclonal rabbit IgG fraction	Amino acids 12–156 of human B-Raf (unique N-terminal region)	Human, mouse, rat, chicken	Santa Cruz Biotechnology	1:1000
Anti-Raf-1 (C-12)	Polyclonal IgG-fraction	Amino acids 637–648 (human)	Human, mouse, rat, chicken	Santa Cruz Biotechnology	1:1000
Anti-phospho-tyrosine 4G10	Murine, monoclonal IgG2bκ	Phosphotyramin-keyhole limpet-hemocyanine conjugate		Upstate Biotechnology	1:1000

205

Table 4
List of Conjugated Secondary Antibodies

Antibody	Origin/Isotype	Specificity	Source	Applied concentration
Goat antirat IgG-HRPO	Goat serum	Rat heavy and light chain	Pierce	1:8000
Goat antimouse IgM-HRPO	Goat serum	Murine heavy and light chain	Pierce	1:8000
Goat antimouse IgG-HRPO	Goat serum	Murine heavy and light chain	Pierce	1:3000–1:10000
Rabbit antigoat IgG-HRPO	Rabbit serum	Goat heavy and light chain	Pierce	1:8000
Goat antirabbit IgG-HRPO	Goat serum	Rabbit heavy and light chain	Pierce	1:8000

The applied working concentration was either provided by the supplier or determined experimentally by titration.

3.7. Analysis of BCR-Mediated Ca^{2+} Mobilization by Flow Cytometry

Apart from other technical approaches, BCR-mediated Ca^{2+} mobilization can be easily and reliably analyzed by flow cytometry using the method described by June and Rabinovitch *(52)*. In brief, this method is based on the dye indo-1 (1-[2-Amino-5-(6-carboxy-2-lindolyl)phenoxy]-2-(2-amino-5-methylphenoxy)ethane-*N,N,N′,N′*-tetraacetic acid). Indo-1 indicates a rise in the cytoplasmic calcium concentration ($[Ca^{2+}]_i$) by an increase in the ratio of a lower to a higher emission wavelength. Indo-1 is nontoxic, is readily taken up by living cells, and is uniformly distributed within the cytoplasm.

The following protocol has been successfully applied to cultured B-cell lines such as DT40 or K46 as well as to primary human and murine B cells (*see* **Fig. 5**).

1. Adjust the cells to a density of 1×10^7/mL in starvation medium (RPMI-1640; 2 m*M* L-glutamine, 100 U/mL penicillin, 100 µg/mL streptomycin, 10 m*M*

Fig. 5. *(see facing page)* Calcium mobilization in different B-cell lines. (**A**) DT40 or K46 B-cell lymphoma cells were loaded with the Ca^{2+}-sensitive fluorescent dye indo-1 for 45 min and then stimulated with 5 µg/mL mouse antichicken IgM Ab M4 or 5 µg/mL goat antimouse κ Abs, respectively . The addition of the stimulants is indicated by a gap in the baseline. The fluorescent properties of the cell suspension were recorded for 511 s at two recordings per second, and the FL4/FL3-ratio (*y*-axis) indicative for the cellular calcium concentration is plotted vs time (*x*-axis) using the

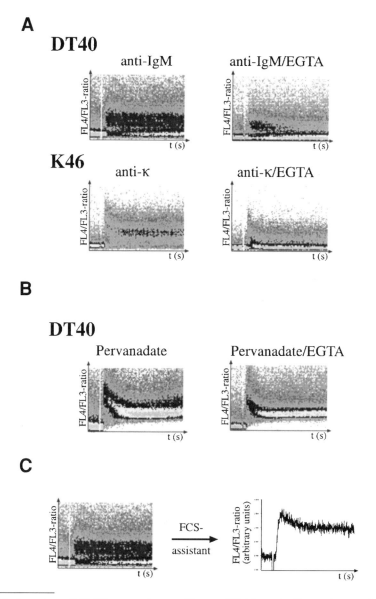

Fig. 5. *(continued)* CellQuest software (Becton Dickinson). Shown is a multicolored density plot with the following color code: green (lowest density) → orange → blue → yellow → light blue (highest density). **(B)** Experiment as described in **(A)** except that the DT40 cells were stimulated with 75 µ*M* PV. Note that addition of 1 m*M* ethylene glycol *bis* (2-aminoethyl ether)-*N,N,N′N′*-tetraacetic acid (EGTA), which chelates extracellular C^{2+}, 2 min prior to the addition of the stimulants strongly attenuates both anti-Ig Ab and PV-mediated calcium mobilization. **(C)** The change in the ratio between FL-4 and FL-3 measured in anti-IgM-stimulated DT40 cells was calculated using the FCSassistant 1.3.1a beta software and is plotted on the *y*-axis against time in seconds on the *x*-axis. Each time-point (0.5 s) represents the integration of the fluorescent parameters of approx 250 cells.

HEPES, 50 μ*M* β-mercaptoethanol, 1% FCS; for DT40 add also 0.25% CS).

2. Subsequently, 1 mL of cell suspension receives 15 μL loading solution (25 μL 2 m*M* indo-1; 25 μL 20% pluronic acid (w/v in DMSO), 113 μL FCS) and is incubated for 45 min at 37°C.

3. Then the cells are washed with 10 mL ice-cold starvation medium, resuspended in 3.3 mL starvation medium, kept on ice, and protected against daylight prior to the analysis.

4. An appropriate volume of the cell suspension is diluted in 2 mL starvation medium (prewarmed to 37°C) so that a flow rate of approx 500 cells/s is achieved.

5. For the evaluation of BCR-mediated Ca^{2+} mobilization, the fluorescence properties of DT40 cells with a forward scatter/side scatter (FSC/SSC) ratio indicative for viable cells are monitored. To record the $[Ca^{2+}]_i$ baseline, indo-1-loaded cells are measured for approx 50 s prior to the addition of the indicated stimulants. As mobilization of Ca^{2+} ions is an energy-dependent and temperature-sensitive process, the temperature of the cell suspension is kept constant at 37°C during the recording.

6. Changes of $[Ca^{2+}]_i$ were evaluated with density plots of the 405 nm/525 nm ratio vs time. For a better demonstration of the changes in $[Ca^{2+}]_i$, the raw data, which were originally acquired with the CellQuest software (Becton Dickinson), can be converted into a standard curve using the FCSassistant software (version 1.3.1a beta; http://www.fcspress.com/).

4. Notes

1. PV represents an unstable complex; to obtain reproducible results, PV must always be prepared immediately before each experiment. In contrast, the 10 m*M* sodium orthovanadate and 30% H_2O_2 stock solutions can be stored at 4°C for a long time.

2. In-vitro immune-complex kinase assays: owing to a plethora of different kinase assay protocols, we cannot provide a general protocol for immune-complex kinase assays. In general, the immune-complex-containing protein G Sepharose beads are washed at least three times with the appropriate kinase buffer to remove detergents and to equilibrate the beads to an environment allowing proper kinase function. Subsequently, a certain amount of the purified kinase is mixed with an appropriate volume of kinase buffer, recombinant substrate protein, and adenosine triphosphate (ATP). In general, tyrosine phosphorylation of substrate proteins can be easily detected by using anti-phospho-tyrosine Abs, which makes the use of radioactive ATP unnecessary and reduces the risk of artifacts caused by copurified Ser/Thr kinases. Likewise, specific phosphorylation events, such as the phosphorylation of recombinant GST-MEK by Raf-kinases, can be quantified by using phospho-specific Abs raised against the serine residues that are phosphorylated by Raf *(53)*. For kinase assays with the BCR-proximal kinase Syk and the Src-family kinase Lyn, the reader is referred to Rolli et al. (2002) *(15)* and Larbolette et al. (1999) *(54)*, respectively.

Acknowledgments

This work was supported by the Deutsche Forschungsgemeinschaft through grant SFB 388 (M. H. and M. R.).

References

1. Rajewsky, K. (1996) Clonal selection and learning in the antibody system. *Nature* **381,** 751– 758.
2. Benschop, R. J. and Cambier, J. C. (1999) B cell development: signal transduction by antigen receptors and their surrogates. *Curr. Opin. Immunol.* **11,** 143–151.
3. Schamel, W. W. and Reth, M. (2000) Monomeric and oligomeric complexes of the B cell antigen receptor. *Immunity* **13,** 5–14.
4. Reth, M., and Wienands, J. (1997) Initiation and processing of signals from the B cell antigen receptor. *Annu. Rev. Immunol.* **15,** 453–479.
5. Reth, M., Wienands, J., and Schamel, W. W. (2000) An unsolved problem of the clonal selection theory and the model of an oligomeric B-cell antigen receptor. *Immunol. Rev.* **176,** 10–18.
6. Wienands, J., Larbolette, O., and Reth, M. (1996) Evidence for a preformed transducer complex organized by the B cell antigen receptor. *Proc. Natl. Acad. Sci. USA* **93,** 7865–7870.
7. Zhang, Y., Wienands, J., Zurn, C., and Reth, M. (1998) Induction of the antigen receptor expression on B lymphocytes results in rapid competence for signaling of SLP-65 and Syk. *EMBO J.* **17,** 7304–7310.
8. Lam, K. P., Kuhn, R., and Rajewsky, K. (1997) In vivo ablation of surface immunoglobulin on mature B cells by inducible gene targeting results in rapid cell death. *Cell* **90,** 1073–1083.
9. Torres, R. M., Flaswinkel, H., Reth, M., and Rajewsky, K. (1996) Aberrant B cell development and immune response in mice with a compromised BCR complex. *Science* **272,** 1802–1804.
10. Reichlin, A., Hu, Y., Meffre, E., et al. (2001) B cell development is arrested at the immature B cell stage in mice carrying a mutation in the cytoplasmic domain of immunoglobulin beta. *J. Exp. Med.* **193,** 13–23.
11. Kraus, M., Saijo, K., Torres, R. M., and Rajewsky, K. (1999) Ig-alpha cytoplasmic truncation renders immature B cells more sensitive to antigen contact. *Immunity* **11,** 537–545.
12. Torres, R. M. and Hafen, K. (1999) A negative regulatory role for Ig-alpha during B cell development. *Immunity* **11,** 527–536.
13. Campbell, K. S. (1999) Signal transduction from the B cell antigen-receptor. *Curr. Opin. Immunol.* **11,** 256–264.
14. Pani, G., Kozlowski, M., Cambier, J. C., Mills, G. B., and Siminovitch, K. A. (1995) Identification of the tyrosine phosphatase PTP1C as a B cell antigen receptor-associated protein involved in the regulation of B cell signaling. *J. Exp. Med.* **181,** 2077–2084.

15. Rolli, V., Gallwitz, M., Wossning, T., et al. (2002) Amplification of B cell antigen receptor signaling by a Syk/ITAM positive feedback loop. *Mol. Cell.* **10,** 1057–1069.

16. Takata, M., Sabe, H., Hata, A., et al. (1994) Tyrosine kinases Lyn and Syk regulate B cell receptor-coupled Ca^{2+} mobilization through distinct pathways. *EMBO J.* **13,** 1341–1349.

17. Jiang, A., Craxton, A., Kurosaki, T., and Clark, E. A. (1998) Different protein tyrosine kinases are required for B cell antigen receptor-mediated activation of extracellular signal-regulated kinase, c-Jun NH2-terminal kinase 1, and p38 mitogen-activated protein kinase. *J. Exp. Med.* **188,** 1297–1306.

18. Kurosaki, T. and Tsukada, S. (2000) BLNK: connecting Syk and Btk to calcium signals. *Immunity* **12,** 1–5.

19. Kurosaki, T., Maeda, A., Ishiai, M., Hashimoto, A., Inabe, K., and Takata, M. (2000) Regulation of the phospholipase C-gamma2 pathway in B cells. *Immunol. Rev.* **176,** 19–29.

20. Gold, M. R. (2002) To make antibodies or not: signaling by the B-cell antigen receptor. *Trends Pharmacol. Sci.* **23,** 316–324.

21. Tuveson, D. A., Carter, R. H., Soltoff, S. P., and Fearon, D. T. (1993) CD19 of B cells as a surrogate kinase insert region to bind phosphatidylinositol 3-kinase. *Science* **260,** 986–989.

22. Okada, T., Maeda, A., Iwamatsu, A., Gotoh, K., and Kurosaki, T. (2000) BCAP: the tyrosine kinase substrate that connects B cell receptor to phosphoinositide 3-kinase activation. *Immunity* **13,** 817–827.

23. Brummer, T., Shaw, P. E., Reth, M., and Misawa, Y. (2002) Inducible gene deletion reveals different roles for B-Raf and Raf-1 in B-cell antigen receptor signalling. *EMBO J.* **21,** 5611–5622.

24. Brazil, D. P. and Hemmings, B. A. (2001) Ten years of protein kinase B signalling: a hard Akt to follow. *Trends Biochem. Sci.* **26,** 657–664.

25. Gold, M. R. (2000) Intermediary signaling effectors coupling the B-cell receptor to the nucleus. *Curr. Top. Microbiol. Immunol.* **245,** 77–134.

26. Hashimoto, A., Okada, H., Jiang, A., et al. (1998) Involvement of guanosine triphosphatases and phospholipase C-gamma2 in extracellular signal-regulated kinase, c-Jun NH2-terminal kinase, and p38 mitogen-activated protein kinase activation by the B cell antigen receptor. *J. Exp. Med.* **188,** 1287–1295.

27. Gugasyan, R., Grumont, R., Grossmann, M., et al. (2000) Rel/NF-kappaB transcription factors: key mediators of B-cell activation. *Immunol. Rev.* **176,** 134–140.

28. Davis, R. E., Brown, K. D., Siebenlist, U., and Staudt, L. M. (2001) Constitutive nuclear factor kappaB activity is required for survival of activated B cell-like diffuse large B cell lymphoma cells. *J. Exp. Med.* **194,** 1861–1874.

29. Karin, M. and Ben-Neriah, Y. (2000) Phosphorylation meets ubiquitination: the control of NF-[kappa]B activity. *Annu. Rev. Immunol.* **18,** 621–663.

30. Bajpai, U. D., Zhang, K., Teutsch, M., Sen, R., and Wortis, H. H. (2000) Bruton's tyrosine kinase links the B cell receptor to nuclear factor kappaB activation. *J. Exp. Med.* **191,** 1735–1744.

31. Petro, J. B. and Khan, W. N. (2001) Phospholipase C-gamma 2 couples Bruton's tyrosine kinase to the NF-kappaB signaling pathway in B lymphocytes. *J. Biol. Chem.* **276,** 1715–1719.

32. Baba, T. W., Giroir, B. P., and Humphries, E. H. (1985) Cell lines derived from avian lymphomas exhibit two distinct phenotypes. *Virology* **144,** 139–151.

33. Masteller, E. L., Larsen, R. D., Carlson, L. M., et al. (1995) Chicken B cells undergo discrete developmental changes in surface carbohydrate structure that appear to play a role in directing lymphocyte migration during embryogenesis. *Development* **121,** 1657–1667.

34. Zhang, Y., Riesterer, C., Ayrall, A. M., Sablitzky, F., Littlewood, T. D., and Reth, M. (1996) Inducible site-directed recombination in mouse embryonic stem cells. *Nucleic Acids Res.* **24,** 543–548.

35. Lanier, L. L. and Warner, N. L. (1981) Cell cycle related heterogeneity of Ia antigen expression on a murine B lymphoma cell line: analysis by flow cytometry. *J. Immunol.* **126,** 626–631.

36. Scott, D. W., Tuttle, J., Livnat, D., Haynes, W., Cogswell, J. P., and Keng, P. (1985) Lymphoma models for B-cell activation and tolerance, II: growth inhibition by anti-mu of WEHI-231 and the selection and properties of resistant mutants. *Cell Immunol.* **93,** 124–131.

37. Hombach, J., Tsubata, T., Leclercq, L., Stappert, H., and Reth, M. (1990) Molecular components of the B-cell antigen receptor complex of the IgM class. *Nature* **343,** 760–762.

38. Venkitaraman, A. R., Williams, G. T., Dariavach, P., and Neuberger, M. S. (1991) The B-cell antigen receptor of the five immunoglobulin classes. *Nature* **352,** 777–781.

39. Klein, G., Giovanella, B., Westman, A., Stehlin, J. S., and Mumford, D. (1975) An EBV-genome-negative cell line established from an American Burkitt lymphoma: receptor characteristics: EBV infectibility and permanent conversion into EBV-positive sublines by in vitro infection. *Intervirology* **5,** 319–334.

40. Reth, M. (2002) Hydrogen peroxide as second messenger in lymphocyte activation. *Nat. Immunol.* **3,** 1129–1134.

41. Brennan, W. A., Jr. and Lin, S.-H. (1996) Solubilization and purification of the rat liver insulin receptor. In *Strategies for Protein Purification and Characterization: A Laboratory Course Manual* (Marshak, D. R., Kadonaga, J. T., Burgess, R. R., Knuth, M. W, Brennan Jr., W. A., and Lin, S.-H., eds.), Cold Spring Harbor Laboratory Press, Cold Spring Harbor, NY.

42. Smith, F. D. and Scott, J. D. (2002) Signaling complexes: junctions on the intracellular information super highway. *Curr. Biol.* **12,** R32–R40.

43. Pawson, T. and Nash, P. (2000) Protein–protein interactions define specificity in signal transduction. *Genes Dev.* **14,** 1027–1047.

44. Leitges, M., Gimborn, K., Elis, W., et al. (2002) Protein kinase C-delta is a negative regulator of antigen-induced mast cell degranulation. *Mol. Cell Biol.* **22,** 3970–3980.

45. Laemmli, U. K. (1970) Cleavage of structural proteins during the assembly of the head of bacteriophage T4. *Nature* **227,** 680–685.

46. Fleming, H. E. and Paige, C. J. (2001) Pre-B cell receptor signaling mediates selective response to IL-7 at the pro-B to pre-B cell transition via an ERK/MAP kinase-dependent pathway. *Immunity* **15,** 521–531.

47. Richards, J. D., Dave, S. H., Chou, C. H., Mamchak, A. A., and DeFranco, A. L. (2001) Inhibition of the MEK/ERK signaling pathway blocks a subset of B cell responses to antigen. *J. Immunol.* **166,** 3855–3864.

48. Flemming, A., Brummer, T., Reth, M., and Jumaa, H. (2003) The adaptor protein SLP-65 acts as a tumor suppressor that limits pre-B cell expansion. *Nat. Immunol.* **4,** 38–43.

49. Glynne, R., Ghandour, G., Rayner, J., Mack, D. H., and Goodnow, C. C. (2000) B-lymphocyte quiescence, tolerance and activation as viewed by global gene expression profiling on microarrays. *Immunol. Rev.* **176,** 216–246.

50. Davies, S. P., Reddy, H., Caivano, M., and Cohen, P. (2000) Specificity and mechanism of action of some commonly used protein kinase inhibitors. *Biochem. J.* **351,** 95–105.

51. Leitges, M., Elis, W., Gimborn, K., and Huber, M. (2001) Rottlerin-independent attenuation of pervanadate-induced tyrosine phosphorylation events by protein kinase C-delta in hemopoietic cells. *Lab. Invest.* **81,** 1087–1095.

52. June, C. H. and Rabinovitch, P. S. (1990) Flow cytometric measurement of intracellular ionized calcium in single cells with indo-1 and fluo-3. *Methods Cell Biol.* **33,** 37–58.

53. Bondzi, C., Grant, S., and Krystal, G. W. (2000) A novel assay for the measurement of Raf-1 kinase activity. *Oncogene* **19,** 5030–5033.

54. Larbolette, O., Wollscheid, B., Schweikert, J., Nielsen, P. J., and Wienands, J. (1999) SH3P7 is a cytoskeleton adapter protein and is coupled to signal transduction from lymphocyte antigen receptors. *Mol. Cell Biol.* **19,** 1539–1546.

15

Isolation of Lipid Rafts From B Lymphocytes

**Anu Cherukuri, Shiang-Jong Tzeng, Arun Gidwani,
Hae Won Sohn, Pavel Tolar, Michelle D. Snyder,
and Susan K. Pierce**

Summary

Recent advances in cell biology have provided evidence that the plasma membrane is not a homogeneous lipid bilayer but rather contains within it sphingolipid- and cholesterol-rich membrane microdomains, termed *lipid rafts*, which serve as platforms for both receptor signaling and trafficking. In B lymphocytes lipid rafts appear to play a key role in the initiation of B-cell antigen receptor (BCR) signaling. Current methods to isolate lipid rafts rely on the relative detergent insolubility of lipid rafts as compared to the nonraft, glycerophospholipid bilayer. Here a method to isolate and characterize lipid rafts from B lymphocytes is described. Particular emphasis is given to the potential artifacts inherent in current procedures that rely on detergents to isolate lipid rafts and alternative technologies that may circumvent these.

Key Words

B cell; lipid rafts; B-cell antigen receptor; sphingolipids; cholesterol.

1. Introduction

Recently evidence has been provided that sphingolipid- and cholesterol-rich membrane microdomains, termed *lipid rafts*, play a role in the initiation of signaling by the B-cell antigen receptor (BCR) (reviewed in **ref. *1***). The sphingolipids in the outer leaflet of the plasma membrane pack tightly into gel-like microdomains owing to the saturated nature of their acyl chains (reviewed in **refs. *2* and *3***). The sphingolipids preferentially bind cholesterol, promoting the formation of a liquid ordered phase. The sphingolipids and cholesterol partition out of the glycerophospholipid bilayer, which exists in a liquid disordered phase attributable to the unsaturated, kinked acyl chains of the glycerophospholipids. Thus, the lipid rafts are relatively ordered domains that float in a sea of disor-

From: *Methods in Molecular Biology, vol. 271: B Cell Protocols*
Edited by: H. Gu and K. Rajewsky © Humana Press Inc., Totowa, NJ

dered glycerophospholipids *(2)*. The inner leaflet of the lipid raft that is coupled to the outer leaflet is less well characterized but is presumably composed of saturated phospholipids. The rafts in resting cells, often referred to as *elemental rafts*, are estimated to be small, submicroscopic, highly dynamic domains possibly containing only thousands of lipids *(2)*. The plasma membrane of immune cells is estimated to be composed of approx 40% raft membranes *(4)*, and thus the cell surface may be viewed as a dynamic mosaic of raft and nonraft domains.

A key feature of lipid rafts is that they provide a mechanism for the lateral segregation of proteins within the plasma membrane (reviewed in **ref. 2**). Most membrane proteins are excluded from lipid rafts but a few preferentially partition into rafts. These include cell-surface proteins that associate with the plasma membrane through glycosylphosphatidylinositol (GPI) linkage and cytoplasmic proteins that associate with the inner leaflet of the membrane owing to their dual acylation by saturated fatty acids (*N*-myristoylation and *S*-palmitoylation). A small number of transmembrane proteins constitutively associate with rafts, but such associations are usually dependent on palmitoylation of the protein. GPI-linked proteins and dually acylated proteins, such as the Src family kinase Lyn, that partition into B-cell rafts can be used as markers for rafts as can glycosphingolipids such as GM1, the ligand for the B subunit of cholera toxin *(5)*. The small size of elemental rafts suggests that these will not be homogenous and that each elemental raft may contain only a subset of all raft-resident proteins.

Lipid rafts are differentially soluble in certain nonionic detergents at low temperatures, and most isolation procedures are based on this property of rafts (reviewed in **refs. 2** and **3**). In the method provided here, rafts are isolated at 4°C from 1% Triton X-100 detergent lysates of B cells by their buoyant density on sucrose gradients *(5)*. The tightly packed acyl chains of the sphingolipids are likely responsible for the detergent insolubility of rafts because in a tightly packed state, lipid–lipid interactions can be more stable than lipid–detergent interactions *(3)*. The solubility of rafts can be influenced by a number of factors that include the choice of detergent, the temperature at which the procedure is carried out, and the composition of the particular cells from which rafts are isolated. Thus, understanding that at present rafts are only operationally defined by their relative detergent insolubility demands that caution be taken in the interpretation of results using the isolation procedure that follows.

The use of detergent solubility to isolate rafts has the potential to introduce a number of artifacts, including the possibility of inducing the formation of membrane structures that do not exist in living cells. Consequently, significant effort is directed toward detecting rafts in intact and/or living cells. Techniques that are being applied include chemical crosslinking *(6)* and fluorescence resonance energy transfer (FRET) *(7)* to detect the proximity of two proteins in the membrane, photonic force microscopy *(8)* to measure the local diffusion of single-membrane proteins, and single-fluorophore tracking microscopy *(9)* to monitor the dynamics of individual proteins and lipids in the plasma membrane. However, these techniques are technically sophisticated and most require expensive instrumentation that may not be generally available. However, when tested the identification of proteins in lipid rafts by such techniques in living cells has correlated with their detergent solubility *(10)* indicating that detergent solubility is a useful tool to study rafts.

The evidence to date indicates that in resting B cells, Lyn is detergent insoluble and the BCR is detergent soluble, as are a variety of other B-cell membrane proteins, including CD45, CD22, CD19/CD21, FcγRIIB, and CD20 (reviewed in **ref. *1***). On BCR crosslinking either by antigens or Abs specific for the BCR, the BCR becomes insoluble as do a number of components of the BCR signaling cascade. In addition, the insoluble BCR is phosphorylated, as are components of the signaling cascade. It has been proposed that the crosslinked BCR initially partitions into submicroscopic elemental lipid rafts *(1)*. These elemental rafts may cluster to form larger, more stable rafts with time, following BCR crosslinking, and may be large enough to be detected by light microscopy. The small elemental rafts are likely to be more detergent soluble as compared to larger, clustered rafts, and thus the BCR that resides in elemental rafts may be more soluble as compared to the BCR in clustered rafts. This possibility should be taken into account when interpreting results from experiments using detergent solubility to isolate rafts from B cells at different states of activation.

Following crosslinking, the BCR forms microscopic patches and ultimately polarized caps. Recently, the polarized caps have been reported to be ordered structures, analogous to the immunological synapse described for T cells *(11)*. The relationship between lipid rafts and the patched and capped B-cell structures has not been established and it is not known if rafts play a role in establishing the spatial order of B-cell receptors described in the polarized cap.

2. Materials

1. Beckman (Fullerton, CA) swinging bucket SW41 or SW55 rotors; 12 mL or 6 mL polyallomer Beckman centrifuge tubes.
2. CH27 mouse B lymphoma cells in log- or late log-phase growth with 85% or greater viability.
3. Phosphate-buffered saline (PBS), pH 7.4.
4. TNE: Tris-HCl, NaCl, ethylenediaminetetraacetic acid (EDTA) buffer, pH 7.5, prepared from 10X TNE stock: 5 mL 1 *M* Tris-HCl, pH 7.5, 50 mL 1.5 *M* NaCl, 5 mL 0.5 *M* EDTA, and 440 mL ddH$_2$O. Store at 4°C.
5. 10% Triton X-100: Triton X-100 (Sigma, St. Louis, MO) in 1X TNE. Stocks are stored at –20°C. Once thawed, they are stored at 4°C.

6. 1000X CLAP protease inhibitors: 2.5 mg/mL each chymostatin, leupeptin, antipain, and pepstatin A in dimethyl sulfoxide (DMSO). All reagents are purchased from Sigma. Stock is stored at –20°C in 20-μL aliquots. Once thawed it is stored at 4°C.

7. 100X phosphatase inhibitors: 200 mM Na$_3$VO$_4$, 1 M NaF (Sigma) in DMSO. Stock is stored at –20°C in 200-μL aliquots. Once thawed it is stored at 4°C.

8. 85% (w/v) Sucrose solution: stock solution of ultrapure sucrose (ICN Biomedicals, Aurora, OH) in 1X TNE buffer. Preparing the stock requires heating the sucrose solution. To avoid caramelizing the sucrose heat indirectly in a water bath. Stock is stored at –20°C in 15-mL aliquots. Once thawed it can be frozen again.

9. 35 and 5% (w/v) Sucrose solutions: prepared fresh from 85% sucrose in 1X TNE buffer.

10. Loose-fit Wheaton-33 borosilicate glass Dounce homogenizer (Wheaton Science Products, Milville, NJ).

11. 1% (w/v) Bovine serum albumin (BSA) (Sigma) in PBS (BSA/PBS).

12. Affinity-pure F(ab′)$_2$ goat antimouse IgM + IgG (Jackson ImmunoResearch, West Grove, PA). Store at 4°C.

13. 15-mL and 50-mL Sterile conical tubes (Corning Life Sciences, Acton, MA). Eppendorf microfuge tubes (Brinkmann, Westbury, NY), 5-mL, 10-mL, and 25-mL pipets (BD Biosciences Labware, Bedford, MA), Pasteur pipets (Kimble Glass Co., Vineland, NJ).

14. Beckman L8-80M ultracentrifuge, Sorvall RC-3C Plus centrifuge (Kendrove Inc, Newton, CT), orbital shaker (VWR Scientific, Westchester, PA).

15. Methyl-β-cyclodextrin (MβCD) (Sigma).

16. Sodium dodecyl sulfate-polyacrylamide gel electrophoresis (SDS-PAGE) and immunoblotting apparatus.

17. Cholera toxin B subunit (CTB) conjugated to horseradish peroxidase (HRP) (Sigma); rat Abs specific for CD45R (PharMingen, San Diego, CA), rabbit Abs specific for Lyn (Santa Cruz Biotechnology, Santa Cruz, CA), HRP-coupled goat Abs specific for rat Ig and HRP-coupled goat Abs specific for rabbit Ig (Jackson ImmunoResearch).

18. DiIC$_{16}$(3) and Fast DiI (Molecular Probes, Eugene, OR).

3. Methods

3.1. Preparation of Cells

CH27 cells, a mouse B-cell lymphoma, were grown in Dulbecco's modified Eagle's medium (DMEM) growth media containing fetal calf serum (FCS) in T-flasks at 37°C in a humidified incubator supplied with 5–7% CO$_2$ and harvested in log-phase growth (*see* **Note 1**). Each sucrose gradient requires 1×10^8 cells; therefore, each experimental condition will require 1×10^8 cells (*see* **Note 2**). Cells were transferred from T-flasks into 50-mL conical tubes using a 25-mL pipet and centrifuged in a Sorvall centrifuge at 1250 rpm for 5 min at 4°C. The supernatant media was discarded, and the cell pellet was washed three times in cold PBS buffer by resuspending the pellet each time using a 10-mL pipet in 10 mL of PBS and centrifuging at 458g for 5 min at 4°C after each resuspension. After the third wash an aliquot of the cells was counted for number and viability. Cells (1×10^8) were suspended in 1 mL of cold 1% BSA/PBS after the final wash and transferred to a 15-mL conical tube for activation. If the experimental design does not require activation in vitro, remove all supernatant from the pelleted cells and proceed to **Subheading 3.3.**

3.2. Activation

To activate CH27 cells to induce the BCR to partition into lipid rafts by BCR crosslinking, 1×10^8 cells were incubated in 1 mL 1% BSA/PBS containing 15 µg/mL affinity pure F(ab')$_2$ goat antimouse IgM + IgG on ice for 30 min on an orbital shaker. The activation reaction was stopped by adding 10 mL cold 1% BSA/PBS; the cells were washed twice by centrifuging at 458g for 5 min at 4°C in a Sorvall centrifuge, then resuspended in 1% BSA/PBS. After the final wash is completed care should be taken to aspirate all the supernatant from the cell pellet.

3.3. Lysis

Fresh lysis buffer was prepared by diluting 10% Triton X-100 in 1X TNE at 1:10 (*see* **Note 3**). CLAP was added to the lysis buffer at a dilution of 1:1000, and the phosphatase inhibitor was added at a dilution of 1:100. It is essential that the cells and lysis buffer be cold and that all subsequent procedures be carried out on ice. Warming the sample even briefly can result in solubilization of the rafts. The cell pellet (approx 1×10^8 cells/sample) was resuspended in 1 mL ice-cold lysis buffer in a 15-mL conical tube and pipetted up and down to resuspend. Samples were incubated on ice for 30 to 45 min. Shaking is not necessary for this step.

3.4. Homogenization of Lysates

Lysates were homogenized with 10 strokes of a loose-fit Dounce homogenizer. The bottom of the homogenizer was submerged in ice while homogenizing. Between samples the homogenizer was rinsed with ddH$_2$O and properly drained. The lysates were transferred to 15-mL conical tubes and centrifuged in the Sorvall centrifuge at 900g for 11 min at 4°C to remove mitochondrial, nuclei, and other large cellular debris. The supernatant lysates were carefully transferred with a 1-mL pipetman into Beckman SW41 polyallomer 12-mL centrifuge tubes (*see* **Note 4**) and placed on ice, with care to avoid pulling up any of the cell pellet. The pellet containing cell debris was discarded.

3.5. Discontinuous Sucrose Gradient (see Note 5)

Stock sucrose (85%) in TNE was thawed at room temperature. The cleared lysates were mixed 1:1 with 85% sucrose in the Beckman centrifuge tubes.

The solution, which is viscous, was pipetted up and down with a 1-mL nega-tive pressure pipetman to mix thoroughly, taking care to avoid air bubbles. Complete mixing at this step is absolutely essential. The 35 and 5% sucrose solutions were prepared fresh in 1X TNE buffer from the 85% sucrose stock, and phosphatase inhibitors were added to both solutions at a final concentration of 1:100. Keeping the tubes on ice, the diluted lysates were overlaid drop-by-drop using a 5-mL pipet with 6-mL of 35% sucrose in TNE followed by 3.5 mL of 5% sucrose in TNE. Care must be taken to add the sucrose solutions slowly so as not to disturb the discontinuous gradient formed. The sample tubes were balanced by weighing them on a digital balance for precise weight, and drops of 5% sucrose in TNE were added carefully at the top of the tube using a Pasteur pipet to bring all tubes to the same weight.

3.6. Ultracentrifugation and Collection of Fractions

Balanced sample tubes were placed in the tube holders of the prechilled SW41 Beckman rotor, caps tightened and the rotor with the tubes was placed carefully inside the precooled Beckman ultracentrifuge with care to avoid any shaking motion of the rotor buckets containing the sample tubes. To precool the ultracentrifuge it was turned on an hour before use, the temperature set to 4°C, and the vacuum turned on. All six buckets of the rotor need to be loaded even if all six chambers are not used. Samples were centrifuged at 200,000g for 16–20 h at 4°C.

After centrifugation was completed the tubes were carefully removed from the holders using forceps and immediately placed on ice. Generally a white cloudy or turbid material is visible at the interface of the 5% and 35% sucrose gradients indicating lipid raft membranes. Using a 1-mL pipetman, 1-mL frac-tions (*see* **Note 6**) were collected from the top of the gradient into Eppendorf tubes with care to discard the pipetman tip after every fraction was dispensed to avoid carryover between fractions. In total 12 fractions were collected from each sample with the last 2 fractions being more dense and viscous than the rest. Sometimes a small pellet may be visible at the bottom of the centrifuge tube (*see* **Note 7**). All gradient fractions were frozen at –20°C for long-term storage of several months to a year for future analysis. Fractions may also be stored at 4°C for short-term storage (1 wk).

3.7. Characterization of Lipid Rafts

Lipid rafts are operationally defined as cholesterol-dependent sphingolipid membrane microdomains that concentrate GPI-linked and dually acylated pro-teins and exclude most other membrane proteins. To verify that rafts have been isolated, the fractions from the sucrose density gradient are analyzed for

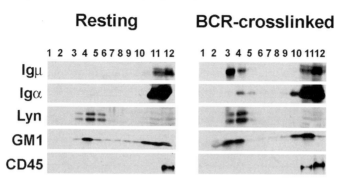

Fig. 1. Fractions from sucrose gradients on which the Triton X-100 lysates of 1×10^8 resting CH27 cells (*left panel*) or BCR-crosslinked CH27 cells (*right panel*) were applied were analyzed by SDS-PAGE and immunoblotting. The immunoblot was probed for Igμ, Igα, Lyn, and CD45 using specific Abs detected with HRP-labeled secondary Abs, or for GM1 using cholera toxin B subunit coupled to HRP.

the presence of the raft components, the Src-family kinase Lyn, and the glycosphingolipid GM1, and for the nonraft component CD45 (*see* **Note 8**). The sucrose gradient fractions are analyzed by SDS-PAGE and immunoblotting, probing the immunoblots for Lyn using Lyn-specific Abs, for GM1 using the B subunit of cholera toxin, and for CD45 using CD45-specific Abs. An example of the immunoblots of sucrose gradient fractions on which lysates from resting B cells and B cells activated by the crosslinking of the BCR were separated is shown (**Fig. 1**). In resting cells the BCR Igμ and Igα chains are entirely in the soluble fractions (fractions 11 and 12). Lyn is predominantly in the insoluble fractions (fractions 3–5). The presence of a significant amount of Lyn in the soluble fractions indicates that the rafts have been solubilized. GM1 is concentrated in rafts, although a variable amount also appears in the soluble fractions. CD45 is entirely in the soluble fraction, and its presence in the insoluble fractions suggests that the membranes have not been adequately solubilized (*see* **Note 9**).

The integrity of lipid rafts is dependent on cholesterol and the presence of the Lyn and GM1 markers in the light fractions of the sucrose gradient should be demonstrated to be dependent on cholesterol. To do so, harvest cells and wash twice in culture medium without serum. Resuspend cells at $1–2 \times 10^7$/mL in serum-free culture medium supplemented with 12.5 mM MβCD, and incubate 20 min at 37°C in a CO_2 incubator (*see* **Note 10**). Wash cells twice in serum-free culture medium. To replete cholesterol, incubate 3 h in culture medium with serum at 37°C in a CO_2 incubator. Proceed to raft isolation procedure.

4. Notes

1. In addition to CH27 cells the protocol given has been used to isolate rafts from several mouse (A20, WEHI), human (Ramos, BJAB, Daudi) and chicken (DT40) cell lines and from mouse splenic B cells and human peripheral blood B cells.

2. The minimum number of cells from which lipid rafts have been isolated and characterized is 10×10^6 in 0.2 mL detergent using a 6-mL sucrose gradient. The maximum number is 200×10^6 in 1 mL detergent using a 12-mL sucrose gradient. It may be necessary to adjust the concentration of the detergent depending on the number of cells to be solubilized. In general, for a given volume less detergent is required to solubilize fewer cells. The maximal and minimal cell numbers apply to cultured cell lines. For primary cells it is generally advisable to double the number of cells because the cell size is significantly smaller.

3. In addition to 1% Triton X-100 detergent at 4°C, Triton X-100 at 4°C can be used at lower concentrations (to 0.25%) to isolate rafts from B cells (*see* **Fig. 2**). In addition, rafts have been isolated from immune cells at 4°C using several different nonionic detergents including 1% Tween-20, 1% Brij-58, 1% Brij-98, 0.5% Brij-96, 0.5% Lubrol, 1% CHAPS, and 1% NP40 *(12)*. Lastly, the isolation of rafts at 37°C in Brij-98 has been described *(13)*. It is likely that the optimal type and concentration of detergent required for efficient raft isolation will need to be established for each cell type *(14)*.

4. For the 6-mL gradients the cell pellets should be solubilized in 0.5 mL of 1% Triton X-100 containing lysis buffer. Because the volume of the lysis buffer is halved for 6-mL gradients, the volume of the 85, 35, and 5% sucrose solutions recommended for the 12-mL gradient should also be halved.

5. In addition to sucrose gradients, B-cell lipid rafts have been isolated on Nycodenz gradients *(15)* and on OptiPrep gradients *(16)*.

6. 450-µL Fractions should be collected from the top of a 6-mL gradient using a 1-mL pipetman.

7. This high-density pellet may contain BCR that has become attached to the actin cytoskeleton, and the pellet should be saved for analysis. Receptors that become associated with the actin cytoskeleton following crosslinking may become insoluble and pellet in the sucrose gradients if they are not associated with sufficient lipid to remain buoyant *(17)*.

8. Not all human B-cell lines contain detectable amounts of GM1. It is also possible to assay for other sphingolipids, including GM3 or the heterotrimeric G protein subunit Gαi. In addition to CD45, the transferrin receptor can be used as a marker for soluble membranes. An alternative method to characterize the raft preparation relies on the incorporation of fluorescent lipids with either saturated acyl chains—$DiIC_{16}(3)$—that when added to cells partition into lipid rafts, or unsaturated acyl chains —FastDiI—that partition into the nonraft membranes *(18)*. Cells (1×10^8/mL) are incubated with either $DiIC_{16}(3)$ (5 µ*M*) or FastDiI (5 µ*M*) at room temperature for 5 min prior to the raft isolation procedure. The fluorescence intensity of the resulting sucrose density gradient fractions is measured. **Figure 2** shows the fluorescence intensity of sucrose density

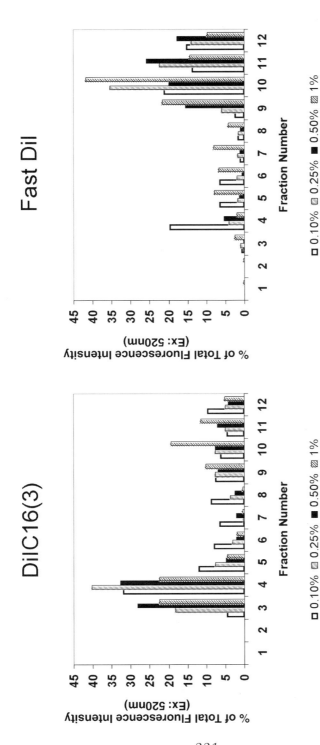

Fig. 2. The fluorescence intensity of the fractions from sucrose gradients to which the lysates of 1×10^8 CH27 cells, which had incorporated $DilC_{16}(3)$ (*left panel*) or FastDil (*right panel*), were applied was measured using a fluorometer. The fluorescence intensity is expressed for each fraction as a percentage of the total fluorescence intensity in the gradient. Lysates were prepared using 0.10%, 0.25%, 0.50%, or 1.0% Triton X-100.

gradient fractions on which lysates prepared from CH27 cells using different concentrations of Triton X-100 (0.1–1.0%) were separated. At 0.25–0.5% Triton X-100 the optimal separation of $DiIC_{16}(3)$ into insoluble membranes, and FastDiI into soluble membranes, was achieved. In choosing a detergent concentration to isolate rafts that were detected using the $DiIC_{16}(3)$ and Fast DiI, it is advisable to verify the separation using the membrane markers described in the protocol.

9. Detection of proteins in the intermediate fractions (fractions 7–9) might suggest that there has been inadvertent mixing of the sucrose gradient owing to bumping the gradient or mixing of the gradient during fraction collection. However, it is also possible that a receptor has partitioned into a microdomain that is more dense than most lipid rafts *(19,20)*.

10. The MβCD concentration and extraction time may need adjustment for a given cell type. Generally, concentrations of MβCD ranging from 5–20 mM and incubation times of 15 to 60 min have been used. MβCD at higher concentrations can be toxic to cells. If this problem is encountered, it is possible to achieve complete cholesterol extraction using two to three rounds of depletion with lower concentrations of MβCD. For a more precise manipulation of cholesterol content, especially for cholesterol repletion, a mixture of MβCD with cholesterol may be used in a defined molar ratio *(21)*. However, measurements of cellular cholesterol content may then be required to ensure that the repletion restores normal cholesterol levels. Certain caution should be observed when interpreting the consequences of MβCD treatment on complex cellular functions. It has been described that, depending on concentrations and incubation times, MβCD treatment can lead to ligand-independent activation of signaling pathways *(22)* and disturb calcium homeostasis *(23)*. Currently, it is not clear whether all of the consequences of MβCD treatment are mediated by cholesterol removal and disruption of lipid rafts, or whether some are less specific effects. Thus, controls should always be performed, including cholesterol repletion and the use of alternative methods to target raft-associated functions. These include polyene antifungal agents filipin and nystatin *(24)*, cholesterol oxidase *(13)*, dipalmitoyl phosphatidylethanolamine *(25)*, and polyunsaturated fatty acids *(26)*.

References

1. Dykstra, M. L., Cherukuri, A., Sohn, H. W., Tzeng, S.-J., and Pierce, S. K. (2003) Location is everything: lipid rafts and immune cell signaling. *Annu. Rev. Immunol.* **21,** 457–481.
2. Simons, K. and Toomre, D. (2000) Lipid rafts and signal transduction. *Nat. Rev. Mol. Cell Biol.* **1,** 31–41.
3. Brown, D. A. and London, E. (2000) Structure and function of sphingolipid- and cholesterol-rich membrane rafts. *J. Biol. Chem.* **275,** 17,221–17,224.
4. Gidwani, A., Holowka, D., and Baird, B. (2001) Fluorescence anisotropy measurements of lipid order in plasma membranes and lipid rafts from RBL-2H3 mast cells. *Biochemistry* **40,** 12,422–12,429.

5. Cheng, P. C., Dykstra, M. L., Mitchell, R. N., and Pierce, S. K. (1999) A role for lipid rafts in BCR signaling and antigen targeting. *J. Exp. Med.* **190,** 1549–1560.

6. Friedrichson, T. and Kurzchalla, T. V. (1998) Microdomains of GPI-anchored proteins in living cells revealed by crosslinking. *Nature* **394,** 802–805.

7. Varma, R. and Mayor, S. (1998) GPI-anchored proteins are organized in submicron domains at the cell surface. *Nature* **394,** 798–801.

8. Pralle, A., Keller, P., Florin, E.-L., Simons, K., and Horber, J. K. H. (2000) Sphingolipid-cholesterol rafts diffuse as small entities in the plasma membrane of mammalian cells. *J. Cell Biol.* **148,** 997–1007.

9. Schutz, G. J., Kada, G., Pastushenko, V. P., and Schindler, H. (2000) Properties of lipid microdomains in a muscle cell membrane visualized by single molecule microscopy. *EMBO J.* **19,** 892–901.

10. Zacharias, D. A., Violin, J. D., Newton, A. C., and Tsien, R. Y. (2002) Partitioning of lipid-modified monomeric GFPs into membrane microdomains of live cells. *Science* **296,** 913–916.

11. Batista, F. D., Iber, D., and Neuberger, M. S. (2001) B cells acquire antigen from target cells after synapse formation. *Nature* **411,** 489–494.

12. Pike, L. J. (2003) Lipid rafts: bringing order to chaos. *J. Lipid Res.* **44,** 655–667.

13. Drevot, P., Langlet, C., Guo, X.-J., et al. (2002) TCR signal initiation machinery is pre-assembled and activated in a subset of membrane rafts. *EMBO J.* **21,** 1899–1908.

14. Schuck, S., Honsho, M., Ekroos, K., Shevchenko, A., and Simons, K. (2003) Resistance of cell membranes to different detergents. *Proc. Natl. Acad. Sci. USA* **100,** 5795–5800.

15. Hostager, B. S., Catlett, I. M., and Bishop, G. A. (2000) Recruitment of CD40 and tumor necrosis factor receptor-associated factors 2 and 3 to membrane microdomains during CD40 signaling. *J. Biol. Chem.* **275,** 15,392–15,398.

16. Sproul, T. W., Malapati, S., Kim, J., and Pierce, S. K. (2000) B cell antigen receptor signaling occurs outside lipid rafts in immature B cells. *J. Immunol.* **165,** 6020–6023.

17. Higuchi, M., Izumi, K. M., and Kieff, E. (2001) Epstein-Barr virus latent-infection membrane proteins are palmitoylated and raft-associated: protein 1 binds to the cytoskeleton through TNF receptor cytoplasmic factors. *Proc. Natl. Acad. Sci. USA* **98,** 4675–4680.

18. Seveau, S., Eddy, R. J., Maxfield, F. R., and Pierini, L. M. (2001) Cytoskeleton-dependent membrane domain segregation during neutrophil polarization. *Mol. Biol. Cell* **12,** 3550–3562.

19. Lindwasser, O. W. and Resh, M. D. (2001) Multimerization of human immuno-deficiency virus type 1 gag promotes its localization to barges, raft-like membrane microdomains. *J. Virol.* **75,** 7913–7924.

20. Holowka, D., Sheets, E. D., and Baird, B. (2000) Interactions between FcεRI and lipid raft components are regulated by the actin cytoskeleton. *J. Cell Sci.* **113,** 1009–1019.

21. Christian, A. E., Haynes, M. P., Phillips, M. C., and Rothblat, G. H. (1997) Use of cyclodextrins for manipulating cellular cholesterol content. *J. Lipid Res.* **38,** 2264–2272.
22. Kabouridis, P. S., Janzen, J., Magee, A. L., and Ley, S. C. (2000) Cholesterol depletion disrupts lipid rafts and modulates the activity of multiple signaling pathways in T lymphocytes. *Eur. J. Immunol.* **30,** 954–963.
23. Pizzo, P., Giurisato, E., Tassi, M., Benedetti, A., Pozzan, T., and Viola, A. (2002) Lipid rafts and T cell receptor signaling: a critical re-evaluation. *Eur. J. Immunol.* **32,** 3082–3091.
24. Xavier, R., Brennan, T., Li, Q., McCormack, C., and Seed, B. (1998) Membrane compartmentation is required for efficient T cell activation. *Immunity* **8,** 723–732.
25. Legler, D. F., Doucey, M.-A., Cerottini, J.-C., Bron, C., and Luescher, I. F. (2001) Selective inhibition of CTL activation by a dipalmitoyl-phospholipid that prevents the recruitment of signaling molecules to lipid rafts. *FASEB J.* **15,** 1601–1603.
26. Stulnig, T. M., Berger, M., Sigmund, T., Raederstorff, D., Stockinger, H., and Waldhausl, W. (1998) Polyunsaturated fatty acids inhibit T cell signal transduction by modification of detergent-insoluble membrane domains. *J. Cell Biol.* **143,** 637–644.

16

Molecular Single-Cell PCR Analysis of Rearranged Immunoglobulin Genes As a Tool to Determine the Clonal Composition of Normal and Malignant Human B Cells

Ralf Küppers

Summary

Owing to the nearly limitless diversity of immunoglobulin (Ig) variable-region gene rearrangements, such rearrangements represent ideal clonal markers for B-lineage cells. This chapter describes an approach to isolate single cells from frozen tissue sections by microdissection using a hydraulic micromanipulator and the subsequent amplification of rearranged IgH and Igκ genes from the cells in a seminested polymerase chain reaction (PCR) approach. The amplification of *a priori* unknown V-gene rearrangements is made possible by the usage of a collection of V-gene family-specific primers recognizing nearly all V-gene segments together with primer mixes for the J-gene segments. By sequence comparison of V-gene amplicates from distinct cells, the clonal relationship of the B-lineage cells can unequivocally be determined. As a large part of the V-gene rearrangements is amplified, the approach is also useful to address additional issues, such as V-, D-, and J-gene usage and the presence and pattern of somatic mutations.

Key Words

B cells; B-cell lymphoma; clonality; Hodgkin's lymphoma; micromanipulation; V-gene recombination; single-cell PCR; somatic hypermutation; *Taq* deoxyribonucleic acid (DNA) polymerase errors.

1. Introduction

During B-cell development, gene segments coding for the variable region of immunoglobulin (Ig) heavy and light chains are assembled by a somatic recombination process. Heavy chain V-region genes are composed of three gene segments, V_H, D_H, and J_H, whereas those of light chains are composed of V and J segments. There are two types of light chains, κ and λ. As the human

From: *Methods in Molecular Biology, vol. 271: B Cell Protocols*
Edited by: H. Gu and K. Rajewsky © Humana Press Inc., Totowa, NJ

genome contains about 50 functional V_H genes (depending on the haplotype), 27 D_H genes and 6 J_H genes *(1–3)*, a large diversity of heavy chain V-gene rearrangements can be generated. This diversity is further increased by variable loss of nucleotides at the ends of the rearranging gene segments and by the addition of nongermline encoded nucleotides (N nucleotides) between the gene segments and the variable opening of a hairpin structure generated as a recombination intermediate (P nucleotides). In the light chain loci, the diversity is more limited, as D segments are missing. Nevertheless, as loss of nucleotides and the addition of N nucleotides and P nucleotides between the V and J segments happens also during light chain gene rearrangements, and as a large number of functional V- and J-gene segments can be used for V–J recombination (30–35 V_κ genes, 5 J_κ genes, 30–37 V_λ genes, and 4 J_λ genes *[4–7]*), a large diversity can also be generated for light chain V-region genes. Even further diversity is generated at later stages of B-cell differentiation, when V-region genes of antigen-activated B cells acquire somatic mutations at a high rate in the course of T-cell-dependent immune responses *(8,9)*. This process is called *somatic hypermutation* and takes place in histological structures of secondary lymphoid organs (such as lymph nodes), the germinal centers.

Owing to the nearly limitless diversity of V-region genes, each mature B cell is equipped with a unique B-cell receptor. Hence, V-region genes represent ideal markers for members of a B-cell clone. This can be used to trace B-cell developmental processes and the clonal relationship of B-lineage cells. For example, by combining microdissection of single germinal center B cells from histological sections of human lymph nodes with a polymerase chain reaction (PCR) analysis of the rearranged V genes of these cells, the oligo-clonality of the germinal center B-cell population and the occurrence of somatic hypermutation in the course of the clonal expansion of these B cells was shown *(10)*. An analysis of rearranged Ig V genes can also be used for the study of malignant B-lineage cells. In our own work, we showed that the putative malignant Hodgkin and Reed-Sternberg (HRS) cells of Hodgkin's lymphoma carry clonal Ig gene rearrangements in a given case *(11,12)*. This proved the B-cell origin of these cells in the large majority of cases and demonstrated that HRS cells represent monoclonal cell populations, a hallmark of tumor cells.

A large number of different protocols have been established to analyze normal and malignant human B cells for clonality (*see* for example **refs. 13–17**). In some situations, PCR amplification from cell populations can be sufficient to identify clonal B-cell populations. Such approaches have often been used to study B-cell non-Hodgkin's lymphomas, where usually most of the B cells in the tissue belong to the lymphoma clone. However, for a more detailed analysis of B cells that are, for example, defined by their location in the histological environment, a single cell analysis is needed. Moreover, also for a characterization of rare cells, such as the HRS cells in Hodgkin's lymphoma, which usually account for less than 1% of cells in the tissue, analysis of single HRS cells is mandatory for a reliable characterization of the cells.

For the analysis of rearranged human Ig V genes from single cells, we established a PCR protocol to amplify these genes from genomic DNA using primers binding to the 5' end of the V-gene segments in framework region I and primers binding to the J-gene segments in framework region IV *(10)*. As the human V_H and V_L genes are grouped into families based on sequence homology (members of a family show at least 80% deoxyribonucleic acid (DNA)-sequence homology to each other), it was possible to design V-gene family-specific primers that bind to all (or at least most) members of a given V-gene family. Because one round of PCR is not sufficient to obtain enough amplificate in nanogram (ng) amounts starting from a single target molecule, the PCR protocol includes a seminested amplification, that is, a second round of PCR is performed with an aliquot of the first round product using the same V-gene primers, but using J-primer mixes that are located 5' of the J primers used in the first round of PCR (i.e., nested J primers). By using framework region I primers in combination with J-gene primers, PCR products of about 300–350 bp in length are obtained that cover most of the V-region gene. Hence, this approach is not only suitable to determine the clonal relationship of distinct cells but also to identify the V-, (D-), and J-gene segments used in the rearrangements and analyze the sequences for the presence of somatic mutations.

We used this single-cell PCR approach for the analysis of single cells isolated from tissue sections by micromanipulation, as described in this chapter. However, the PCR protocol can also be used to amplify rearranged Ig V genes from cells isolated by flow cytometry *(18,19)*.

2. Materials
2.1. Histological Stainings

1. Acetone.
2. Antibodies: for example, CD30 (Ber-H2, mouse antihuman) (DAKO, Hamburg, Germany), 1:20 diluted, or CD20 (L26, mouse antihuman) (DAKO), 1:50 diluted.
3. Biotinylated F(ab')$_2$ fragment of rabbit antimouse immunoglobulins (DAKO), 1:200 diluted.
4. Avidin–biotin alkaline phosphatase complex (ABComplex/AP) (DAKO) or streptavidin–biotin alkaline phosphatase complex (StreptABComplex/AP) (DAKO).
5. Fast Red with levamisole (DAKO).
6. Tris-buffered saline (TBS): dissolve 8 g of NaCl, 0.2 g of KCl, and 3 g of Tris-base in approx 800 mL of water, adjust pH to 7.4 with HCl, and add water to 1 L.

7. TBS/bovine serum albumin (BSA): TBS supplemented with 1 or 5% (w/v) BSA.
8. Mayer's hemalum.

2.2. Microdissection

1. Inverse microscope equipped with a 40-fold long-distance lens (IMT-2; Olympus, Hamburg, Germany).
2. TV camera and TV monitor, Reflex camera.
3. Vertical pipet puller (Bachhofer, Reutlingen, Germany).
4. Grinding machine (Bachhofer).
5. Glass capillaries (GD1; 1 mm × 90 mm; Narishige, Japan).
6. TBS.
7. PCR buffer from Expand High Fidelity PCR system (Roche, Mannheim, Germany).

2.3. Proteinase K Digestion

1. Proteinase K, PCR grade (Roche).
2. 10 mM Tris-HCl, pH 7.6.
3. Mineral oil.

2.4. First Round of PCR

1. High-fidelity polymerase mixture, 3.5 U/µL (Roche) with high-fidelity PCR buffer and MgCl$_2$.
2. Deoxynucleotide triphosphates (dNTPs) (Pharmacia, Freiburg, Germany).
3. V_H/V_κ primer mix: mix same volumes of 2.5 µM working dilutions (V_H 1–6, 3' J_H primers, V_κ 1–6, and 3' J_κ primers) (**Table 1**).

2.5. Second Round of PCR

2.5.1. V_H PCR

1. *Taq* DNA polymerase (5 U/µL) with PCR buffer and MgCl$_2$.
2. dNTPs (Pharmacia).
3. Mineral oil.
4. V_H primers (**Table 1**); make a 2.5-µM working dilution.
5. J_H primer mix: mix same vol of 2.5 µM working dilutions of the four 5' J_H primers (0.625 µM per 5' J_H primer; **Table 2**).

2.5.2. V_κ PCR

1. *Taq* DNA polymerase (5 U/µL) with PCR buffer and MgCl$_2$.
2. dNTPs (Pharmacia).
3. Mineral oil.
4. V_κ primers (**Table 1**); make a 2.5-µM working dilution.
5. J_κ primer mix: 2.5 µM per 5' J_κ primer (**Table 2**).

Table 1
Primers for the First Round of PCR

Name	Sequence
V_H1	5′-CAG-TCT-GGG-GCT-GAG-GTG-AAG-A-3′
V_H2	5′-GTC-CTR-CGC-TGG-TGA-AAC-CCA-CAC-A-3′
V_H3	5′-GGG-GTC-CCT-GAG-ACT-CTC-CTG-TGC-AG-3′
V_H4	5′-GAC-CCT-GTC-CCT-CAC-CTG-CRC-TGT-C-3′
V_H5	5′-AAA-AAG-CCC-GGG-GAG-TCT-CTG-ARG-A-3′
V_H6	5′-ACC-TGT-GCC-ATC-TCC-GGG-GAC-AGT-G-3′
$3′J_H1.2.4.5$	5′-ACC-TGA-GGA-GAC-GGT-GAC-CAG-GGT-3′
$3′J_H3$	5′-ACC-TGA-AGA-GAC-GGT-GAC-CAT-TGT-3′
$3′J_H6$	5′-ACC-TGA-GGA-GAC-GGT-GAC-CGT-GGT-3′
$V_\kappa1$	5′-GAC-ATC-CRG-WTG-ACC-CAG-TCT-CCW-TC-3′
$V_\kappa2$	5′-CAG-WCT-CCA-CTC-TCC-CTG-YCC-GTC-A-3′
$V_\kappa3$	5′-TTG-TGW-TGA-CRC-AGT-CTC-CAG-SCA-CC-3′
$V_\kappa4$	5′-AGA-CTC-CCT-GGC-TGT-GTC-TCT-GGG-C-3′
$V_\kappa5$	5′-CAG-TCT-CCA-GCA-TTC-ATG-TCA-GCG-A-3′
$V_\kappa6$	5′-TTT-CAG-TCT-GTG-ACT-CCA-AAG-GAG-AA-3′
$3′J_\kappa1.2.4$	5′-ACT-CAC-GTT-TGA-TYT-CCA-SCT-TGG-TCC-3′
$3′J_\kappa3$	5′-GTA-CTT-ACG-TTT-GAT-ATC-CAC-TTT-GGT-CC-3′
$3′J_\kappa5$	5′-GCT-TAC-GTT-TAA-TCT-CCA-GTC-GTG-TCC-3′

R, A&G; W, A&T; Y, C&T; S, G&C.

2.6. PCR Product Purification and Direct Sequence Analysis

1. Agarose.
2. 1 mg/mL Ethidium bromide; dilute 1:2000.
3. 3 *M* Sodium acetate, pH 5.0–5.5.
4. Ethanol 100%, 70%.
5. TAE: 40 m*M* Tris-acetate, 1 m*M* ethylenediaminetetraacetic acid (EDTA).
6. QIAEX gel extraction kit (Qiagen, Hilden, Germany).
7. DNA mass ladder (Invitrogen, Karsruhe, Germany).
8. Big Dye sequencing kit (PE Biosystems, Weiterstadt, Germany).
9. Sodium acetate/dextran blue: 3 *M* sodium acetate, pH 5.0–5.5, with 12.5 mg/mL dextran blue.

3. Methods
3.1. Histological Staining

1. Dry frozen sections overnight on glass slides. Fix in acetone for 10 min at –20°C. Dry sections for 30–60 min. Incubate for 30–60 min with TBS/5% BSA.

Table 2
Sequences of J_H and $J_κ$ Primers for the Second Round of PCR

Name	Sequence
5'J_H1.4.5	5'-GAC-GGT-GAC-CAG-GGT-KCC-CTG-GCC-3'
5'J_H2	5'-GAC-AGT-GAC-CAG-GGT-GCC-ACG-GCC-3'
5'J_H3	5'-GAC-GGT-GAC-CAT-TGT-CCC-TTG-GCC-3'
5'J_H6	5'-GAC-GGT-GAC-CGT-GGT-CCC-TTK-GCC-3'
5'$J_κ$1.2	5'-TTG-ATY-TCC-ASC-TTG-GTC-CCY-TGG-C-3'
5'$J_κ$3	5'-TTG-ATA-TCC-ACT-TTG-GTC-CCA-GGG-C-3'
5'$J_κ$4	5'-TTG-ATC-TCC-ACC-TTG-GTC-CCT-CCG-C-3'
5'$J_κ$5	5'-TTA-ATC-TCC-AGT-CGT-GTC-CCT-TGG-C-3'

K, G&T; Y, C&T; S, G&C.

2. Dilute primary antibody in TBS/1% BSA. Incubate section with primary antibody for 1 h at room temperature. Wash three times with TBS.
3. Dilute secondary antibody in TBS/1% BSA. Incubate for 30–60 min at room temperature. Wash three times with TBS.
4. Incubate for 30 min with ABComplex/AP prepared according to the instructions. Wash three times with TBS.
5. Incubate for 1–15 min with Fast Red prepared according to the instructions of the supplier. Wash three times with TBS.
6. ounterstain with Mayer's hemalum (*see* **Note 1**).

3.2. Microdissection

3.2.1. Preparation of Glass Capillaries

1. Glass capillaries are prepared on a capillary puller and then polished on a grinding machine. After this procedure, the tip of the capillary should have a diameter of 3–5 mm for the capillary designated to "scratch out" the cells.
2. Another group of capillaries (designated for aspiration of the mobilized cells) has to be polished more extensively to yield a tip width of 10–15 mm (e.g., aspiration of B cells) or 30–40 mm (for aspiration of HRS cells).
3. After grinding, the capillaries are sterilized at 220°C for 5 h resulting in destruction of any DNA molecules possibly adsorbed to the glass capillaries.

3.2.2. Microdissection

1. The stained tissue sections (*see* **Subheading 3.1.**) are covered by TBS buffer (*see* **Note 2**).
2. Single cells are mobilized at ×600 magnification with the help of a micropipette fixed to a hydraulic micromanipulator (*see* **Note 3**). The hydraulic micromanipu-

lator itself is fixed to a 3D manipulator and operated by a joystick. A TV camera fixed to the microscope allows control of the micromanipulation on a monitor. The capillaries should be operated at an angle of <45°.

3. Mobilized cells are aspirated into a glass capillary fixed to a second micromanipulator with as little buffer as possible.

4. Isolated cells are released into a PCR tube containing 20 µL 1X PCR buffer (without MgCl$_2$). To make sure that the cell was transferred into the PCR tube, the tip of the capillary may be broken off and dropped into the tube.

5. The PCR tubes are centrifuged at full speed for 1 min to spin down the micromanipulated cells and are stored at –20°C or –80°C.

6. Before and after micromanipulation a photograph is taken to allow the exact localization of the micromanipulated cells within the histological microenvironment.

7. To monitor possible contamination by DNA from fragmented neighboring cells, every 5th to 10th cell should be followed by one sample that contains aspirated buffer from the tissue section but no mobilized cells (these samples are used as negative controls for micromanipulation). Micromanipulation can be carried out for about 24 h without a detectable decrease of PCR efficiency. In regular time intervals, fresh sterile TBS buffer is added to the section to avoid drying out (*see* **Note 4**).

3.3. Proteinase K Digestion

1. Dilute proteinase K 1:1 with 10 mM Tris-HCl, pH 7.6.
2. Add 0.5 µL of this enzyme dilution to the 20 µL PCR buffer containing the single cell and overlay with mineral oil.
3. Incubate for 2 h at 50°C. Inactivate the enzyme by incubation at 95°C for 10 min. Cool down to 4°C.

3.4. First Round of PCR

1. Prepare a master mix consisting of 2.5 µL dNTP solution (2 mM), 3 µL 10X high-fidelity PCR, 5 µL primer mix, 5 µL 25 mM MgCl$_2$, and 13.5 µL water per reaction.
2. Add 29 µL master mix to the reaction mixture from the proteinase K digestion (20.5 µL).
3. Add 0.5 µL Expand high-fidelity enzyme mix after the first denaturation (*see* first round of PCR in **Subheading 3.4., step 4**), and carry out PCR.
4. The PCR program consists of: 95°C for 2 min; 65°C, pause (add enzyme); 72°C for 1 min; 34 × (95°C for 50 s; 59°C for 30 s; 72°C for 60 s); 72°C for 5 min; 10°C, pause (*see* **Notes 5–8**).

3.5. Second Round of PCR

3.5.1. V$_H$ PCR

1. Prepare on ice six master mixes (one for each of the six V$_H$ family-specific primers) each consisting of 5 µL dNTPs (2 mM), 5 µL 10X PCR buffer, 2.5 µL

of the respective V_H primer, 2.5 µL 5′ J_H primer mix, 3 µL 25 mM MgCl$_2$ (5 µL for V_H1), 30.7 µL water (28.7 µL for V_H1), and 0.3 µL *Taq* DNA polymerase.
2. Pipet 49 µL master mix in fresh PCR tube. Overlay with mineral oil.
3. To master mix add 1 µL of the first round reaction.
4. Start PCR program, but take PCR tubes from ice to heating block only after block has reached 95°C.
5. The program consists of: 95°C for 1 min, 45 × (95°C for 50 s, 61°C for 30 s, 72°C for 60 s), 72°C for 5 min, 10°C, pause.

3.5.2. V_κ PCR

1. Prepare on ice six master mixes (one for each of the six V_κ family-specific primers) each consisting of 5 µL dNTPs (2 mM), 5 µL 10X PCR buffer, 2.5 µL of the respective V_κ primer, 2.5 µL J_κ primer mix, 5 µL 25 mM MgCl$_2$, 28.7 µL water and 0.3 µL *Taq* DNA polymerase.
2. Pipet 49 µL master mix in fresh PCR tube. Overlay with mineral oil.
3. To master mix add 1 µL of the first round reaction.
4. Start PCR program, but take PCR tubes from ice to heating block only after block has reached 95°C.
5. The program consists of: 95°C for 1 min, 45 × (95°C for 50 s, 61°C for 30 s, 72°C for 60 s), 72°C for 5 min, 10°C, pause (*see* **Note 9**).

3.6. PCR Product Purification and Direct Sequence Analysis

Because primer dimers and additional bands are often observed when PCR products are analyzed by agarose gel electrophoresis, it is recommended to purify the PCR products from agarose gels before sequencing. The amount of purified product is estimated from an agarose gel using a DNA mass ladder to estimate the DNA concentration (*see* **Note 10**).

1. PCR products are first analyzed on a 2% analytical agarose gel containing ethidium bromide and visualized under ultraviolet (UV) light.
2. For samples with a PCR product, the second round of PCR (*see* **Subheading 3.5.**) is repeated with 2–3 reactions in 50 or 100 µL vol. These reactions are then mixed and precipitated by adding 1/10 vol of 3 M Na-acetate and 2 vol 100% ethanol. After centrifugation and washing of the pellet with 70% ethanol, the pellet is dissolved in 20 µL Tris-EDTA (TE) buffer and loaded on a 2% preparative agarose gel.
3. The PCR product is cut out of the gel under UV light and isolated with the QIAEX gel extraction kit, as recommended by the supplier.
4. To estimate the concentration of the isolated DNA, one tenth of the DNA is run on a 2% agarose gel together with a DNA mass ladder, which allows estimation of the DNA concentration by comparing the strength of the PCR product bands to the strength of the DNA fragments of the mass ladder.
5. Sequence reaction: mix the following components on ice: 30–40 ng PCR product, 1.5 µL primer (2.5 µM—usually the V primer), 2 µL Big Dye reaction mix plus × µL water to obtain a final volume of 10 µL. Overlay with 1 drop of mineral oil.

Run PCR: 96°C for 2 min; 34 × (96°C for 30 s; 50°C for 15 s; 60° for C 4 min); cool to 10°C.

6. Precipitation (at room temperature): Add 40 μL water, remove oil, transfer reaction mixture to fresh tube, add 5 μL 3 mM sodium acetate/dextran blue and 125 μL 100% ethanol. Mix by vortexing. Directly centrifuge at approx 12,000g for 30 min. Remove supernatant and wash the pellet with 400 μL 70% ethanol. Centrifuge at approx 12,000g for 5 min; remove supernatant and let the pellet air-dry. Store at –20°C if not run on a sequencing gel on the same day.

7. Analyze sequence reaction on an ABI sequencer.

3.7. Sequence Evaluation

V-gene sequences can be evaluated for the identity of the gene segments used in the rearrangement and the presence of somatic mutations by comparing to GenBank entries (www.ncbi.nlm.nih.gov/blast/). However, a more convenient way is to use the ImMunoGeneTics (IMGT) database and the DNA plot software (www.dnaplot.de/input/human_v.html) or (http://imgt.cines.fr/textes/vquest/). With this tool, rearranged human Ig genes are analyzed for sequences homologous to V, D-, and J-gene segments, and the five most homologous germline gene sequences are given (*see* **Note 11**).

4. Notes

1. Although the protocol described here uses alkaline phosphatase for immuno-staining, the procedure can also be performed using horseradish peroxidase as enzyme for the color reaction. Moreover, PCR also works efficiently with cells labeled by *in situ* hybridization. In our own work, we labeled Epstein-Barr virus (EBV)-infected cells by *in situ* hybridization for EBV-encoded small non-polyadenylated RNAs (EBER) transcripts *(20)*.

2. Sometimes cells are "sticky" for example, because of fibrosis in the tissue. In that case, some collagenase H (Roche, Mannheim, Germany) can be added to the TBS covering the stained section, and the section should be incubated for 30 min at room temperature before the buffer covering the section is replaced with fresh TBS buffer. Sometimes it is worthwhile to discard the tissue sections and to start from the beginning with new sections.

3. The microdissection of the cells can be done in two ways: (a) Scratch off all surrounding cells. This has the advantage that the cell can be recovered entirely without risking fragmentation, and the procedure works relatively fast. However, this increases the risk of contamination by floating cell fragments and leads to loss of material and the destruction of the histologic environment. (b) Enucleate the target cell only. This has the advantage that the histologic environment is preserved, and it reduces the risk of buffer contamination. However, this technique requires more experience in handling to avoid loss of nuclear fragments.

4. As an alternative to the hydraulic micromanipulation with glass capillaries, a laser-based microdissection apparatus may also be used. With the PALM laser

(Bernried, Germany) sections are usually mounted on a membrane before staining. Cells are first microdissected by a laser beam, ablating the tissue directly adjacent to the cell that is to be isolated, and then the cell on a piece of supporting membrane is catapulted by a brief laser pulse into a PCR cap that is placed above the section. The usage of the PALM laser is described in detail in **refs. 21,22**. The advantage of the laser-based system is that it is a completely nontouching method and is faster than the hydraulic system. However, the section has to be dried after staining, which reduces the quality of the microscopic picture.

5. The protocol described here uses V_H and V_κ family-specific primers binding to the framework region I of the respective genes together with combinations of J-segment primers. We also established PCR strategies for the coamplification of V_H and V_λ gene rearrangements from single cells *(23)*, and a strategy to coamplify V_H, V_κ, and V_λ gene rearrangements *(23)*. Also in these approaches, V-gene family-specific primers binding to sequences in the framework regions I are used together with primers for the J-gene segments. For the amplification of V_H gene rearrangements, V_H family-specific primers binding to sequences in the leader region may be used instead of the framework region I primers *(24)*. Similarly, we also established a protocol for the amplification of V_κ gene rearrangements from single cells using V_κ family-specific leader primers *(25)*. For a detailed characterization of the IgH and Igκ loci of human B-lineage cells, there are also protocols available to amplify $D_H J_H$ joints or fragments indicative of germline configuration of the IgH locus *(26)*, and to detect rearrangements of the kappa deleting element *(18)*.

6. Because 60% of human B cells use kappa light chains, the V_H and V_κ primer combination is the preferred one to analyze human B cells for clonality and the presence of somatic mutations. However, one has to keep in mind that most λ light chain expressing B cells carry $V_\kappa J_\kappa$ joints, and these joints are usually inactivated by rearrangements of a kappa-deleting element (KDE) that is located downstream of the C_κ gene *(27–29)*. Rearrangement of the KDE to a sequence in the J_κ–C_κ intron deletes not only the C_κ gene but also the two kappa enhancers. However, $V_\kappa J_\kappa$ joints remain on the chromosome and can be amplified. As the Ig enhancers are needed for somatic hypermutation activity, such $V_\kappa J_\kappa$ joints remain unmutated when the other Ig loci in the cell acquire mutations *(18)*.

7. The protocol described here allows one to coamplify V_H and V_κ gene rearrangements from single human cells. By first performing a whole-genome amplification with the DNA of a single cell, it is possible to analyze the cells in parallel for multiple genes *(30)*. Using this strategy, single cells can be analyzed in parallel for IgH, Igκ and Igλ gene rearrangements *(31)* and also for other features, such as mutations in oncogenes or tumor-suppressor genes, or infection by viruses *(20,32)*. A similar approach to combine whole-genome amplification with PCR for rearranged human Ig genes has also been described by Brezinschek and colleagues *(33,34)*.

8. PCR contamination is a major concern when performing single-cell PCR. Therefore, a number of precautions should be followed to obtain reliable results: Pre-

and post-PCR work should be performed in separate rooms. Gloves should be worn and frequently changed. Separate pipets, reaction tubes, aerosol-resistant pipet tips, and other equipment should be dedicated only for the pre-PCR work of the first round of PCR. First round reaction tubes should not be opened in the room where the pre-PCR work is done.

9. Because V_κ genes of the $V_\kappa 5$ and $V_\kappa 6$ families are only very rarely used, the V_κ analysis may be restricted to families 1–4.

10. The PCR products amplified with the V-gene primers binding to the framework region I of human V_H and V_κ genes and the corresponding J-segment primers are usually 300–350 bp long. However, occasionally longer or shorter PCR products are also obtained. These should not be disregarded, as they also may represent V-gene rearrangements. On the one hand, one has to keep in mind that the J-gene segments are located close to each other in the IgH and Igκ loci (200–600-bp distances; *see* **refs.** *3,4*), and J-primer mixes are used that bind to all J_H and J_κ segments. Hence, longer PCR products may represent amplificates ranging from the V-primer binding site to a J-gene segment downstream of the one that is actually rearranged. If a somatic mutation in the rearranged J-gene segments prevents successful amplification from this J segment, the product from the next downstream J-gene segment may be the predominant one. On the other hand, somatic hypermutation not only introduces point mutations, but also deletions or duplications. These events account for approx 4% of somatic mutations in nonselected (e.g., out-of-frame) rearrangements *(35)*. Overall, about 40% of mutated out-of-frame rearrangements from human B cells have been found to carry deletions and/or duplications, which may range in length from single nucleotides to several hundred basepairs *(35)*.

11. Although the single cell approach is technically demanding and laborious, it has several advantages compared with PCR analysis of cell populations: (i) PCR products can be directly sequenced without cloning. This is not only fast, but also has the important advantage that errors owing to *Taq* DNA polymerase mistakes do not pose a problem for the reliable identification of somatic mutations, as a "consensus" sequence is obtained. Even if a *Taq* DNA polymerase error happens in the first cycle of amplification, it should principally be present only in a quarter of the final molecules of the PCR (assuming that both strands of the target molecule are replicated in that cycle and that the mutated sequence does not have an amplification advantage in the further cycles). (ii) PCR hybrid artifacts, which may represent a considerable fraction of sequences if PCR products obtained from DNA of cell populations are cloned before sequencing, do not play a role in single-cell PCR. (iii) V_H and V_L genes can be coamplified from a single cell, so that they can be assigned to distinct cells, which is not possible if heavy and light chain genes are amplified from cell populations. (iv) The sizes of B-cell clones can be estimated when single cells from defined histological locations are analyzed. If PCR is performed with DNA from cell populations, it is difficult to discern whether identical sequences derive from a single cell or from an expanded clone.

Acknowledgments

This work was supported by the Deutsche Forschungsgemeinschaft through SFB502 and a Heisenberg Award. I am grateful to all present and previous members of the group and coworkers who were involved in the establishment of the protocols described here. I thank Andreas Bräuninger for helpful comments on the manuscript.

References

1. Cook, G. P. and Tomlinson, I. M. (1995) The human immunoglobulin VH repertoire. *Immunol. Today* **16**, 237–242.
2. Corbett, S. J., Tomlinson, I. M., Sonnhammer, E. L., Buck, D., and Winter, G. (1997) Sequence of the human immunoglobulin diversity (D) segment locus: a systematic analysis provides no evidence for the use of DIR segments, inverted D segments, "minor" D segments or D-D recombination. *J. Mol. Biol.* **270**, 587–597.
3. Ravetch, J. V., Siebenlist, U., Korsmeyer, S., Waldmann, T., and Leder, P. (1981) Structure of the human immunoglobulin mu locus: characterization of embryonic and rearranged J and D genes. *Cell* **27**, 583–591.
4. Hieter, P. A., Maizel, J. V., and Leder, P. (1982) Evolution of human immunoglobulin kappa J region genes. *J. Biol. Chem.* **257**, 1516–1522.
5. Kawasaki, K., Minoshima, S., Nakato, E., et al. (1997) One-megabase sequence analysis of the human immunoglobulin lambda gene locus. *Genome Res.* **7**, 250–261.
6. Schäble, K. F. and Zachau, H. G. (1993) The variable genes of the human immunoglobulin kappa locus. *Biol. Chem. Hoppe Seyler* **374**, 1001–1022.
7. Vasicek, T. J. and Leder, P. (1990) Structure and expression of the human immunoglobulin lambda genes. *J. Exp. Med.* **172**, 609–620.
8. Neuberger, M. S., and Milstein, C. (1995) Somatic hypermutation. *Curr. Opin. Immunol.* **7**, 248–254.
9. Rajewsky, K. (1996) Clonal selection and learning in the antibody system. *Nature* **381**, 751–758.
10. Küppers, R., Zhao, M., Hansmann, M. L., and Rajewsky, K. (1993) Tracing B cell development in human germinal centers by molecular analysis of single cells picked from histological sections. *EMBO J.* **12**, 4955–4967.
11. Kanzler, H., Küppers, R., Hansmann, M. L., and Rajewsky, K. (1996) Hodgkin and Reed-Sternberg cells in Hodgkin's disease represent the outgrowth of a dominant tumor clone derived from (crippled) germinal center B cells. *J. Exp. Med.* **184**, 1495–1505.
12. Küppers, R., Rajewsky, K., Zhao, M., et al. (1994) Hodgkin disease: Hodgkin and Reed-Sternberg cells picked from histological sections show clonal immunoglobulin gene rearrangements and appear to be derived from B cells at various stages of development. *Proc. Natl. Acad. Sci. USA* **91**, 10,962–10,966.
13. Deane, M., McCarthy, K. P., Wiedemann, L. M., and Norton, J. D. (1991) An improved method for detection of B-lymphoid clonality by polymerase chain reaction. *Leukemia* **5**, 726–730.

14. Küppers, R., Zhao, M., Rajewsky, K., and Hansmann, M. L. (1993) Detection of clonal B cell populations in paraffin-embedded tissues by polymerase chain reaction. *Am. J. Pathol.* **143**, 230–239.
15. McCarthy, K. P., Sloane, J. P., and Wiedemann, L. M. (1990) Rapid method for distinguishing clonal from polyclonal B cell populations in surgical biopsy specimens. *J. Clin. Pathol.* **43**, 429–432.
16. Ohno, T., Stribley, J. A., Wu, G., Hinrichs, S. H., Weisenburger, D. D., and Chan, W. C. (1997) Clonality in nodular lymphocyte-predominant Hodgkin's disease. *N. Engl. J. Med.* **337**, 459–465.
17. Trainor, K. J., Brisco, M. J., Wan, J. H., Neoh, S., Grist, S., and Morley, A. A. (1991) Gene rearrangement in B- and T-lymphoproliferative disease detected by the polymerase chain reaction. *Blood* **78**, 192–196.
18. Goossens, T., Bräuninger, A., Klein, U., Küppers, R., and Rajewsky, K. (2001) Receptor revision plays no major role in shaping the receptor repertoire of human memory B cells after the onset of somatic hypermutation. *Eur. J. Immunol.* **31**, 3638–3648.
19. Klein, U., Rajewsky, K., and Küppers, R. (1998) Human immunoglobulin (Ig)M+IgD+ peripheral blood B cells expressing the CD27 cell surface antigen carry somatically mutated variable region genes: CD27 as a general marker for somatically mutated (memory) B cells. *J. Exp. Med.* **188**, 1679–1689.
20. Kurth, J., Spieker, T., Wustrow, J., et al. (2000) EBV-infected B cells in infectious mononucleosis: viral strategies for spreading in the B cell compartment and establishing latency. *Immunity* **13**, 485–495.
21. Becker, I., Becker, K. F., Rohrl, M. H., Minkus, G., Schütze, K., and Höfler, H. (1996) Single-cell mutation analysis of tumors from stained histologic slides. *Lab. Invest.* **75**, 801–807.
22. Böhm, M., Wieland, I., Schütze, K., and Rubben, H. (1997) Microbeam MOMeNT: non-contact laser microdissection of membrane-mounted native tissue. *Am. J. Pathol.* **151**, 63–67.
23. Bräuninger, A., Küppers, R., Spieker, T., et al. (1999) Molecular analysis of single B cells from T-cell-rich B-cell lymphoma shows the derivation of the tumor cells from mutating germinal center B cells and exemplifies means by which immunoglobulin genes are modified in germinal center B cells. *Blood* **93**, 2679–2687.
24. Bräuninger, A., Küppers, R., Strickler, J. G., Wacker, H. H., Rajewsky, K., and Hansmann, M. L. (1997) Hodgkin and Reed-Sternberg cells in lymphocyte predominant Hodgkin disease represent clonal populations of germinal center-derived tumor B cells. *Proc. Natl. Acad. Sci. USA* **94**, 9337–9342 (correction appeared in: 94, 14,211).
25. Müschen, M., Küppers, R., Spieker, T., Bräuninger, A., Rajewsky, K., and Hansmann, M. L. (2001) Molecular single-cell analysis of Hodgkin- and Reed-Sternberg cells harboring unmutated immunoglobulin variable region genes. *Lab. Invest.* **81**, 289–295.
26. Küppers, R., Bräuninger, A., Müschen, M., Distler, V., Hansmann, M. L., and Rajewsky, K. (2001) Evidence that Hodgkin and Reed-Sternberg cells in Hodgkin disease do not represent cell fusions. *Blood* **97**, 818–821.

27. Bräuninger, A., Goossens, T., Rajewsky, K., and Küppers, R. (2001) Regulation of immunoglobulin light chain gene rearrangements during early B cell development in the human. *Eur. J. Immunol.* **31,** 3631–3637.

28. Hieter, P. A., Korsmeyer, S. J., Waldmann, T. A., and Leder, P. (1981) Human immunoglobulin kappa light-chain genes are deleted or rearranged in lambda-producing B cells. *Nature* **290,** 368–372.

29. Korsmeyer, S. J., Hieter, P. A., Sharrow, S. O., Goldman, C. K., Leder, P., and Waldmann, T. A. (1982) Normal human B cells display ordered light chain gene rearrangements and deletions. *J. Exp. Med.* **156,** 975–985.

30. Zhang, L., Cui, X., Schmitt, K., Hubert, R., Navidi, W., and Arnheim, N. (1992) Whole genome amplification from a single cell: implications for genetic analysis. *Proc. Natl. Acad. Sci. USA* **89,** 5847–5851.

31. Kanzler, H., Küppers, R., Helmes, S., Wacker, H. H., Chott, A., Hansmann, M. L., and Rajewsky, K. (2000) Hodgkin and Reed-Sternberg-like cells in B-cell chronic lymphocytic leukemia represent the outgrowth of single germinal-center B-cell-derived clones: potential precursors of Hodgkin and Reed-Sternberg cells in Hodgkin's disease. *Blood* **95,** 1023–1031.

32. Müschen, M., Re, D., Bräuninger, A., et al. (2000) Somatic mutations of the CD95 gene in Hodgkin and Reed-Sternberg cells. *Cancer Res.* **60,** 5640–5643.

33. Brezinschek, H. P., Brezinschek, R. I., and Lipsky, P. E. (1995) Analysis of the heavy chain repertoire of human peripheral B cells using single-cell polymerase chain reaction. *J. Immunol.* **155,** 190–202.

34. Brezinschek, H. P., Foster, S. J., Dorner, T., Brezinschek, R. I., and Lipsky, P. E. (1998) Pairing of variable heavy and variable kappa chains in individual naive and memory B cells. *J. Immunol.* **160,** 4762–4767.

35. Goossens, T., Klein, U., and Küppers, R. (1998) Frequent occurrence of deletions and duplications during somatic hypermutation: implications for oncogene translocations and heavy chain disease. *Proc. Natl. Acad. Sci. USA* **95,** 2463–2468.

17

Analysis of B-Cell Immune Tolerance Induction Using Transgenic Mice

Helen Ferry and Richard J. Cornall

Summary

Over the past 15 yr, the use of transgenic mice has led to significant advances in our understanding of immunological tolerance. In a normal repertoire the number of B cells with a single antigen receptor specificity is very small, making the study of their fate difficult. In contrast, animals that carry transgenes encoding rearranged immunoglobulin genes generate large numbers of B cells that, by the process of allelic exclusion, have an identical specificity. Exploitation of this effect has enabled the mechanisms involved in B-cell tolerance to be explored in some detail.

In this review we use the hen egg lysozyme (HEL) model system to illustrate the generation and preparation of a transgene. In our example, we describe the generation of mice expressing HEL as a systemic, intracellular, membrane-bound self-antigen. The same principles and methods apply to immunoglobulin transgenes. We briefly discuss the techniques that could be used to explore mechanisms of tolerance to systemic intracellular antigens in these mice.

Key Words

B lymphocytes; hen egg lysozyme; autoimmunity; transgenics; self-tolerance; antigen; immunoglobulin.

1. Introduction

Several immunoglobulin transgenic models have been used to study the mechanisms of B-cell tolerance. These experiments have revealed how self-tolerance depends on B-cell receptor affinity and the form and abundance of self-antigen. Self-reactive B cells that encounter abundant low valency and soluble self-antigens, such as single-stranded deoxyribonucleic acid (DNA) *(1)* and soluble hen egg lysozyme (HEL) *(2,3)*, are functionally inactivated. These anergic B cells persist in the repertoire for about 3 d before dying in the secondary lymphoid organs, but they can be reactivated by multivalent foreign antigens.

From: *Methods in Molecular Biology, vol. 271: B Cell Protocols*
Edited by: H. Gu and K. Rajewsky © Humana Press Inc., Totowa, NJ

Self-reactive B cells that bind multivalent self-antigens including MHC class I antigens, double-stranded DNA, and membrane-bound HEL are eliminated in the bone marrow by deletion *(4–6)* or receptor editing *(7,8)*. Other self-reactive B cells are not tolerized by antigen, particularly those specific for tissue-specific antigens *(9)*. In immunoglobulin transgenic models, mice have been made with specificities for both endogenous *(1,6,10)* and neoantigens. Neo-self-antigen transgenes may themselves have deleterious effects, which can complicate the interpretation of results; however, the advantage of using a neo-self-antigen is that the same antigen can be manipulated to express in different tissues and forms. Significantly, results from immunoglobulin/antigen double-transgenic mice can be compared with antigen negative controls.

When starting a new transgenic experiment, the construct design will be the most important factor determining whether or not the transgene will be successfully expressed. First, it is necessary to consider how the choice of promoter will affect the expression of the protein. Whenever possible it is advisable to use promoters and their genetic control elements that have already been proved in vivo. Remember the literature gives a false impression of success as the number of failed constructs with abnormal, little, or no transgene expression in vivo is undocumented; our own experience runs into more than 10 unproductive constructs. In the case of HEL, expression is usually robust, and the gene has been successfully expressed under a variety of ubiquitous (e.g., major histocompatibility complex [MHC] class I H-2K^b promoter *[11]*) and tissue-specific promoters (e.g., interphotoreceptor retinoid-binding protein [IRBP; *12*; and in our lab, manuscript in preparation]). HEL has also been expressed under an inducible promoter. In ML-transgenic mice, soluble HEL is under the control of the mouse liver metallothionein I promoter (ML), which is transcriptionally active during embryonic, fetal, and adult life and can be induced to higher transcription levels, using zinc or other heavy metals *(13)*.

When swapping the promoter in a construct it is important that the Kozak *(14)* consensus sequence and transcription start site are not disrupted, and it is preferable to leave these unchanged. We take care to sequence the boundary between the 3′ end of the promoter and the 5′ end of the lysozyme gene, as well as any new ligation points or polymerase chain reaction (PCR) products within the construct.

Splicing and polyadenylation of the transcribed product is also important for efficient expression *(15)*, and for this reason we used the HEL genomic sequence, which includes both introns and exons. Genomic DNA is also likely to include associated enhancer elements, which, if identified, should be included whenever possible. As a rule it is best to avoid using complementary DNA (cDNA) in constructs, but if necessary include artificial introns 5′ (preferably) or 3′ of the coding sequence. Finally, as bacterial sequences present in the plasmid vector adversely effect the expression of transgenes *(16,17)*, it is advantageous to design the construct so these elements can be removed easily by restriction digest prior to microinjection. This digest will also linearize the transgene, which is essential for efficient integration into the host genome.

2. Materials

1. Oligonucleotide primers.
2. Restriction enzymes, shrimp alkaline phosphatase (Boehringer Mannheim), T4 DNA ligase, *Taq* polymerase, VentR® polymerase (New England Biolabs, or other high-fidelity, thermostable polymerase), and deoxynucleotide-triphosphates (dNTPs).
3. Agarose gel electrophoresis equipment and DNA markers.
4. ElectroMAX DH10B™ competent cells (Gibco-BRL).
5. MERmaid® and GENECLEAN II® spin kits (Anachem).
6. QIAquick® PCR purification kit and plasmid mini and mega kits (Qiagen).
7. 10X Phosphate-buffered saline (PBS): 80 g NaCl, 2 g KCl, 14.4 g Na_2HPO_4, 2.4 g KH_2PO_4, triple-distilled water to 1 L.
8. Gene Pulser® (Bio-Rad).
9. Fetal calf serum (FCS), European origin (First Link). Heat inactivate at 55°C for 40 min, then store in 10- and 50-mL aliquots at –20°C.
10. RPMI plus extras: RPMI plus 2% FCS, 20 mM HEPES (Gibco-BRL), 10 mM $NaHCO_3$ (Sigma), 0.05 mM 2-mercaptoethanol (Sigma), 2 mM l-glutamine (Sigma), 10^5 U/L penicillin (Sigma), and 10 mg/L streptomycin (Sigma).
11. Transfection selection media: RPMI plus extras plus 2 µM HCl (BDH), 250 µg/mL xanthine (Sigma), 15 µg/mL hypoxanthine (Sigma), and 1 µg/mL mycophenolic acid (Gibco-BRL). A 30X stock can be made up as a solution in 0.1 M NaOH (only neutralize with HCl just prior to use) and stored at –70°C.
12. Antibodies (Abs) for histology and flow cytometry, and normal serum blocking reagent.
13. Histology staining equipment (e.g., humid chamber, slide racks, slide baths, slides, and cover slips).
14. Aqueous (BDH), and VECTASHIELD® (Vector Laboratories) mounting media; hematoxylin (Sigma).
15. Tris-buffered saline (TBS): 50 mM Tris-HCl, pH 7.6, 150 mM NaCl.
16. Fast Red/napthol AS-MX: Dissolve 1 mg naphthol AS-MX phosphate, free acid (Sigma) in 1 mL *N,N*-dimethylformamide (Sigma) in a glass universal. Add 100 mM Tris-HCl, pH 8.2, to 50 mL. Split into 5 × 10-mL aliquots, and add 10 µL 1 M levamisole (Sigma) to each. Store at –20°C for up to 6 mo. Add 50 mg Fast Red TR salt (Sigma).
17. Cryo-M-Bed (BDH).
18. Elutip-d® minicolumns (Schleicher and Schuell).
19. OptiSeal™ centrifuge tubes (Beckman Coulter).
20. Low-salt buffer (LSB): 0.2 M NaCl, 20 mM Tris-HCl, pH 7.4, 1 mM ethylene-diaminetetraacetic acid (EDTA).
21. High salt buffer: 1 M NaCl, 20 mM Tris-HCl, pH 7.4, 1 mM EDTA.

22. CsCl gradient: 29 g ultrapure CsCl (Gibco-BRL) in 22 mL clean, fresh Tris-EDTA (TE) buffer, pH 8.0.
23. Injection buffer: 5 mM Tris-HCl, pH 7.4, 5 mM NaCl, 0.1 mM EDTA in water for embryo transfer (Sigma).
24. UltraPURE™ dialysis tubing, 3/4-inch diameter, 12–14 kDa exclusion limit (Gibco-BRL) and dialysis clips (BDH).
25. Proteinase K: make up fresh in triple-distilled water at 10 mg/mL.
26. TE, pH 8.0: 10 mM Tris-HCl, 1 mM EDTA, pH 8.0.
27. Tail punch buffer: 50 mM Tris-HCl, pH 8.0, 100 mM EDTA, 0.5% sodium dodecyl sulfate (SDS).
28. Phenol (BDH).
29. Sevag: chloroform 40:1 isoamyalchol.
30. Gene Images™ random prime labeling module (Amersham Biosciences).
31. Hybond™ N+ (Amersham Biosciences).
32. Sodium dodecyl sulfate (SDS) stock: 10% (w/v).
33. 20X sodium saline citrate (SSC): 0.3 M sodium citrate, 3 M NaCl.
34. Denaturing buffer: 1.5 M NaCl, 0.5 M NaOH.
35. Neutralizing buffer: 1.5 M NaCl, 0.5 M Tris-HCl, pH adjusted to 7.5.
36. Hybridization buffer: 5X SSC, 1 in 20 dilution liquid block, 0.1% (w/v) SDS, 5% (w/v) dextran sulphate.
37. Gene Images CDP-Star detection module (Amersham Biosciences).
38. Buffer A: 100 mM Tris-HCl, 300 mM NaCl, pH 9.5. Autoclave in aliquots; do not use for more than 1 d once opened.
39. BioMax MR-1 X-ray film (Kodak) and an X-ray cassette.
40. Ear punch buffer: 50 mM Tris-HCl, pH 8.0, 2 mM NaCl, 10 mM EDTA, 1% SDS.
41. Flow cytometry staining media: Hank's balanced salt solution (BSS) plus 2% FCS and 0.1% sodium azide.
42. ICE-T for enzyme-linked immunosorbent assay (ELISA) protocol: 1% FCS, 1% milk powder (w/v), 0.1% Tween-20, and 0.1% sodium azide in PBS.
43. AMP buffer: 150 mg MgCl$_2$ • 6H$_2$O, 100µL Triton X-405, 1 g sodium azide, 95.8 mL 1 M 2-amino-2-methyl-1-propanol (AMP) buffer solution (1.5 mol/L, pH 10.3 at 25°C; (Sigma), in 900 mL triple-distilled water. Adjust pH to 10.25 with conc. HCl and make volume up to 1 L.
44. ICE-T for enzyme-linked immunosorbent spot-forming cell (ELISPOT) protocol: 1% FCS, 1% milk powder (w/v), 0.05% Tween-20 in PBS.
45. *p*-Nitrophenyl phosphate (NPP) buffer, pH 9.8: 50 mM Na$_2$CO$_3$, 0.5 mM MgCl$_2$.

3. Methods

3.1. Generation of HEL Transgenes

Subheadings 3.1.1.–3.1.5. describe the modification of the cell surface mHEL construct (KLK), to one in which HEL will be retained intracellularly by the addition of a dilysine endoplasmic reticulum (ER) retention motif to the c-terminal cytoplasmic tail (mHEL-KK). Included are the design and preparation of the modified insert, cloning procedures involved, confirmation of the construct's sequence, and confirmation of in vitro expression.

3.1.1. The HEL Constructs

Figure 1 illustrates the HEL constructs used to generate transgenic mice. To generate the KLK *(18)* transgene Hartley et al. modified the ML *(2,3)* construct by inserting a complementary DNA fragment encoding the extracellular spacer sequence and cytoplasmic tail of the H-2K^b MHC class I molecule. The metallothionein I promoter was replaced with the H-2K^b gene promoter *(11)*. To model tolerance to intracellular self-antigens, the KLK construct was modified by adding a dilysine ER retention motif to the C-terminal cytoplasmic tail (mHEL-KK). Proteins carrying dilysine motifs bind to cytosolic COP I proteins, which causes their continuous and avid retrieval from the Golgi to the ER *(19)*.

3.1.2. Preparation of Modified Insert

The modified H-2K^b transmembrane region including a dilysine-retention motif was produced by a PCR amplification of the original KLK, introducing the new coding sequence (*see* **Note 1**). The forward primer (5' to 3': ccagggtcgcctggcgcaacc) encodes the junction between the 3' sequence of the lysozyme gene and the 5' H-2K^b transmembrane sequence. The reverse primer (5' to 3': ccatgcggccgctgtcttcaatccc tcttacccttttttaccagtccttctcatcttcatcacaaaagcc) encodes a *Not*I site and the dilysine retention motif derived from human uridine diphosphate (UDP)-glucuronosyltransferase 2B4 *(20)* as a 5' tag, with the 3' end priming off the H-2K^b transmembrane sequence. The PCR reactions were cleaned up using a QIAquick PCR purification kit (Qiagen) and assayed by relative fluorescence intensity on an agarose gel.

3.1.3. Cloning

All cloning steps were performed using standard molecular biology techniques that will not be discussed here *(21)*. All fragments, unless stated otherwise, were cleaned after gel electrophoresis using a GENECLEAN II spin kit (Anachem). The 265-bp PCR product was digested with *Xho*I and *Not*I, cleaned after gel electrophoresis with a MERmaid spin kit designed for isolating small fragments (Anachem), subcloned into pBluescript, and checked for fidelity by sequencing. To allow directional cloning, the pKLK H-2K^b transmembrane 3' *Xho*I restriction site was destroyed and a *Not*I site created via the introduction of an oligonucleotide linker (pKLK+linker). The final construct was obtained by digesting pKLK+linker with *Xho*I and *Not*I, and ligating it to the 196-bp *Xho*I/*Not*I PCR fragment.

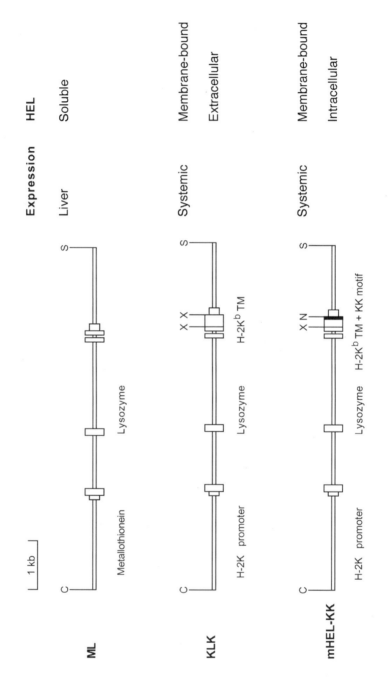

3.1.4. Restriction Digest

A panel of restriction digests of the original and modified constructs was set up to demonstrate that the various elements of the new construct were correctly orientated and of the right size (**Fig. 2**). Having confirmed that the construct was correct, a Qiagen plasmid mega kit was used to prepare DNA for future experiments (>1 mg) from a 500-mL culture of ElectroMAX DH10B competent cells (Gibco-BRL) transformed with the final plasmid (mHEL-KK). ElectroMAX *Escherichia coli* DH10B competent cells were used in all subcloning steps.

3.1.5. In Vitro Transfection

Stable transfectants of mHEL-KK in Sp2/0 myeloma cells were made to determine if the construct was able to both express HEL and retain the protein intracellularly. Mycophenolic acid resistance was conferred by the *Eco*-gpt gene present in the sequence of the pSVG-gpt vector used.

1. Linearize 10 µg of plasmid DNA by cutting with *Cla*I in a total volume of 50 µL. Harvest Sp2/0 cells (kind gift from M. Puklavec, Sir William Dunn School of Pathology) midlog phase, wash in sterile PBS, and resuspend at 1×10^7/mL. Keep cells on ice prior to transfection. Add the DNA digest to 700 µL of cells, gently mix, and transfer to a prechilled 4-mm electroporation cuvet (Bio-Rad); then allow to stand on ice for 15 min. Electroporate samples using a Bio-Rad Gene Pulser with 250 V, 960 µF, and return to ice for a further 15 min. Gently resuspend the cells in 12 mL RPMI + extras, then plate 500-µL aliquots in 24-well flat-bottom tissue-culture plates, and culture at 37°C, 5% CO_2.
2. After 24 h add 500 µL of 2X selection media to each well. Replace the medium with 1 mL 1X selection media on every other day (*see* **Note 2**). Split colony positive wells into two tissue-culture flasks, and culture until enough cells are available for testing (*22*; *see* **Note 3**).
3. Resuspend cells at 5×10^5/mL, cytospin onto slides, and allow to air-dry overnight. Fix slides in acetone for 10 min, then air-dry for the same period. Add the primary Ab, HyHEL9-Tc, to one of the two samples on each slide and incubate at room temperature for 30 min. Wash for 5 min in PBS. Add the secondary Ab, mouse anti-IgG-fluorescein isothiocyanate (FITC), to both samples on each slide and incubate for 2 h. Wash slides in PBS and mount using a nonfluorescing mounting media containing the antinuclear stain 4,6-diamidino 2-phenylindole (DAPI; VECTASHIELD, Vector Laboratories).

Fig. 1. *(see opposite page)* Schematic diagram of three different HEL transgenes (ML, KLK, and mHEL-KK), illustrating the differences in both design and expression between the constructs.

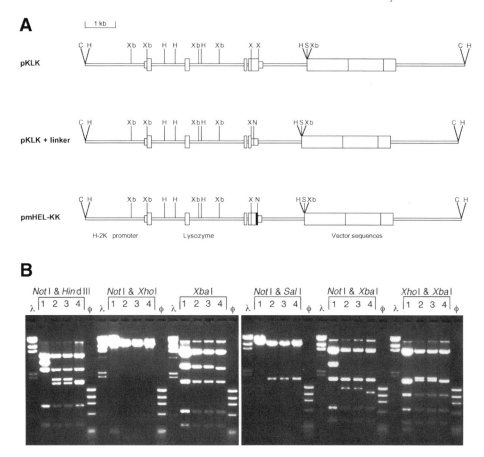

Fig. 2. Restriction maps of pKLK, pKLK+linker, and pmHEL-KK (**A**). Restriction enzymes: C, *Cla*I; H, *Hin*dIII; Xb, *Xba*I; X, *Xho*I; S, *Sal*I. Agarose gel (1%) showing pKLK (**1**), two alternative pmHEL-KK constructs (**2**) and (**3**), and pKLK+linker (**4**), cut with the restriction enzymes shown (**B**). All fragments produced were as expected.

3.2. Preparation of the Transgene for Microinjection (23)

Microinjection is a costly technique that yields relatively low numbers of positive founders (typically 10–30% efficiency in our hands). Careful preparation of the transgene greatly improves yields. Dust can block the needle during the process of microinjection, and detergent can kill the oocytes. Therefore to maximize the chance of success, it is very important that great care is taken when preparing the transgene to ensure that the risk of contaminating the sample with either dust or detergent is minimized at every step. These issues are critical from the point of ultracentrifugation onward (*see* **Notes 4–10**).

1. Digest 50–100 µg of construct with *Sal*I and *Cla*I to excise the 7.15-kb transgene from plasmid DNA.
2. Run the digested DNA with the marker λ *Hin*dIII on a 0.8% agarose Tris-acetate-EDTA (TAE) gel without ethidium bromide at 30 V for 24 h. Stain the gel with 0.5 µg/mL ethidium bromide in TAE for 30 min on a slowly rocking platform. Wash the gel in TAE for 1 h with gentle rocking. Repeat wash with fresh TAE. Quickly view the gel with a long-wave ultraviolet (UV) light box. Cutting completely through the gel, cut a 3-mm-long trough in front of the transgene band using a ruler and a clean scalpel blade. Return the gel to the electrophoresis tank, and fill with fresh TAE until the level of the buffer is just below the top of the gel. Fill the trough with TAE. To elute the DNA into the trough run at 100 V for 2 min intervals, collect the buffer from the trough, and replace with clean buffer. Run for another 2 min, collect buffer from trough, and replace with fresh buffer. Repeat this procedure until approx 25 mL has been collected into a single tube. Reexamine the gel under UV to determine if the entire transgene band has been electroeluted or if further collections are required. Save all of the elutant; this may be as much as 50 mL in total.
3. Remove the tip from an Elutip-d minicolumn (Schleicher and Schuell) with a clean scalpel blade. Equilibrate the column with 5 mL LSB at a rate of approx 1 drop/s. TAE is a LSB, and the elute can be loaded directly onto the column. Aspirate the gel eluant into a suitable volume syringe; load onto the column at a rate of 1–2 drops/s. Wash the column twice with 5 mL LSB. Elute the DNA from the column with 500 µL of high-salt buffer (HSB).
4. Add 450µL of the Elutip-d minicolumn elutant to an 8.9-mL OptiSeal tube (Beckman Coulter), and add the equivalent volume of HSB to a balance tube. Top up the tubes with CsCl and balance to within 0.01 g. Seal the tubes and place on caps required for ultracentrifugation. Place the tubes in an 80-Ti rotor and run at 200,000*g* for 36 h at 20°C, with maximum acceleration and no brake.
5. Carefully remove the ultracentrifuge tubes from the rotor, and attach securely to an upright with tape. Insert a large gage needle at the top of the tube and a second needle approx 1 cm from the bottom of the tube. Collect 25 fractions of 10 drops each in clean, dust-free 1.5-mL Eppendorf tubes (rinse them first with sterile water). Run 10 µL of each fraction on a 1% ethidium containing agarose gel to identify the fractions containing DNA. Pool the fractions containing the majority of the transgene: Typically there will be two to three.
6. Rinse two clean 4-L conical flasks, dialysis clips, and a magnetic flea with water 10 times. Make up 4 L of injection buffer in each flask using triple-distilled water straight from the source using a clean, rinsed measuring cylinder and pipets. Chill the buffer in a cold room or refrigerator.
7. Rinse a sterile beaker so it is free of detergent, and fill it with sterile distilled water. Gently pull the dialysis tubing into the beaker of clean water. Cut a length of approx 10 inches using a sterile blade, and then rinse the tubing with several changes of water using gentle agitation. Clip at one end with two dialysis clips. Using a rinsed 2-mL pipet transfer the combined fractions from **step 5** into the

end of the tubing; clip the other end of the tube a sufficient distance from the first clip to ensure that the membrane is not pulled taut by the sample. Gently drop the tubing into one of the 4-L flasks containing injection buffer (it will float) plus a magnetic flea; set to dialyze on a stirrer in a cold room. Change the buffer daily for 5 d; each time pour off the spent dialysate, rinse the flask with 500 mL of the fresh injection buffer, and then replace it with the remaining 3.5 L. Make up fresh dialysis buffer in the empty flask for the next time.

8. At the end of the dialysis aspirate the sample with a rinsed sterile pipet. Assay the sample by optical density $(OD)_{260nm}$ and by comparison with known standards on a gel (50, 25, 20, 15, 10, and 5 ng per lane of the same fragment, from an unpurified digest of the original plasmid cut with the same enzymes as used in **step 1**, and DNA standard **Fig. 3**). There should be good agreement between the two assays.

9. Dilute into 50-µL aliquots at 2, 1, and 0.5 ng/µL using injection buffer. The samples are now ready for microinjection and can be stored frozen alongside the remaining stock.

3.3. Microinjection

The transgene was injected into the male pronuclei of fertilized oocytes at a concentration of 2 ng/µL using standard techniques, which we will not detail here (Robert Sumner and Richard Corderoy, Transgenic facility, BMSU, Oxford University). A male C57BL/6 was crossed with a superovulated female CBA/J mouse to produce fertilized (C57BL/6 × CBA/J)$_{F1}$ oocytes. It is possible to inject C57BL/6 embryos and some facilities are very successful with this approach; but, in our hands, it has proved unreliable owing to the poor quality and number of oocytes. After injection the embryos were transferred to a pseudopregnant female to gestate. In this experiment, after weaning a total of 82 mice, potential founders were generated from two rounds of injections.

3.4. Screening Founders

Subheadings 3.4.1. and **3.4.2.** describe the process of screening potential founders generated from microinjection of the transgene. Included are notes on breeding strategies to determine if the transgene is transmitted, techniques used to test for the integration of the transgene, and testing for the expression of the transgene.

3.4.1. Screening for Transgene Integration and Transmission

Potential founders were screened by both Southern blot analysis, using a fluorescently labeled probe, and PCR. Eight were positive for the transgene. Positive founders were then crossed with C57BL/6, and as the PCR results replicated those obtained by Southern blot analysis, all offspring were screened

mHEL-KK *Sal* I & *Cla* I

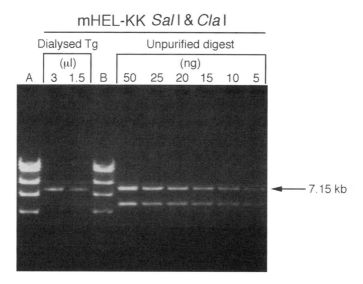

Fig. 3. Agarose assay gel (1%) comparing the fluorescent intensity of the dialyzed transgene (Tg) with varying amounts of the same fragment from an unpurified plasmid digest (*arrow*) and λ *Hind*III (**A**, 250 ng; **B**, 125 ng). The dialyzed transgene OD_{260nm} equivalent to 5.75 ng/μL. The 3 μL transgene band has a fluorescent intensity equivalent to 15–20 ng of the 7.15-kb digest fragment, thus, the dialyzed transgene calculated to be at a concentration of 5.83 ng/μL (17.5 ng per 3 μL).

by PCR alone. If a breeding pair produced 20 offspring that were all nontransgenic, it was deemed that the transgene was not transmitting and that the positive parent was presumably mosaic for the transgene (i.e., the transgene integrated after the one-cell stage).

3.4.1.1. SOUTHERN BLOT ANALYSIS

DNA Preparation *(23)*.

1. Cut off 1–2 cm from the tip of the tail; transfer to a clean 1.5-mL Eppendorf tube.
2. Add 700 μL of tail punch buffer and 35 μL of proteinase K solution to each tube, and incubate at 55°C overnight.
3. Centrifuge for 5 min.
4. Remove the supernatant, aspirating as little debris as possible, and transfer to a clean tube with 700 μL phenol. Cap tube and shake vigorously until the phases are completely mixed.
5. Centrifuge for 5 min.
6. Transfer the upper, aqueous phase to a clean tube containing 350 μL phenol and 350 μL Sevag. Cap tube and shake vigorously until the phases are completely mixed.

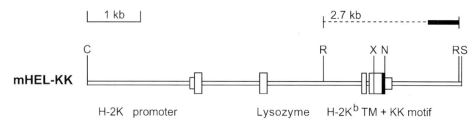

Fig. 4. The HEL probe (indicated by the *thick line*) consists of a 550-bp *Sma*I DNA fragment obtained from pKLK.

7. Centrifuge for 5 min.
8. Transfer the upper, aqueous phase to a clean tube containing 700 µL Sevag. Cap tube and shake vigorously until the phases are completely mixed.
9. Centrifuge for 5 min.
10. Again transfer the upper, aqueous phase to a clean tube. Add 50 µL 3 *M* sodium acetate, pH 5.2, and 700 µL ice-cold absolute ethanol (*see* **Note 11**). Mix gently until DNA precipitate becomes visible, then invert five times to mix thoroughly.
11. Centrifuge for 2 min.
12. Remove and discard the supernatant. Add 1 mL of 70% ethanol, and flick the tube to ensure the pellet lifts off the bottom of the tube. Invert several times to wash the pellet.
13. Centrifuge for 2 min.
14. Remove and discard the supernatant. Allow the pellet to air-dry.
15. Add 100 µL of TE, pH 8.0, to dissolve the pellet (*see* **Note 12**).

Generation of the Nonradioactive Probe

Figure 4 shows the probe used for screening mHEL-KK transgenics. The probe consisted of a 550-bp *Sma*I DNA fragment obtained from the original KLK construct. The probe was labeled using the Gene Images random prime-labeling module (Amersham Biosciences) in accordance with the manufacturer's instructions. Briefly the probe DNA was diluted to 5 ng/µL and denatured. The labeling reaction was set up using 50 ng of probe DNA, incubated at 37°C for 1 h, then stopped by the addition of EDTA to a final concentration of 20 m*M*. A rapid labeling assay was used to check that the labeling process was successful.

Southern Blotting

The concentration of each DNA extraction was determined by optical density (OD$_{260nm}$ 1 = 50 µg/mL single-stranded DNA), and 10 µg of each sample digested with *Eco*RI plus 4 m*M* spermidine. The digests plus λ *Hin*dIII were electrophoresed at 50 V on a 10-cm-long 1% agarose gel, until the dye front reached the end of the gel. A photograph of the gel under UV fluorescence was taken to ascertain the size of the hybridized band (*see* **Note 13**). The Southern blot transfer *(21)* was set up using Hybond N+ (Amersham Biosciences). After disassembling the capillary apparatus, the blot was fixed via UV cross-linking.

Depurination

This step was not performed as the target sequence is smaller than 10 kb (depurination buffer: 250 m*M* HCl).

Hybridization and Stringency Washing

Hybridization buffer (0.125 mL/cm^2) was preheated to 60°C, placed on the blots in a hybridization bottle (Techne or similar), and incubated for 30 min at 60°C in a hybridization oven. After denaturing the probe, 25 µL was pipetted into the buffer in the hybridization bottle, taking care not to pipet it directly onto the membrane. Hybridization was performed at 60°C overnight. The blots were washed for 20 min in preheated 1X SSC, 0.1% SDS (w/v), at 60°C with agitation. A second wash was performed at 60°C using preheated 0.5X SSC and 0.1% SDS (w/v) with agitation.

Developing Blot

The hybridized probe was detected using the Gene Images CDP-Star detection module (Amersham Biosciences) in accordance with the manufacturer's instructions, discussed briefly here.

The blot was incubated with gentle agitation for 1 h at room temperature in 1 mL/cm^2 of a 1 in 10 dilution of liquid blocking agent in buffer A fo r 1 h. The blot was transferred to antifluorescein–alkaline phosphatase conjugate diluted 5000-fold in 0.5% bovine serum albumin (BSA)/buffer A, incubated at room temperature for 1 h with gentle agitation, and then washed three times with 4 mL/cm^2 0.3% Tween-20/buffer A before draining off the excess buffer. Sterile detection reagent (35 µL/cm^2) was pipetted over the blot and left for 5 min. BioMax MR-1 X-ray film (Sigma) was exposed to the blot in an X-ray cassette for an initial period of 5 min before developing (**Fig. 5**).

3.4.1.2. PCR

Ear punch DNA preparation (*see* **Note 14**):

1. Add 150 µL of proteinase K solution to 3 mL of ear punch buffer.
2. Incubate the ear punch in a 0.5-mL Eppendorf tube plus 20 µL of the enzyme mix for 20 min at 55°C.
3. Remove samples from the water bath and vortex well; return to the water bath for a further 20 min.
4. At the end of the incubation period, pulse spin the tubes, and add 180 µL of triple-distilled water.

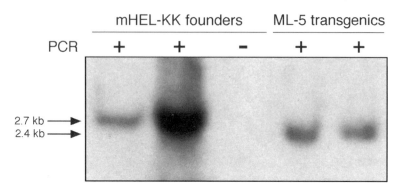

Fig. 5. Southern blot of tail DNA from mHEL-KK founders and ML-5 transgenic mice was digested with *Eco*RI; the sizes of the resulting hybridizing bands are indicated (also refer to **Fig. 1**).

5. Incubate the tubes at 100°C for 5 min to heat-inactivate the proteinase K.
6. Store the prepared DNA at 4°C if screening will be performed within a few days, or long-term at –20°C.
7. Use 1 µL of ear punch DNA for each PCR reaction in a total volume of 25 µL (*see* **Note 15**).

Figure 6 illustrates the presence or absence of the HEL transgene, detected using the following primers:

HEL3F   5′   GAGCGTGAACTGCGCGAAGA   3′
HEL4R   5′   TCGGTACCCTTGCAGCGGTT   3′

3.4.2. Screening for Expression

Once a transgenic line has been established, it is necessary to determine that the transgene is expressing as intended in vivo. There are many techniques that can be used for this purpose including reverse transcriptase polymerase chain reaction (RT-PCR; sensitive, straightforward, and quantifiable; it does not test for correct translation but is usually sufficient in addition to in vitro assays), immunohistochemistry (site-specific but insensitive), or Western blots (gives information on size and protein associations but is relatively insensitive and may require denaturation). The actual technique chosen depends on the expected pattern and to some extent the level of expression.

3.4.2.1. IMMUNOHISTOCHEMISTRY

Histological examination of the spleen was used in the preliminary screening for expression of this systemic transgene. The spleen was chosen because it has

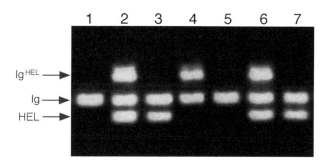

Fig. 6. Example of an agarose PCR gel (2%) showing the products generated using both the HEL-specific and Ig^HEL-specific primers in the same reaction. The HEL transgene is detected by HEL3F and HEL4R (160 bp PCR product). The immunoglobulin transgene is detected by IgH$_{F1}$ and IgH$_{R1}$ (Ig^HEL, 430 bp), and IgH$_{F2}$ and IgH$_{R1}$ generate the endogenous immunoglobulin band (Ig, 264 bp). Nontransgenic (*lanes 1* and *5*), mHEL-KK transgenic (*lanes 3* and *7*), Ig^HEL-transgenic (*lane 4*), and Ig^HEL/mHEL-KK double transgenic mice (*lanes 2* and *6*) are shown.

high levels of MHC class I expression, it is an easy organ to handle, and it yields plenty of sections, which are often required to refine the staining protocol.

Perform all incubations at room temperature.

1. Snap-freeze organs in liquid nitrogen, and store at –80°C until sectioning.
2. Mount organs in Cryo-M-Bed and cut 7 µm cryostat sections, pick sections up onto 8-place poly-L-lysine slides and air-dry overnight.
3. Store the sections in pairs, wrapped back-to-back in aluminium foil until ready to stain (*see* **Note 16**).
4. Incubate the slides in acetone for 10 min (this fixes and permeablizes the section), then air-dry for 10 min (*see* **Note 17**).
5. Block slides with 5% normal goat serum in TBS for 30 min.
6. Wash slides twice in TBS for 5 min.
7. Incubate sections with the primary Ab, unpurified polyclonal rabbit anti-HEL serum diluted in TBS, for 30 min (*see* **Note 18**).
8. Wash slides in TBS for 5 min.
9. Incubate sections with the secondary Ab, alkaline-phosphatase-conjugated goat antirabbit IgG, for 30 min.
10. Wash slides in TBS for 5 min.
11. Add the Fast Red salt to the napthol AS-MX mix and filter directly onto the slides; leave for 15 min (*see* **Note 19**).
12. Wash briefly in TBS and then water.
13. Counterstain with hematoxylin, and mount using an aqueous mounting media.

3.4.2.2. FLOW CYTOMETRY

Flow cytometry was used to demonstrate that the introduced dilysine retention motif did indeed result in the intracellular sequestration of HEL. The basic procedure used to stain samples for flow cytometric analysis is outlined in **Subheading 3.5.2.** Briefly, bone marrow and spleen suspensions from nontransgenic, KLK (surface expression) and mHEL-KK mice were stained with the anti-HEL monoclonal Ab HyHEL9-tricolor (Tc) and HyHEL9-FITC to stain for the presence of surface HEL. All analysis was performed using a FACSCalibur flow cytometer and CellQuest™ software (Becton Dickinson).

3.4.2.3. QUANTIFICATION OF SERUM HEL CONCENTRATION

This bioassay was used to determine the amount of HEL present in the serum of the mHEL-KK transgenic mice.

1. Incubate splenocytes (1×10^5) from an Ig^{HEL} transgenic mouse on ice for 45 min, with either HEL serially diluted in normal mouse serum or serum alone.
2. Stain with HyHEL9-Tc and B220-FITC as for flow cytometry.
3. Plot a standard curve of the mean HyHEL9-Tc fluorescence of positively staining B cells, vs the concentration of HEL.
4. Determine the concentration of serum HEL in the KK transgenic lines by incubating Ig^{HEL} cells with mHEL-KK serum and comparing the mean fluorescence with the standard curve.

3.5. Analysis

Subheadings 3.5.1.–3.5.4. describe selected aspects of functional studies, which may be important to consider in other experiments, focusing on breeding, flow cytometry, and quantitative assays.

3.5.1. Breeding With Immunoglobulin (Ig^{HEL}) Transgenic Mice

As the transgenics were founded in (C57BL/6 × CBA/J)$_{F1}$ mice, each line was backcrossed at least 6 generations to C57BL/6 (99.22% C57BL/6) to remove major effects from background CBA genes before analysis. Full backcrossing is a relative concept, but is considered to require at least 10 generations (>99.95% C57BL/6), so we would not generally intercross the mice until after this point. For analysis, the mHEL-KK transgenic mice were crossed with Ig^{HEL} transgenic mice to study the effect of the self-antigen on the autoreactive B cells. It has been noted that female Ig^{HEL} transgenic mice do not always breed as well as nontransgenic litter mates; therefore, when setting up crosses we always ensure that the male breeder is positive for the Ig^{HEL} transgene and the female breeder carries the mHEL-KK transgene. This strategy also ensures that only hemizygous Ig^{HEL} transgenic mice are bred. It is essential to have an unambiguous PCR screen. The presence of both transgenes can be detected by a single PCR reaction using the HEL-specific primers described in **Subheading 3.4.1.2.** and the following primers for the Ig^{HEL} transgene and endogenous immunoglobulin gene (**Fig. 6**):

IgH$_{F1}$	5′	GCGACTCCATCACCAGCGAT	3′
IgH$_{F2}$	5′	CTGGAGCCCTAGCCAAGGAT	3′
IgH$_{R1}$	5′	ACCACAGACCAGCAGGCAGA	3′

3.5.2. Flow Cytometry

The μ and δ genes used in the Ig^{HEL} transgenic were derived from a Balb/c hybridoma (IgH^a allotype) *(2)*. Thus by using monoclonal Abs that are allotype specific, it is possible to distinguish between the transgenic B cells and those that derive from the endogenous C57BL/6 (IgH^b allotype) immunoglobulin genes. For example, by comparing B cells from Ig^{HEL} transgenics and mHEL-KK/Ig^{HEL} double transgenics stained with IgM^a and IgD^a Ab, it is possible to observe the effects of antigen on developmental blocks, or receptor modulation. Because the antigen in this system is commercially available, it is also possible to add exogenous antigen and stain with an anti-HEL monoclonal Ab that is specific for an epitope that does not overlap with the epitope recognized by the Ig^{HEL} transgenes. This sandwich staining technique allows the amount of receptors bound to antigen to be determined and is unaffected by previously bound self-antigen. It is also the basis for the bioassay outlined in **Subheading 3.4.2.2.** When setting up bone marrow chimeras it is useful to use mice that also express allotypic cell markers, as this allows donor and recipient B cells to be distinguished—for example, CD45.1 (Ly-5.1).

Basic staining protocol (*see* **Notes 20–22**):

1. Incubate 5×10^5 cells in a round-bottom 96-well plate with 25 μL of Ab in the dark, for 20 min on ice.
2. Wash twice using flow cytometry staining media.
3. Add 25 μL of secondary Ab if required and incubate as in **step 1**.
4. Wash twice using flow cytometry staining media.
5. Fix using 200 μL 1% paraformaldehyde in PBS.

3.5.3. IgM^a Anti-HEL ELISPOT

The following protocol is used to ascertain the number of HEL-specific IgM^a-secreting plasma cells per spleen or other tissue.

1. Coat 96-well flat-bottom plates with 125 μL/well of 1 mg/mL HEL in carbonate buffer, pH 9.8 (HEL binds at high pH). Incubate overnight at 4°C, or for 3 h at 37°C.
2. Wash plates twice with 0.05% Tween-20/PBS.

3. Block with 125 μL/well of 1% BSA/PBS. Incubate overnight at 4°C, or for 1 h at 37°C.
4. Wash plates three times with PBS.
5. Add a unicellular splenocyte suspension to the wells in graded fractions in duplicate. Incubate for 4 h at 37°C, 5% CO_2.
6. Wash plates with PBS, twice with 0.05% Tween-20/PBS, then fill the wells with 0.05% Tween-20/PBS and allow to stand for 3 min; finally repeat the 3-min rinse.
7. Add 125 μL/well of ICE-T that has been prewarmed to 37°C. Incubate at 37°C for 30 min.
8. Wash plates twice with 0.05% Tween-20/PBS.
9. Add 60 μL/well of anti-IgM[a] (clone DS-1) diluted in ICE-T. Incubate overnight at 4°C, or for 2 h at 37°C.
10. Wash plates three times with 0.05% Tween-20/PBS.
11. Add 60 μL/well of avidin alkaline phosphatase (Sigma) diluted in 0.1% BSA/PBS/Tween-20. Incubate for 1 h at 37°C.
12. Wash five times with 0.05% Tween-20/PBS, then twice with triple-distilled water.
13. Add 100 μL/well of 1 mg/mL 5-bromo-4-chloro-3-indoyl phosphate disodium salt (BCIP)/0.2% agarose/AMP buffer (*see* **Note 23**).
14. Leave plate at room temperature for a few hours for the color to develop, then count using an inverted microscope.

3.5.4. IgM[a] Anti-HEL ELISA

1. Coat 96-well ELISA plates with 100 μL/well of 10 μg/mL HEL in carbonate buffer, pH 9.8. Incubate overnight at 4°C, or for 2 h at 37°C.
2. Wash plates three times with 0.05% Tween-20/PBS.
3. Block with 200 μL/well of 1% BSA/PBS. Incubate overnight at 4°C, or for 2 h at 37°C.
4. Wash plates three times with 0.05% Tween-20/PBS.
5. Dilute serum samples, including a standard of known concentration, in 0.1% BSA/PBS, and serially dilute across plate using a multichannel pipet. Incubate for 2 h at 37°C.
6. Wash plates five times with 0.05% Tween-20/PBS.
7. Add 60 μL/well of anti-IgM[a] (clone DS-1) diluted in ICE-T. Incubate overnight at 4°C, or for 2 h at 37°C.
8. Wash plates five times with 0.05% Tween-20/PBS.
9. Add 50 μL/well of avidin alkaline phosphatase (Sigma) diluted in 0.1% BSA/PBS/Tween-20. Incubate for 1 h at 37°C.
10. Wash five times with 0.05% Tween-20/PBS, then twice with PBS, once with triple-distilled water, and finally once with NPP buffer.
11. Add 100 μL/well of 1 mg/mL Sigma[104®]/NPP buffer. Read at 405 nm.

4. Notes

1. When amplifying a sequence by PCR for use in cloning it is critical that no errors are introduced into the sequence. To minimize the possibility of introducing such

an error, use a high-fidelity polymerase (e.g., VentR polymerase) and titrate down the number of PCR cycles to the minimum required to give enough product to manipulate. Set up any PCR reactions, and perform all subsequent subcloning steps, in tandem. Check the final results by sequencing; if there is an unexpected base change present in both constructs it is unlikely that the same error has occurred in both PCR reactions and is more likely owing to a mistake in the published sequence. However, if only one sequence differs from the published data, then the error has been introduced during the PCR process and that construct should be discarded.

2. Colonies should usually start growing out by 2 wk; however, this has taken up to 6 wk to occur.

3. When culturing cells in preparation for testing by immunohistochemistry it is important that the cells are healthy. The proportion of viable cells in a culture can be calculated by staining with trypan blue (excluded by living cells).

4. The percentage of agarose gel used to resolve the transgene fragment and the choice of marker depends on the size of the fragments to be separated.

5. When cutting the trough for the transgene elution it is helpful to cut the trough a few millimeters narrower than the width of the transgene band; this makes it easy to ascertain that all of the transgene band has been eluted, as the edges of the band can be seen clearly on either side of the trough.

6. The eluted transgene should be purified using the Elutip-d minicolumns as soon as possible, but it can be stored overnight at 4°C if necessary. When loading the eluant onto the column a blue band usually appears at the top of the column; this is perfectly normal.

7. When running the gel to identify the fractions containing the transgene after ultra-centrifugation, it is important to note that the DNA will run abnormally, owing to the CsCl, and will only just penetrate the gel; however, it is helpful to run the gel so that the dye front has run a short distance away from the wells and does not obscure the positive wells.

8. To prevent the concentration of the transgene becoming too dilute, combine only the fractions that contain the majority of the DNA. Usually there will be a group of two to three wells with bright intensity, with a lane of weaker intensity on either side.

9. If a narrow band of DNA-containing fractions is not observed when the fractions collected after ultracentrifugation are run out on a gel, it may be that the transgene band in the CsCl gradient has been disrupted. Likely causes of this are knocking the tube after centrifugation, or inserting the needles into the tube in the wrong order, disrupting the band by air bubbles. If this is the case, the wells may all appear to have a faint fluorescence under UV light compared to neighboring empty wells.

10. It is possible to dialyze more than one sample at the same time; just clip the tubes in a different pattern so that they can be distinguished. It should not be necessary to increase the dialysis period.

11. Ice-cold absolute ethanol is best achieved by keeping a clean stock at –20°C until required.

12. Using TE that has been prewarmed to 55°C can help when hydrating the pellet at the end of the tail extraction procedure. If the pellet is difficult to dissolve it is probably attributable to overdrying. This can be overcome by incubating at to 55°C (overnight if necessary) and/or by the addition of more buffer.

13. Measure the distance from the wells to the hybridizing band on the film developed from the Southern blot. Look at where this distance falls within the DNA ladder to determine the size of the fragment hybridized by the probe.

14. Ensure that the ear punch tool is thoroughly rinsed and dried between mice; this minimizes DNA carryover, which could otherwise lead to false-positive PCR results, especially when screening transgenic mice with high transgene copy numbers. We have not found contamination to be a problem using this technique.

15. When pipetting ear punch DNA for the PCR reaction, be sure to aspirate the sample from just through the meniscus, avoiding the debris at the bottom of the tube as this can lead to poor-quality results.

16. When removing stored cryostat section slides from the freezer, it is important to allow the slides to come up to room temperature before unwrapping. Patchy staining will result if slides that have been exposed to condensation are returned to the freezer for use at a later date.

17. Acetone fixation may result in poor cellular morphology. If organelle structure is important, improved morphology can be achieved using different fixation methods. For example, fix sections with 4% paraformaldehyde/250 mM HEPES for 10 min, then 8% paraformaldehyde/HEPES for 50 min, and wash with PBS. Quench sections with 50 mM NH_4Cl, 5 min, and wash. Finally permeabilize sections with 0.1% Triton X-100/PBS for 10 min, and wash before staining as before. We used this method prior to staining for confocal microscopy of thymocytes (as they have larger cytoplasm than splenocytes) and were able to show that mHEL-KK is retained within the endoplasmic reticulum, as expected.

18. We always include a "no Ab" and "secondary Ab only" control on each slide. A hydrophobic Pap pen (BioGenex) can be used to prevent solutions from running inappropriately from one section to another on the same slide.

19. Always ensure that fresh levamisole is used when making up napthol AS-MX, or high background staining from endogenous alkaline phosphatase can result. The napthol AS-MX mix can be made in batches and stored at –20°C for up to 6 mo.

20. Plate out the volume of cell suspension required to give 5×10^5 cells, and add flow cytometry staining media to 200 µL. Pulse centrifuge the staining plates to 800g. Flick off the media, and tap the plate to resuspend the cells before adding staining Ab. At the end of the incubation, add 200 µL of staining media, then spin, flick off, and tap as previously. Repeat as required by the protocol.

21. All Abs are diluted in flow cytometry staining medium. RPMI containing phenol red must not be used as staining media because this can result in autofluorescence.

22. It is critical that all Abs used to stain cells for flow cytometry are titrated. This will not only save you money but will also reduce nonspecific staining, which can occur when the concentration of Ab used is too high. A starting dilution of 1 in 10, serially diluted twofold to give 10 dilution factors should be sufficient to

determine which concentration is best. Buying several vials of a commonly used in Ab (e.g., B220) will ensure that they are all from the same batch and thus only one titration need be performed.

23. Fifteen minutes prior to the end of the final incubation step, make up a solution of 1.25 mg/mL BCIP/AMP buffer, heat to 40°C for 10 min, then filter-sterilize (0.2 μM), and split into 4-mL aliquots. Make up 50 mL of 1% agarose in triple-distilled water, and microwave it just before adding it to the plates so that it doesn't set too early. Wash the plates as described, add 1 mL of the hot agarose to the BCIP solution, mix well, and pipet into the wells using a multichannel pipet.

Acknowledgments

We acknowledge the excellent facilities at the Biomedical Services Unit at the John Radcliffe Hospital. This work was supported by the Wellcome Trust.

References

1. Erikson, J., Radic, M. Z., Camper, S. A., Hardy, R. R., Carmack, C., and Weigert, M. (1991) Expression of anti-DNA immunoglobulin transgenes in nonautoimmune mice. *Nature* **349,** 331–334.
2. Goodnow, C. C., Crosbie, J., Adelstein, S., et al. (1988) Altered immunoglobulin expression and functional silencing of self-reactive B lymphocytes in transgenic mice. *Nature* **334,** 676–682.
3. Goodnow, C. C., Crosbie, J., Jorgensen, H., Brink, R. A., and Basten, A. (1989) Induction of self-tolerance in mature peripheral B lymphocytes. *Nature* **342,** 385–391.
4. Hartley, S. B., Cooke, M. P., Fulcher, D. A., et al. (1993) Elimination of self-reactive B lymphocytes proceeds in two stages: arrested development and cell death. *Cell* **72,** 325–335.
5. Nemazee, D. A. and Burki, K. (1989) Clonal deletion of B lymphocytes in a trans-genic mouse bearing anti-MHC class I antibody genes. *Nature* **337,** 562–566.
6. Okamoto, M., Murakami, M., Shimizu, A., et al. (1992) A transgenic model of autoimmune hemolytic anemia. *J. Exp. Med.* **175,** 71–79.
7. Tiegs, S. L., Russell, D. M., and Nemazee, D. (1993) Receptor editing in self-reactive bone marrow B cells. *J. Exp. Med.* **177,** 1009–1020.
8. Gay, D., Saunders, T., Camper, S., and Weigert, M. (1993) Receptor editing: an approach by autoreactive B cells to escape tolerance. *J. Exp. Med.* **177,** 999–1008.
9. Akkaraju S., Canaan K., and Goodnow C. C. (1997) Self-reactive B cells are not eliminated or inactivated by autoantigen expressed on thyroid epithelial cells. *J. Exp. Med.* **186,** 2005–2012.
10. Hayakawa, K., Asano, M., Shinton, S. A., et al. (1999) Positive selection of natural autoreactive B cells. *Science* **285,** 113–116.
11. Morello, D., Moore, G., Salmon, A. M., Yaniv, M., and Babinet, C. (1986) Studies on the expression of an H-2K/human growth hormone fusion gene in giant trans-genic mice. *EMBO J.* **5,** 1877–1883.

12. Liou, G. I., Matragoon, S., Yang, J., Geng, L., Overbeek, P. A., and Ma, D. P. (1991) Retina-specific expression from the IRBP promoter in transgenic mice is conferred by 212 bp of the 5′-flanking region. *Biochem. Biophys. Res. Commun.* **181,** 159–165.

13. Palmiter, R. D. and Brinster, R. L. (1986) Germ-line transformation of mice. *Annu. Rev. Genet.* **20,** 465–499.

14. Kozak, M. (1987) At least six nucleotides preceding the AUG initiator codon enhance translation in mammalian cells. *J. Mol. Biol.* **196,** 947–950.

15. Brinster, R. L., Allen, J. M., Behringer, R. R., Gelinas, R. E., and Palmiter, R. D. (1988) Introns increase transcriptional efficiency in transgenic mice. *Proc. Natl. Acad. Sci. USA* **85,** 836–840.

16. Chada, K., Magram, J., Raphael, K., Radice, G., Lacy, E., and Costantini, F. (1985) Specific expression of a foreign beta-globin gene in erythroid cells of transgenic mice. *Nature* **314,** 377–380.

17. Townes, T. M., Lingrel, J. B., Chen, H. Y., Brinster, R. L., and Palmiter, R. D. (1985) Erythroid-specific expression of human beta-globin genes in transgenic mice. *EMBO J.* **4,** 1715–1723.

18. Hartley, S. B., Crosbie, J., Brink, R., Kantor, A. B., Basten, A., and Goodnow, C. C. (1991) Elimination from peripheral lymphoid tissues of self-reactive B lymphocytes recognizing membrane-bound antigens. *Nature* **353,** 765–769.

19. Teasdale, R. D. and Jackson, M. R. (1996) Signal-mediated sorting of membrane proteins between the endoplasmic reticulum and the Golgi apparatus. *Annu. Rev. Cell. Dev. Biol.* **12,** 27–54.

20. Jackson, M. R., McCarthy, L. R., Harding, D., Wilson, S., Coughtrie, M. W., and Burchell, B. (1987) Cloning of a human liver microsomal UDP-glucuronosyl-transferase cDNA. *Biochem. J.* **242,** 581–588.

21. Sambrook, J., Fritsch, E. F., and Maniatis, T. (1989) *Molecular Cloning : A Laboratory Manual,* 2nd ed. Cold Spring Harbor Laboratory Press, Cold Spring Harbor, NY.

22. Gascoigne, N. R., Goodnow, C. C., Dudzik, K. I., Oi, V. T., and Davis, M. M. (1987) Secretion of a chimeric T-cell receptor-immunoglobulin protein. *Proc. Natl. Acad. Sci. USA* **84,** 2936–2940.

23. Hogan, B. (1994) *Manipulating the Mouse Embryo: A Laboratory Manual,* 2nd ed. Cold Spring Harbor Laboratory Press, Cold Spring Harbor, NY.

18

Analysis of B-Cell Signaling
Using DT40 B-Cell Line

Tomoharu Yasuda and Tadashi Yamamoto

Summary

A chicken DT40 B-cell line provides an excellent model system for the analysis of the function of genes related to B-cell antigen receptor (BCR) signaling. Crosslinking of BCR by μ-chain-specific antibody stimulates DT40 cells to undergo apoptosis. Activation of protein-tyrosine kinases (PTKs) and intracellular calcium signaling are essential for BCR-induced apoptosis. Because DT40 B cells are highly proliferative and readily integrate exogenous DNA by homologous recombination, gene targeting is conveniently employed to inactivate genes of interest. This chapter describes procedures for (1) generation of knockout DT40 cell line, (2) analysis of surface BCR expression, (3) detection of tyrosine-phosphorylated proteins, (4) detection of intracellular calcium mobilization, and (5) reconstitution of depleted gene by transfection.

Key Words

DT40; homologous recombination; gene targeting; BCR; *cbl*; calcium.

1. Introduction

Chicken B-cell line DT40, which integrates transfected deoxyribonucleic acid (DNA) constructs at high frequency by homologous recombination, is a convenient model for analysis of phenotypes caused by disruption of one or more genes. DT40 cells express surface B-cell antigen receptor (BCR; IgM isotype) and respond to BCR crosslinking, which leads to calcium mobilization and apoptosis. The DT40 system has been used to study effects of signaling molecules downstream of BCR and the functional and physical relations between these molecules. Here, we describe methods for generating DT40 cells carrying target deletions of specific genes and for analysis of signaling events regulated by BCR, using experiments with *cbl*-deficient DT40 cells as an example *(1)*.

From: *Methods in Molecular Biology, vol. 271: B Cell Protocols*
Edited by: H. Gu and K. Rajewsky © Humana Press Inc., Totowa, NJ

2. Materials

1. DT40 cell (chicken B-cell line; *see* **Note 1**).
2. *cbl* complementary DNA (cDNA).
3. Chicken genomic library (Clontech Laboratories, Palo Alto, CA).
4. pBluescript vector.
5. Drug resistance cassettes.
6. Blasticidin.
7. Histidinol.
8. Puromycin.
9. G418.
10. Zeocin.
11. Genepulser II (Bio-Rad, Hercules, CA).
12. 0.4-cm Cuvet.
13. Anti-*cbl* antibody (Santa Cruz Biotechnology, Santa Cruz, CA).
14. Fluorescein isothiocyanate (FITC)-conjugated antichicken IgM antibody (Bethyl Laboratories, Montgomery, TX).
15. Antichicken IgM antibody, M4 (Southern Biotechnology, Birmingham, AL).
16. Antiphosphotyrosine antibody, 4G10 (Upstate Biotechnology, Lake Placid, NY).
17. FACS Calibur (Becton Dickinson, Franklin Lakes, NJ).
18. HBSS/CaCl$_2$: Hank's balanced salt solution (HBSS) supplemented with 10 mM HEPES, pH 7.2, 0.025% bovine serum albumin (BSA), and 1 mM CaCl$_2$.
19. HBSS/ethylene glycol *bis*-(2-aminoethyl ether)-$N,N,N'N'$-tetraacetic acid (EGTA): HBSS supplemented with 10 mM HEPES, pH 7.2, 0.025% BSA, and 1 mM EGTA.
20. 3 µM Fura2 AM solution: suspend Fura2 AM (Molecular Probes, Eugene, OR) in HBSS/CaCl$_2$ with 0.006% Cremophore EL (Sigma-Aldrich, St. Louis, MO), and sonicate before loading into cells.
21. CAF110 spectrofluorometer (JASCO, Tokyo, Japan).
22. pAzeo expression vector.

3. Methods

We describe methods for generation of knockout DT40 cells, analysis of surface BCR expression, detection of tyrosine-phosphorylated proteins, detection of intracellular calcium mobilization, and reconstitution of a disrupted gene by transfection.

3.1. Generation of Knockout DT40 Cells

A knockout construct for DT40 cells should contain homologous sequences from the gene of interest flanking a marker gene. The marker gene, which typically confers drug resistance, can either be inserted into or replace part of the coding sequence of the targeted gene. The drug resistance cassette is typically oriented opposite to the direction of transcription of the target gene. It should be noted that although most genes have two alleles in DT40 cells, there are genes with only one allele (*Syk [2]*), and in some cases there are three (*Lyn [2]*) or four

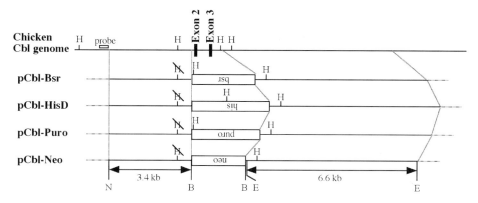

Fig. 1. Structure of the chicken *cbl* allele and the targeting vector. Restriction sites for *Hind*III (**H**), *Not*I (**N**), *Bam*HI (**B**), and *Eco*RI (**E**) are indicated. The 3.4-kb *cbl* genomic fragment on the targeting vectors, which isolated from the chicken genomic library, did not contain the *Hind*III site.

(*cbl [1]*) alleles (*see* **Notes 2** and **3**). In such cases, multiple targeting events may be necessary.

3.1.1. Construction of Targeting Vectors to Disrupt cbl

Chicken *cbl* cDNA was isolated by reverse transcriptase polymerase chain reaction (RT-PCR) from ribonucleic acid (RNA) from chicken DT40 B cells. This cDNA was used as a probe to identify *cbl* genomic DNA fragments in a chicken genomic library. A 3.4-kb *Not*I–*Bam*HI fragment and a 6.6-kb *Eco*RI–*Eco*RI fragment of the chicken *cbl* gene were subcloned into a pBlue-script vector with a *Bam*HI site between the fragments. The targeting vectors pCbl-Bsr, pCbl-HisD, pCbl-Puro, and pCbl-Neo were constructed by replacing the region that corresponds to amino acids 250–336 of human Cbl with *bsr*, *hisD*, *puro*, and *neo* cassettes, respectively (**Fig. 1**), inserted as *Bam*HI fragments in reverse orientation relative to the *cbl* coding sequence. The resistance markers can be readily changed, if two, three, or four targeting events are required.

3.1.2. Electroporation

pCbl-Bsr was linearized and introduced into wild-type DT40 cells by electroporation as described here. Transfectants were selected in the presence of 50 µg/mL blasticidin S, and resistant clones were screened by Southern blotting analysis. We found that appropriate targeting events occurred in 10 of 24 blasticidin-resistant clones. The hybridization signal of the targeted allele was two- or threefold weaker than that of the wild-type allele, suggesting that

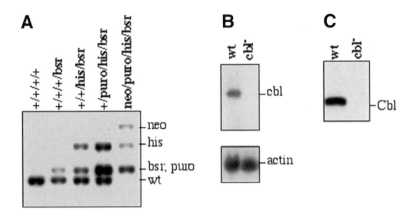

Fig. 2. Disruption of the *cbl* gene in chicken DT40 B Cells. (**A**) Southern blot analysis of wild-type and various targeted DT40 cells. *Hin*dIII-digested genomic DNA separated on agarose gel was blotted and hybridized with a chicken *cbl* cDNA probe. Wild-type and *bsr*-, *his*D-, *puro*-, *neo*-targeted alleles showed 4.9, 5.4, 7.4, 5.4, or 7.9-kb fragments, respectively. (**B**) Northern blot analysis using a chicken cDNA probe for *cbl* (*top*) or β-*actin* (*bottom*). (**C**) Protein expression analysis of Cbl in wild-type and *cbl*-deficient DT40 cells. Total cell lysate (2×10^6 cells) were prepared and analyzed by Western blotting using anti-Cbl antibody.

DT40 cells contain more than two *cbl* alleles (**Fig. 2A**). Disruption of the remaining *cbl* alleles was accomplished by sequential transfection of three other targeting constructs, pCbl-HisD, pCbl-Puro, and pCbl-Neo. pCbl-HisD was transfected into a *bsr*-targeted clone, and integration was selected with both blasticidin S (50 μg/mL) and histidinol (1 mg/mL). pCbl-Puro was transfected into a *bsr/his*-targeted clone and selected with blasticidin S (50 μg/mL), histidinol (0.5 mg/mL), and puromycin (0.5 μg/mL). pCbl-Neo was then transfected into a *bsr/his/puro*-targeted clone and selected with blasticidin S (50 μg/mL), histidinol (0.5 mg/mL), puromycin (0.5 μg/mL), and G418 (2 mg/mL; **Fig. 2A**). Hybridization with *bsr*, *his*D, *puro*, and *neo* probes indicated that the final clone contained a single copy of each construct. Lack of Cbl was verified by Northern blot and Western blot analysis (**Fig. 2B,C**).

1. Spin down 1×10^7 cells and wash with 10 mL of phosphate-buffered saline (PBS) immediately.
2. Resuspend cells in 0.5 mL PBS and transfer to a 0.4-cm cuvet.
3. Add 25–30 μg of linearized DNA.
4. Incubate on ice for 10 min.
5. Electroporate with Genepulser II at the condition of 550 V and 25 μF.
6. Incubate on ice for 10 min.

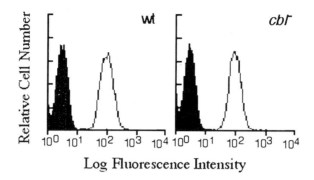

Fig. 3. Cell-surface expression of BCR. BCR expression on the surface of wild-type and *cbl*-deficient DT40 cells was confirmed by flow cytometry. Unstained cells were used as negative controls (*shaded histogram*).

7. Transfer cells to a T75 flask containing 20 mL of fresh media without drugs.
8. Incubate cells at 37°C, 5% CO_2 for 24 h.
9. Resuspend transfected cells in 85–90 mL of selection media. Drug concentrations used for selection are 0.05 mg/mL blasticidin, 1 mg/mL histidinol, 0.5 µg/mL puromycin, 2 mg/mL G418, 1 mg/mL zeocin, 2 mg/mL hygromycin B, and 0.3 mg/mL phleomycin.
10. Transfer 200 µL of transfected cells into each well of four 96-well plates.
11. In approx 5–7 d, drug-resistant clones may be visible through the bottom of the plate. When colonies reach approx 2 mm in diameter, clones should be transferred to 24-well plates. Cultures can be expanded to T25 flasks approx 2 d later.
12. Once stable clones are established, they can be grown in media without selection.

3.2. Flow Cytometric Analysis of Surface BCR Expression

Because the surface level of BCR is a critical factor for the determination of signaling threshold, expression of BCR should be confirmed prior to analysis of BCR-mediated signaling events in *cbl*-deficient cells. BCR expression on the surface of wild-type and *cbl*-deficient cells (**Fig. 3**) was evaluated as follows.

3.2.1. Preparation of Cells for FACS Analysis of BCR Expression

1. Spin down 1×10^6 cells and wash with 0.2 mL of PBS immediately.
2. Resuspend cells in 50 µL of PBS with or without 1 µg of FITC-conjugated antichicken IgM antibody.
3. Incubate for 30 min at 4°C.
4. Wash twice with 0.2 mL of PBS.
5. Resuspend in 0.5% formaldehyde/PBS.
6. Analysis with FACS Calibur.

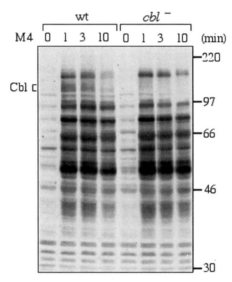

WCL, anti-phosphotyrosine (4G10)

Fig. 4. BCR-induced tyrosine phosphorylation in wild-type and *cbl*-deficient DT40 cells. At the indicated time-points after addition of M4 (4 µg/mL), whole-cell lysates prepared from 2×10^6 cells were loaded onto an 8% sodium dodecyl sulfate-polyacrylamide gel electrophoresis (SDS-PAGE) gel. The blotted membrane was incubated with antiphosphotyrosine antibody (4G10).

3.3. Detection of Tyrosine-Phosphorylated Proteins
Following BCR Stimulation

Stimulation of BCR initiates signaling cascades beginning with the activation of cytoplasmic PTKs. Three distinct families of nonreceptor PTKs, Src family PTKs, Syk, and Btk, are activated on BCR engagement and contribute to the activation of multiple downstream effectors including Ras-MAPK, phosphatidylinositol 3-kinase, and PLC-γ2 (*3–6*). Comparison of BCR-induced tyrosine phosphorylation between wild-type and mutant cells provides initial information about PTK activity and tyrosine-phosphorylated proteins downstream of BCR. In *cbl*-deficient cells, there were no significant changes in tyrosine phosphorylation patterns, except that the band corresponding to Cbl itself was absent (**Fig. 4**).

3.3.1. BCR Stimulation of DT40 Cells
and Antiphosphotyrosine Western Blotting

1. Spin down 2×10^6 cells and wash with 0.5 mL of PBS immediately.
2. Resuspend in 0.5 mL of PBS.

3. Add 4 µg of antichicken IgM (M4) and incubate at 37°C.
4. Spin down and quickly freeze cells on dry ice.
5. Suspend in 80 µL of sodium dodecyl sulfate (SDS) sample buffer and boil for 8 min.
6. Separate 20 µL of whole-cell lysates on 8% polyacrylamide gel and transfer to polyvinylidene difluoride (PVDF) membrane.
7. Western blotting with antiphosphotyrosine antibody (4G10) after blocking membrane with 5% BSA/TBS-T.

3.4. Calcium Measurements

One of the hallmarks of BCR-induced signaling is calcium mobilization. The rise of intracellular Ca^{2+} ($[Ca^{2+}]_i$) induced by BCR or ionomycin stimulation can be measured with Fura2 loading of DT40 cells (**Fig. 5A**).

3.4.1. Intracellular Calcium Mobilization

1. Spin down 2.5×10^6 cells and wash with 0.5 mL of PBS immediately.
2. Resuspend cells in 3 µM Fura2 AM solution.
3. Incubate for 30 min at 30°C.
4. Wash twice with HBSS/CaCl$_2$.
5. Suspend 5×10^5 cells in 0.5 mL of HBSS/CaCl$_2$, transfer to a cuvet, and put into the CAF110 apparatus.
6. Incubate cells at 30°C for a few minutes to stabilize the fluorescence ratio detected at 500 nm with excitation at 340 nm and 380 nm.
7. Begin recording fluorescence intensity and add 1 µg of antichicken IgM (M4).
8. Add 25 µL of 10% Triton X-100 to determine R_{max} and F_{max} values.
9. Add 6 µL of 0.5 M EGTA to determine R_{min} and F_{min} values.
10. The following equation can be used to relate the intensity ratio to Ca^{2+} levels:

$$[Ca^{2+}]_i = K_d * Q * (R - R_{min})/(R_{max} - R)$$
$K_d = 224$ nm and is the Ca^{2+} dissociation constant of Fura2.
R represents the fluorescence intensity ratio F_{340nm}/F_{380nm}.
Q is the ratio of F_{min} to F_{max} at 380 nm.

3.4.2. Calcium Release From Intracellular Calcium Store

Calcium release from intracellular calcium stores was measured in 0.5 mL HBSS/EGTA in conjunction with the method described in **Subheading 3.4.1.** To evaluate extracellular calcium influx, Ca^{2+} was restored to the buffer, adding CaCl$_2$ to a final concentration of 2 mM (**Fig. 5B**).

3.5. Generation of Stable Clones Expressing Transfected Gene

Expression vectors that harbor drug resistance genes can be used to examine exogenous gene expression. To reconstitute Cbl expression in *cbl*-deficient cells, we introduced the *cbl* cDNA—cloned into the pAzeo expression vector—

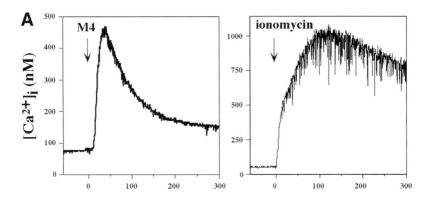

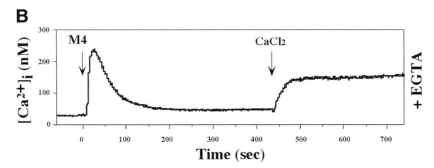

Fig. 5. Ca^{2+} mobilization in DT40 cells. **(A)** [Ca^{2+}]$_i$ were monitored by spectrophotometer after stimulation with M4 (2 μg/mL) or ionomycin (0.5 μg/mL). **(B)** Ca^{2+} release from intracellular Ca^{2+} store was monitored after stimulation with M4 (2 μg/mL) in the presence of 1 m*M* EGTA (+EGTA). Ca^{2+} was restored to the media to evaluate extracellular Ca^{2+} influx. *Arrows* indicate the time-point of addition of M4, ionomycin, and CaCl$_2$.

into *cbl*-deficient cells by electroporation. Stable clones that produce the Cbl protein were identified by screening for zeocin resistance. Electroporation and screening were carried out following the method described in **Subheading 3.1.2.** pAzeo expression vector was generated by replacing the CMV promoter of pcDNA3.1/Zeo (Invitrogen, Carlsbad, CA) with the chicken actin promoter from the pApuro expression vector *(2)* (**Fig. 6A**). Expression vector was transfected into *cbl*-deficient DT40 cells by electroporation at 550 V, 25 μF. Selection was performed in the presence of zeocin (1 mg/mL). Expression of the transfected gene was confirmed by Western blotting (**Fig. 6B**).

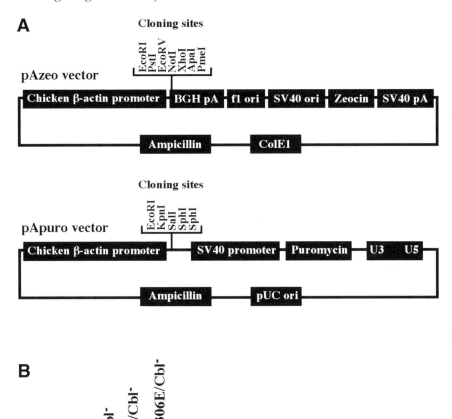

Fig. 6. (A) Schematic representation of pAzeo and pApuro expression vectors. (B) Expression of wild-type or G306E Cbl in *cbl*-deficient cells. Total lysates (2×10^6 cells) were prepared and analyzed by Western blotting using anti-Cbl antibody.

4. Notes

1. DT40 B cells were maintained in RPMI-1640 supplemented with 10% fetal calf serum (FCS), 1% chicken serum, 50 μM β-mercaptoethanol, 2 mM L-glutamine, penicillin, and streptomycin. Confluency, $1.5–2.0 \times 10^6$ cells/mL; doubling time, less than 12 h; passage, 1:20 every other day; freezing, $1.0–1.5 \times 10^7$ cells in 1 mL of 90% FCS/10% dimethyl sulfoxide (DMSO).

2. A number of articles describing mutant DT40 cells have been published. A list of knockout DT40 cell lines can be found in **ref. 7**. The expressed sequence tag (EST) database with over 7000 ESTs from bursal lymphocytes would facilitate identification of possible targets for gene disruption studies in DT40 cells *(8)*. Details of the bursal EST sequencing project and access to database search forms are available at the DT40 Web site (http://genetics.hpi.uni-hamburg.de/dt40.html).

3. If the gene of interest is essential for cell proliferation, gene disruption in DT40 cells can be done by creating conditional loss-of-function mutations with either Cre-recombinase-mediated excision *(9)*, tetracycline-regulated transcription *(10)*, or tamoxifen-regulated protein transport *(11)*.

Acknowledgments

We would like to thank Dr. W. Y. Langdon for providing human c-Cbl cDNA and T. Kurosaki for excellent collaboration in the study on Cbl function in B cells.

References

1. Yasuda, T., Maeda, A., Kurosaki, M., et al. (2000) Cbl suppresses B cell receptor-mediated phospholipase C (PLC)-γ2 activation by regulating B cell linker protein-PLC-γ2 binding. *J. Exp. Med.* **191,** 641–650.
2. Takata, M., Sabe, H., Hata, A., et al. (1994) Tyrosine kinases Lyn and Syk regulate B cell receptor-coupled Ca2+ mobilization through distinct pathways. *EMBO J.* **13,** 1341–1349.
3. Reth, M. and Wienands, J. (1997) Initiation and processing of signals from the B cell antigen receptor. *Annu. Rev. Immunol.* **15,** 453–479.
4. DeFranco, A. L. (1997) The complexity of signaling pathways activated by the BCR. *Curr. Opin. Immunol.* **9,** 296–308.
5. Tamir, I. and Cambier, J. C. (1998) Antigen receptor signaling: integration of protein tyrosine kinase functions. *Oncogene.* **17,** 1353–1364.
6. Kurosaki, T. (1999) Genetic analysis of B cell antigen receptor signaling. *Annu. Rev. Immunol.* **17,** 555–592.
7. Winding, P. and Berchtold, M. W. (2001) The chicken B cell line DT40: a novel tool for gene disruption experiments. *J. Immunol. Methods* **249,** 1–16.
8. Abdrakhmanov, I., Lodygin, D., Geroth, P., et al. (2000) A large database of chicken bursal ESTs as a resource for the analysis of vertebrate gene function. *Genome Res.* **10,** 2062–2069.
9. Fukagawa, T., Hayward, N., Yang, J., et al. (1999) The chicken HPRT gene: a counter selectable marker for the DT40 cell line. *Nucleic Acids Res.* **27,** 1966–1969.
10. Wang, J., Takagaki, Y., and Manley, J. L. (1996) Targeted disruption of an essential vertebrate gene: ASF/SF2 is required for cell viability. *Genes Dev.* **10,** 2588–2599.
11. Fukagawa, T. and Brown, W. R. (1997) Efficient conditional mutation of the vertebrate CENP-C gene. *Hum. Mol. Genet.* **6,** 2301–2308.

B-Cell Development and Pre-B-1 Cell Plasticity In Vitro

Antonius G. Rolink

Summary

B-lymphopoiesis in vivo is a very complex process that is influenced by cooperation between cells, cytokines, and other receptor-ligand interactions, which developmentally occur at different cellular stages. Various in vitro models have been very useful in unraveling this complex process of B-cell development. Here the protocols for how to grow and to study the differentiation of mouse pre-B-1, as developed in our laboratory, are described. Moreover, the protocols of how to grow and to test the plasticity of *Pax-5*-deficient pre-B-1 cells are also outlined.

Index Entries

Pre-B-1 cells; stromal cells; interleukin-7; cell sorting; retroviral transduction; *Pax-5*-deficient mice; pre-B-1 cell plasticity.

1. Introduction

B-cell development as it takes place in the liver before birth and in the bone marrow thereafter can be subdivided into various developmental stages. The differential expression of cell-surface markers and the subsequent analysis of the rearrangement status of immunoglobulin heavy (IgH) and light (IgL) chain genes has allowed these developmental stages to be placed in sequence. In Chapter 1, Hardy and Shinton describe the cell-surface markers they use to dissect B-cell development.

In our laboratory, monoclonal antibodies (mAbs) against B220, CD19, CD117 (c-kit), CD25, and IgM are used to dissect the various stages of B-cell development *(1–3)*. In **Fig. 1** the scheme of B-cell development using these markers is shown. Moreover, the rearrangement status of IgH and IgL chain genes, as determined by single cell PCR analysis *(4)*, as well as the cell cycle status of various cells is also shown in this figure. The relevant cell type for this

From: *Methods in Molecular Biology, vol. 271: B Cell Protocols*
Edited by: H. Gu and K. Rajewsky © Humana Press Inc., Totowa, NJ

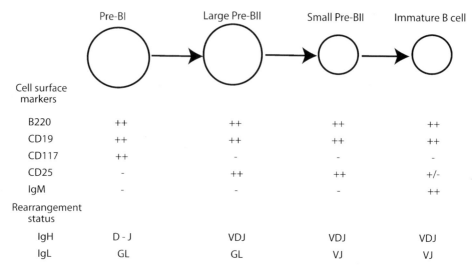

Fig. 1. Scheme of mouse B-cell differentiation in bone marrow based on the differential expression of B220, CD19, CD117, CD25, and IgM and the rearrangement status of the IgH and IgL chain loci as determined by PCR at the single cell level. ++, expression; –, indicates the absence of expression of a given marker. Large cells indicate actively proliferating subpopulations while small cells indicate nonproliferating populations.

protocol is the so-called pre-B-1 cell. This population of cells is characterized by the surface expression of CD19, B220, and CD117 while being negative for CD25 and IgM. In young adult mice the bone marrow pre-B-1 cell compartment comprises 2–4% of all nucleated cells. A characteristic bone marrow B220, CD117 staining is shown in **Fig. 2**. A similar population of pre-B-1 cells is present in the fetal liver.

Analysis of the rearrangement status of Ig genes in single cells revealed that pre-B-1 cells carry predominantly two D_H–J_H rearranged IgH alleles while the IgL chain loci are still in germline configuration (4). Another characteristic of the pre-B-1 cells is their capacity to proliferate long term in vitro on stromal cells in the presence of the growth factor interleukin-7 (IL-7) (5,6).

During this in vitro proliferation, pre-B-1 cells keep their IgH chains D_H–J_H rearranged. However, pre-B-1 cells grown in vitro have not lost the capacity to differentiate. Removal of the growth factor IL-7 almost immediately results in the transcriptional up-regulation of the *RAG* genes and, as a consequence of this, a completion of IgH and IgL chain rearrangements in these cells (6,7). Thus, by d 2–3 after removal of IL-7, a low number of sIgM-positive, LPS-reactive B cells are found in these cultures, showing that these pre-B-1 cells can differentiate into immature sIg-positive B cells *(6,7)*. Another interesting feature of these in vitro grown pre-B-1 cells is that transplantation into *RAG*-deficient mice results in a partial reconstitution of the peripheral B-cell compartment *(8–9)*.

Rather recently, we have shown that pre-B-1 cells isolated from *Pax-5*-deficient mice can also be propagated under the same culture conditions. Since *Pax-5* expression is required for B-cell development beyond the pre-B-1 cell stage, removal of IL-7 from these cultures did not result in the formation of immature B cells. However, and to our great surprise, in vitro grown *Pax-5*-deficient pre-B-1 cells have the potential to differentiate under appropriate cytokine conditions into macrophages, granulocytes, dendritic cells, osteoclasts, and NK cells in vitro, while in vivo these cells can also give rise to T cells *(10,11)*.

2. Materials

2.1. Mice

1. Bone marrow cells of various inbred strains of mice at 3–6 wk of age are used as the source of pre-B-1 cells.
2. *Pax-5*-deficient pre-B-1 cells are isolated from bone marrow of mice at 1.5-2 wk-of-age (the majority of *Pax-5*-deficient mice die at 2–3 wk after birth).
3. Fetal liver pre-B-1 cells are isolated from embryos at d15–18 of gestation.
4. Balb/c or C57BL/6 *RAG*-2 deficient mice at 6–12 wk of age are used in trans-plantation experiments.

2.2. Tissue Culture Reagents and Cytokines

1. Iscove's modified Dulbecco's medium (IMDM) powder (Invitrogen).
2. Dulbecco's modified Eagle's medium (DMEM) (Invitrogen).
3. Fetal calf serum (FCS) (Amimed): batches of FCS from different companies are screened. Batches that have low mitogenicity for T and B cells and strongly pro-mote lipopolysaccharide (LPS)-driven proliferation of B cells are selected.
4. Primatone (Quest, Naarden, The Netherlands).
5. Penicillin–streptomycin (Invitrogen).
6. β-Mercaptoethanol (Invitrogen).
7. Trypsin ethylenediaminetetraacetic acid (Trypsin-EDTA; Invitrogen).
8. Trypan blue (Invitrogen).
9. IL-7-conditioned medium of J558 cells transfected with the cDNA encoding IL-7. Titration of IL-7 is done using the IL-7-dependent 5.7-μ^{low} cells (Rolink, unpub-lished data). Proliferation is measured by ^{3}H-thymidine incorporation during the last 12 h of a 3-d incubation period. Five times the highest dilution giving the maximal proliferation is used for the growth of pre-B-1 cells.
10. Macrophage colony-stimulating factor (M-CSF; R&D).
11. Granulocyte-macrophage colony-stimulating factor (GM-CSF; R&D).

2.3. Tissue Culture Plates and Flasks

1. 96-Well flat-bottom plates (Nunc).
2. 24-Well plates (Nunc).
3. Tissue culture flasks: 25 cm^2, 80 cm^2, 175 cm^2 (Nunc).

2.4. Stromal Cells and Retroviral Packaging Cells

1. ST-2 *(12)* and ST-2 cells transfected with the Puromycin resistance gene.
2. PA6 *(13)*.
3. S17 *(14)*.
4. OP-9 *(15)*.
5. GPE-E86 (ATCC).

2.5. Cell Sorting and Flow Cytometric Analyses

1. Cell sorter and analyzer: we have used the FACS Vantage and FACS Aria from Becton Dickinson as well as the Moflo from Cytomation/Dako to sort pre-B-1 cells; FACS Calibur (Becton Dickinson) is used for analysis.
2. Monoclonal antibodies and secondary reagents:
   a. B220-FITC (clone RA3-6B2, Pharmingen/Becton-Dickinson).
   b. CD117-APC (clone 2B8-eBioscience) or ACK4-biotin or ACK4-PE (Pharmingen/Becton-Dickinson).
   c. CD19-FITC, PE, PE-Cy7 or APC-Cy7 (clone ID3, Pharmingen/Becton-Dickinson).
   d. CD25-biotin (clone 7D4; Pharmingen/Becton-Dickinson).
   e. IgM-FITC, -PE or -biotin (clone IB4BI; eBioscience).
   f. Streptavidin-PE (Southern Biotechnology Assoc.).
   g. Streptavidin-PE-Cy7 (Pharmingen/Becton-Dickinson).

2.6. Miscellaneous

1. Irradiator (X-ray or Cesium source).

3. Methods

3.1. Cell Cultures

3.1.1. Stromal Cell Maintenance and Amplification

The stromal cell lines we used are propagated in IMDM supplemented with 5×10^{-5} *M* β-mercaptoethanol, 100 U/mL antibiotics, 0.03% Primatone, and 2% fetal calf serum (FCS) for ST-2, PA6, and S17, and 20% FCS for OP9. The IMDM medium is made fresh every 2–3 wk from IMDM powder dissolved in quartz-distilled water. The use of commercial liquid IMDM is not recommended, since growth of stromal cells and especially pre-B-1 (*see* **Subheadings 3.1.3.** and **3.1.4.**) is retarded. An alternative medium that can be used is DMEM supplemented with 5×10^{-5} *M* β-mercaptoethanol, 100 U/mL antibiotics and 10% FCS.

Table 1
**Number of Stromal Cells Plated Per Well or Per Tissue Culture Flask
at Various Days Before Onset of Pre-B Cell Cultures**

	Day 3	Day 2	Day 1	Day 0
96-Well plate[a]	1.5×10^3	3×10^3	7×10^3	1×10^4
24-Well plate[b]	0.5×10^4	1×10^4	2×10^4	5×10^4
25 cm² Tissue culture flask[c]	5×10^4	10^5	2×10^5	5×10^5
80 cm² Tissue culture flask[d]	1.5×10^5	3×10^5	6×10^5	1.2×10^6

Stromal cells are plated in: [a]0.2 mL medium, [b]1 mL medium, [c]7 mL medium and [d]25 mL medium

We routinely maintain and expand stromal cells in 175-cm² tissue culture flasks. At d 0, 6×10^5 stromal cells are seeded per flask in 50 mL of medium. At d 3 of culture at 37°C in 10% CO_2, the stromal cells are practically confluent and harvested by trypsinization. For this the culture medium is removed and 10 mL of trypsin-EDTA is added. After 10 min of incubation at 37°C, stromal cells have detached and can be harvested. Stromal cells are then spun down, resuspended in fresh IMDM medium, and counted. About $3–4 \times 10^6$ stromal cells can be recovered from a 175-cm² tissue culture flask.

3.1.2. Stromal Cell Density and Treatment for Pre-B-1 Cell Growth

The density of the stromal cell layer used for growing pre-B-1 cells is rather critical. In **Table 1**, the number of stromal cells plated per well or flask at various days before onset of the pre-B cell cultures is given. At the day of seeding pre-B-1 cells (d 0), stromal cell-containing plates or flasks are irradiated at 3000 rad. After irradiation the medium is replaced by fresh IMDM supplemented with IL-7.

When no irradiator for plates or flasks is available, stromal cells can also be irradiated in suspension. After irradiation the cells are washed once and counted. The number of stromal cells to be plated is identical to the number given in **Table 1** for d 0. Before adding the pre-B-1 cells, these cultures should be incubated at 37°C in 10% CO_2 for 2–3 h to give the stromal cells time to attach to the plastic.

3.1.3. Initiation of Pre-B-1 Cell Cultures

For setting up new cultures, pre-B-1 cells from bone marrow or fetal liver are sorted (B220+, c-kit+; **Fig. 2**). About 5×10^3 sorted pre-B-1 cells are then plated in 1 mL IMDM supplemented with IL-7 on stromal cells in wells of a 24-well plate. After 5–7 d, pre-B-1 cells have expanded to an extent that they

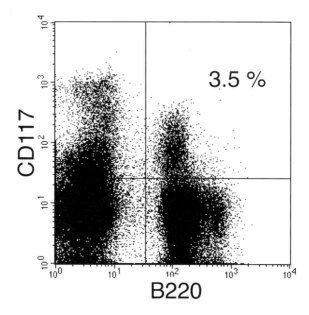

Fig. 2. pre-B-1 cells in wild type bone marrow. Bone marrow cells from 4-wk-old C57BL/6 mice were stained with B220-FITC and CD117-APC. Stained cells were then analyzed using a FACS Calibur. As shown, 3.5% of the cells are B220[+] CD117[+] pre-B-1 cells.

largely cover the stromal cells. Feeding and/or exchange of medium during this period is not necessary. At this point of time, pre-B-1 cells are harvested by strong pipeting using a Pasteur pipet and transferred to a 25-cm^2 tissue culture flask containing stromal cells and 7 mL IMDM medium supplemented with IL-7. After another 3–4 d of culture pre-B-1 cells in some of these small flasks will have grown to almost confluency and should be harvested. It should be noted, however, that for unknown reasons pre-B-1 cells do not grow to confluency in all small flasks seeded. Therefore, we always set up at least five parallel cultures. In our hands, no difference between ST2, PA6, S17, and OP9 stromal cells in promoting growth of sorted pre-B-1 cells has been observed.

3.1.4. Propagation of Pre-B-1 Cell Cultures

Pre-B-1 cells harvested from small flasks should be spun down (100g for 10 min), re-suspended in fresh culture medium containing IL-7 and counted. About 5 × 10^6 pre-B-1 cells are recovered from a confluent 25-cm^2 tissue culture flask. For propagation, either 1 × 10^6 pre-B-1 cells are plated into a 25-cm^2 tissue culture flask containing fresh, irradiated stromal cells and 7 mL culture medium supplemented with IL-7 or 3 × 10^6 pre-B-1 cells are plated into a 80-cm^2

tissue culture flask containing stromal cells and 25 mL culture medium supplemented with IL-7.

When pre-B-1 cell growth is optimal, $5-10 \times 10^6$ pre-B-1 cells can be recovered from the 25-cm^2 tissue culture flask (or $2-4 \times 10^7$ from a 80-cm^2 tissue culture flask) after 3 d. From this point of time on, we transfer the pre-B-1 cells every 3 d onto new stromal cells in fresh IL-7 supplemented medium. A critical point to be noted here is that IL-7 should never become a limiting factor. If it does, the pre-B-1 cells will stop growing and cannot be grown again (*see* also **Subheading 3.1.6.**).

Using this protocol we have been able to establish long-term (>6 mo) growing pre-B-1 cell lines and clones from fetal liver of all mouse strains that have detectable numbers of pre-B-1 cells. Moreover, these established lines and clones can be frozen and thawed using standard protocols. However, and in marked contrast to fetal liver-derived pre-B-1 cells, bone marrow derived pre-B-1 cells have a much more restricted growth potential. The vast majority of bone marrow derived pre-B-1 cells stop proliferating after 3–4 wk in culture. To date we have not been able to find an explanation for this difference between fetal liver and bone marrow derived pre-B-1 cells.

Also for the propagation of pre-B-1 cells, we have not observed any difference between ST2, PA6, S17, and OP9 stromal cells. It should be noted, however, that all four lines, with time in culture, lose their growth promoting activity. Therefore, every 2 mo, we routinely thaw an early passage of stromal cells.

3.1.5. Retroviral Gene Transduction of Cultured Pre-B-1 Cells

Stable introduction of genes into cultured pre-B-1 cells by transfection has not been successful thus far. However, retroviral transduction of genes into pre-B-1 cells is rather efficient. For retroviral transduction, the GPE-E86 packaging cell line stably transfected with the pMSCV puro expressing the gene of interest is used. Transfected GPE-E86 cells can then be used as stromal cell support for the growth of pre-B-1. The number of GPE-E86 cells to be plated, as well as their irradiation, is identical to that described above for the classical stromal cells. Also the number of pre-B-1 cells plated is identical to the numbers used for propagation on normal stromal cells. After 3 d of coculture in the presence of IL-7, about half the control number pre-B-1 cells can be recovered from this type of culture. Transduction efficiency was tested using the *GFP* gene as a model system. Between 20 and 60% of the pre-B-1 cells express the *GFP* gene after 3 d of coculture. Subsequent culturing of these pre-B-1 cells for 3–5 d in IL-7 medium supplemented with 2 µg/mL puromycin on puromycin-resistant ST-2 cells results in pre-B-1 cells of which over 99% express the *GFP* gene. Using the same system, other genes such as the anti-apoptotic *Bcl-2* gene have been successfully introduced into pre-B-1 cells.

3.1.6. Differentiation of In Vitro Grown Pre-B-1 Cells Into Immature B Cells

In the presence of stromal cells and IL-7, pre-B-1 cells proliferate and keep their D_H–J_H rearranged status. When IL-7 is removed, pre-B-1 cells stop growing and the expression of *RAG-1* and 2 genes is quickly up-regulated. By d 3 of culture without IL-7, 5–15% of the cells have differentiated into sIgM-positive immature B cells, which can be further differentiated into Ig-secreting cells upon addition of LPS. Stromal cells are not required for this differentiation into immature B cells. An important point to note is that pre-B-1 cells cultured for 24 h or longer without IL-7 have lost the capacity to grow on stromal cells and IL-7.

During B-cell development in the bone marrow, expression of the pre-BCR, composed of the IgH chain and the surrogate light chain components VpreB and $\lambda 5$, results in a very strong proliferation of large pre-B-2 cells *(8,16)*. This pre-BCR-driven proliferation determines to a large extent the efficiency of immature B-cell production in the bone marrow. In vitro, this pre-BCR-driven proliferation can be mimicked by culturing *ex vivo*-sorted pre-B-1 cells in plain medium without any growth factors *(17)*. However, during the in vitro differentiation of pre-B-1 cells grown on stromal cells in the presence of IL-7, the pre-BCR-driven proliferation does not seem to operate. This conclusion is largely based on the finding that in vitro grown pre-B-1 cells from $\lambda 5$-deficient mice possess a differentiation potential similar to wild type pre-B-1 cells in this assay *(18)*.

A serious drawback of this in vitro B-cell differentiation system is that the vast majority of cells die by apoptosis during this process. By d 3 of culture in the absence of IL-7, only 5% of the initially seeded cells are still viable. The apoptosis can be blocked by the expression of the anti-apoptotic *Bcl-2* gene in the pre-B cells. Thus, pre-B-1 cell lines from *Bcl-2* transgenic mice differentiate with the same efficiency into immature B cells in the absence of IL-7 but without massive apoptosis. A similar inhibition of apoptosis can be achieved by the retroviral introduction of the *Bcl-2* gene into wild type pre-B-1 cells.

3.1.7. Establishment of Pax-5-Deficient Pre-B-1 Cell Lines

The method to generate stromal cell and IL-7-dependent pre-B-1 cell lines from *Pax-5*-deficient mice is identical to the method described for wild type mice. However, and in marked contrast to wild type mice, in *Pax-5*-deficient mice pre-B-1 cells that can be grown are present in the bone marrow but absent in the fetal liver *(19,20)*. Another remarkable difference between wild type and *Pax-5*-deficient bone marrow pre-B-1 cells is that the latter can be grown long term in culture, while wild type-derived ones stop growing after

3–4 wk. Similar to wild type pre-B-1 cells, stable introduction of genes by transfection into *Pax-5*-deficient pre-B-1 cells has not been successful, while retroviral transduction is rather efficient.

3.1.8. In Vitro Plasticity of Pax-5⁻/⁻ Pre-B-1 Cells

Pax-5-deficient mice have an absolute block in B-cell development at the pre-B-1 cell stage. Therefore, withdrawal of IL-7 from *Pax-5*-deficient pre-B-1 cell cultures does not result in the generation of immature B cells. However, and in marked contrast to wild type pre-B-1 cells, *Pax-5*-deficient pre-B-1 cells possess the capacity to differentiate in vitro into various cells of the myeloid lineage. Thus, *Pax-5* pre-B-1 cells cultured in the presence of 10 ng/mL M-CSF (no IL-7) differentiate within 2–3 wk into macrophages while the combination of M-CSF and GM-CSF (10 ng/mL) results in the generation of myeloid dendritic cells. Since *Pax-5*⁻/⁻ pre-B-1 undergo massive apoptosis in the absence of IL-7, the generation of macrophages and dendritic cells is relatively poor. However, and in analogy to wild type pre-B-1 cells giving rise to immature B cells, the efficiency of myeloid differentiation is dramatically improved by retroviral expression of the *Bcl-2* gene in *Pax-5*-deficient pre-B-1 cells.

3.2. In Vivo Developmental Potential of In Vitro Grown Wild-Type and Pax-5-Deficient Pre-B-1 Cells

Long-term in vitro grown wild type pre-B-1 cells are also able to reconstitute, at least in part, the mature B-cell compartment in immunodeficient mice. Therefore C57Bl/6 or Balb/c *RAG*-2-deficient mice are irradiated with 400 rad and 2–4 h thereafter injected iv with $2–10 \times 10^6$ in vitro grown pre-B-1 cells. At around 4 wk after transplantation a low number (2–5%) of IgM-positive B cells of donor origin are found in the spleen. This B cell reconstitution is long-term since a similar number is still detectable after 6 mo. Phenotypically these B cells are similar if not identical to B-1 B cells, i.e., CD5⁺, IgM^high, IgD⁺/⁻, and CD23⁻. Functional analysis reveals that these reconstituted mice mount an almost normal T-cell-independent humoral immune response.

Moreover, *RAG*-deficient mice reconstituted with pre-B-1 cells and 2–3 mo thereafter injected with $1–5 \times 10^5$ sorted CD4 T cells are also able to mount a significant T-cell-dependent humoral immune response. This T-cell dependent immune response is also accompanied by the formation of germinal centers. Thus, in vitro propagated pre-B-1 cells are able to reconstitute a functional competent B-cell compartment in *RAG*-deficient mice. It should be noted however that a reconstitution of the pre- and immature B cell compartments in the bone marrow of these reconstituted mice is never observed. Moreover, we have never observed the development of non-B cells (T cells, myeloid cells) from transferred wild type in vitro-grown pre-B-1 cells.

In marked contrast, transfer of 2–10 × 10^6 in vitro grown *Pax-5*-deficient pre-B-1 cells into sublethal irradiated *RAG*-deficient mice results in complete thymus reconstitution with all thymocyte subpopulations at 3 wk after injection *(10)*. At later time points, mature T cells are found in the periphery. Moreover, in vitro grown *Pax-5*-deficient pre-B-1 cells can also give rise to macrophages, granulocytes and NK cells in vivo *(21)*.

Another remarkable difference between wild type and *Pax-5*-deficient pre-B-1 cells is the fact that upon transfer, *Pax-5*-deficient pre-B-1 cells home back to the bone marrow and stay there as pre-B-1 cells. These pre-B-1 cells can be isolated and can be grown again on stromal cells and IL-7. Moreover, upon transfer these cells can give rise to thymocytes, T cells, macrophages, granulocytes, and NK cells *(22)*.

Acknowledgments

I thank Drs. Jan Andersson and Rod Ceredig for critically reviewing the manuscript. Antonius G. Rolink holds the chair in Immunology endowed by Hoffmann-La Roche Ltd. This work was supported by grants from the Swiss National Science Foundation.

References

1. Rolink, A., Grawunder, U., Winkler, T. H., et al. (1994) IL-2 receptor α chain (CD25, TAC) expression defines a crucial stage in pre-B cell development. *Int. Immunol.* **6,** 1257–1264.
2. Rolink, A., Ghia, P., Grawunder, U., Haasner, D., Karasuyama, H., Kalberer, C., Winkler, T., and Melchers, F. (1995) In-vitro analyses of mechanisms of B-cell development. *Semin. Immunol.* **7,** 155–167.
3. Osmond, D. G., Rolink, A., and Melchers, F. (1998) Murine B lymphopoiesis: towards a unified model. *Immunol. Today* **19,** 65–68.
4. ten Boekel, E., Melchers, F., and Rolink, A. (1995) The status of Ig loci rearrangements in single cells from different stages of B cell development. *Int. Immunol.* **7,** 1013–1019.
5. Rolink, A., Haasner, D., Nishikawa, S., and Melchers, F. (1993) Changes in frequencies of clonable pre B cells during life in different lymphoid organs of mice. *Blood* **81,** 2290–2300.
6. Rolink, A., Kudo, A., Karasuyama, H., Kikuchi, Y., and Melchers, F. (1991) Long-term proliferating early pre B cell lines and clones with the potential to develop to surface Ig-positive, mitogen reactive B cells in vitro and in vivo. *EMBO J.* **10,** 327–336.
7. Rolink, A., Grawunder, U., Haasner, D., Strasser, A., and Melchers, F. (1993) Immature surface Ig+ B cells can continue to rearrange κ and λ L chain gene loci. *J. Exp. Med.* **178,** 1263–1270.
8. Melchers, F., ten Boekel, E., Seidl, T., et al. (2000) Repertoire selection by pre-B-cell receptors and B-cell receptors, and genetic control of B-cell development from immature to mature B cells. *Immunol. Rev.* **175,** 33–46.

9. Reininger, L., Winkler, T. H., Kalberer, C. P., Jourdan, M., Melchers, F., and Rolink, A. G. (1996) Intrinsic B cell defects in NZB and NZW mice contribute to systemic lupus erythematosus in (NZB × NZW)F1 mice. *J. Exp. Med.* **184,** 853–861.

10. Rolink, A. G., Nutt, S. L., Melchers, F., and Busslinger, M. (1999) Long-term in vivo reconstitution of T-cell development by Pax5-deficient B-cell progenitors. *Nature* **401,** 603–606.

11. Nutt, S. L., Heavey, B., Rolink, A. G., and Busslinger, M. (1999) Commitment to the B-lymphoid lineage depends on the transcription factor Pax5. *Nature* **401,** 556–562.

12. Ogawa, M., Nishikawa, S., Ikuta, K., Yamamura, F., Naito, M., and Takahashi, K. (1988) B cell ontogeny in murine embryo studied by a culture system with the monolayer of a stromal cell clone, ST2: B cell progenitor develops first in the embryonal body rather than in the yolk sac. *EMBO J.* **7,** 1337–1343.

13. Kodama, H., Sudo, H., Koyama, H., Kasai, S., and Yamamoto, S. (1984) In vitro hemopoiesis within a microenvironment created by MC3T3-G2/PA6 preadipocytes. *J. Cell Physiol.* **118,** 233–240.

14. Collins, L. S. and Dorshkind, K. (1987) A stromal cell line from myeloid long-term bone marrow cultures can support myelopoiesis and B lymphopoiesis. *J. Immunol.* **138,** 1082–1087.

15. Nakano, T., Kodama, H., and Honjo, T. (1994) Generation of lymphohematopoietic cells from embryonic stem cells in culture. *Science* **265,** 1098–1101.

16. Martensson, I. L., Rolink, A., Melchers, F., Mundt, C., Licence, S., and Shimizu, T. (2002) The pre-B cell receptor and its role in proliferation and Ig heavy chain allelic exclusion. *Semin. Immunol.* **14,** 335–342.

17. Rolink, A. G., Winkler, T., Melchers, F., and Andersson, J. (2000) Precursor B cell receptor-dependent B cell proliferation and differentiation does not require the bone marrow or fetal liver environment. *J. Exp. Med.* **191,** 23–32.

18. Rolink, A., Karasuyama, H., Grawunder, U., Haasner, D., Kudo, A., and Melchers, F. (1993) B cell development in mice with a defective λ5 gene. *Eur J. Immunol.* **23,** 1284–1288.

19. Nutt, S. L., Morrison, A. M., Dorfler, P., Rolink, A., and Busslinger, M. (1998) Identification of BSAP (Pax-5) target genes in early B-cell development by loss- and gain-of-function experiments. *EMBO J.* **17,** 2319–2333.

20. Nutt, S. L., Urbanek, P., Rolink, A., and Busslinger, M. (1997) Essential functions of Pax5 (BSAP) in pro-B cell development: difference between fetal and adult B lymphopoiesis and reduced V-to-DJ recombination at the IgH locus. *Genes Dev.* **11,** 476–491.

21. Schaniel, C., Bruno, L., Melchers, F., and Rolink, A. G. (2002) Multiple hematopoietic cell lineages develop in vivo from transplanted Pax5-deficient pre-B I-cell clones. *Blood* **99,** 472–478.

22. Schaniel, C., Gottar, M., Roosnek, E., Melchers, F., and Rolink, A. G. (2002) Extensive in vivo self-renewal, long-term reconstitution capacity, and hematopoietic multipotency of Pax5-deficient precursor B-cell clones. *Blood* **99,** 2760–2766.

Index

Automated single-cell deposition, flow
cytometry, 183

B-cell receptor (BCR),
crosslinking studies, 215
DT40 B-cell line gene targeting
studies,
calcium flux measurement, 267
flow cytometry of B-cell receptor
expression, 265
homologous recombination for
gene knockout,
alleles, 262, 263, 270
electroporation, 263–265
stable clone generation, 267, 268
Western blotting of
phosphoproteins, 266, 267
lipid rafts in signaling, 213, 214
phosphorylative signaling,
cascade components, 190, 191
protein tyrosine kinases, 190,
266
signal transduction analysis,
calcium flux analysis with flow
cytometry, 206, 208
cell culture, 194
cell lines, 193, 194
glutathione S-transferase pull-
down assay,
fusion protein preparation, 199,
200
precipitation, 200
immunoprecipitation of signaling
proteins, 197
kinase inhibitor studies, 201, 202
stimulation of B cells,

anti-immunoglobulin
antibodies, 194–196
phosphatase-inhibitor-PV, 196,
197, 208
Western blot analysis of
phosphoproteins, 200, 201,
204–206, 208
structure, 189, 190
BCR, see B-cell receptor
BrdU, see Bromodeoxyuridine
Bromodeoxyuridine (BrdU),
B-cell life-span analysis with flow
cytometry,
advantages over tritiated
thymidine determination,
59, 60
bromodeoxyuridine labeling, in
vivo, 62, 63–65
monoclonal antibody staining, 60,
61
staining, 63–65
cell cycle analysis, 21
plasma cell labeling and analysis
after immunization, 112,
115, 117, 118, 121, 122

Calcium flux, B-cell signaling,
flow cytometry analysis, 206, 208
Fura-2 assay, 267
Cell transfer assay, B-cell lineage
potential, 10, 11
Cholesterol rafts, see Lipid rafts
Clonality analysis, see V-region genes
CLP, see Common lymphoid progenitor
Common lymphoid progenitor (CLP),
features, 67

Cre-loxP system,
  conditional gene targeting,
    planning, 93, 94
    rationale, 91, 92
    single-copy, knock-in transgenes,
      97, 98
    targeting vector construction,
      embryonic stem cell lines, 95
      genomic DNA sources, 95, 96
      pEASY-FLIRT, 96, 97
      pEASY-FLOX, 96, 97
  Cre strains and transgenes for B-cell
    development studies, 93
  embryonic stem cells,
    double-resistant colony selection
      and screening, 103–105, 107
    freezing in microtiter plates, 104
    preparation for transfection, 101,
      102, 106
    selection marker deletion from
      targeted locus, 105–108
    thawing in microtiter plates, 105,
      107
    transfection, 102, 102, 107
  mouse embryonic fibroblast isolation
    for transfection of embryonic
    stem cells, 98–101, 106
  Southern blot screening of
    homologous recombinants,
    93, 94
  STOP cassette, 97
Cytoplasmic staining, gene expression
    analysis in B-cell
    development, 20
DT40 cell line, see B-cell receptor

FACS, see Fluorescence-activated cell
    sorting
Fetal thymic organ culture assay, B-cell
    lineage potential,
  culture, 14
  lobe analysis, 14
  plate preparation, 13
  thymic lobe dissection, 13, 14

Flow cytometry,
  B-cell development analysis,
    cell preparation, 5
    8 to 12-color analysis, 9
    four or five-color analysis, 6, 9
    peripheral immune system
      analysis,
      cell suspensions and staining,
        26, 27, 31, 34
      flow cytometry, 27, 28, 30, 31
      four-color staining, 27
    staining, 5, 6, 10, 21, 22
  B-cell differentiation from
      hematopoietic stem cells,
      75, 76
  B-cell life-span analysis with
      bromodeoxyuridine,
    advantages over tritiated
      thymidine determination,
      59, 60
    bromodeoxyuridine labeling, in
      vivo, 62, 63–65
    monoclonal antibody staining, 60,
      61
    staining, 63–65
  B-cell migration assay, 168, 169
  B-cell receptor expression, 265
  B-cell subsets, see Fluorescence-
      activated cell sorting
  calcium flux analysis, 206, 208
  immunoglobulin class switching
      analysis,
    fixation, 155, 156
    intracellular staining, 155–157
    surface staining, 156, 157
  RAG blastocyst complementation
      system B-cell analysis, 84,
      85
  transgene expression screening,
      254
Fluorescence-activated cell sorting
    (FACS),
  B-cell isolation for migration assays,
      165

B-cell subset analysis,
fluorescence compensation, 39,
40, 52, 53
gating and visualization of
subsets, 45, 46, 49, 50
Logicle visualization, 42–45
staining, 42, 50

Germinal center reaction, *see* T-cell-
dependent affinity
maturation
Glutathione *S*-transferase pull-down
assay, B-cell signaling,
fusion protein preparation, 199, 200
precipitation, 200

Hapten-protein conjugates for T-cell-
dependent immunization,
4-hydroxy-3-nitro-phenylacetyl-
coupled proteins for T-cell-
dependent immunization,
memory B cell development,
exclusion adoptive transfers of
single-memory B cell
subsets, 181–182, 186
extended cell-surface phenotype
analysis, 178, 180
flow cytometry of memory B
cells, 176–178, 184–186
plasma cell analysis,
bromodeoxyuridine labeling,
112, 115, 117, 118, 121, 122
germinal center
micromanipulation, 118, 122
immunization, 114
immunohistochemical staining,
115–117, 122
spleen preparation, 115, 122
V-region gene rearrangement,
118–123
repertoire studies with single-cell
resolution,
automated single-cell
deposition, 183

complementary DNA
synthesis, 183
polymerase chain reaction,
184, 186, 187
sequence analysis, 184
2-phenyl-oxazolone-coupled proteins
germinal center analysis, 132–135
immunization, 130
immunohistochemical staining,
131, 132, 135
somatic mutation analysis, 135–137
V–region genes,
polymerase chain reaction, 134
rearrangement, 129
Hematopoietic stem cell (HSC), *see
also* Lymphopoiesis,
B-lymphocyte differentiation
system,
culture conditions, 72, 73
cytokine treatment, 73–75
flow cytometry analysis, 75, 76
materials, 68, 69
stromal cell lines, 70, 72
differentiation potential, 67, 68
High oxygen submersion culture/
multilineage progenitor
(HOS/MLP) assay, B-cell
lineage potential, 14, 15
HOS/MLP assay, *see* High oxygen
submersion culture/
multilineage progenitor assay
HSC, *see* Hematopoietic stem cell
4-Hydroxy-3-nitro-phenylacetyl-coupled
T-cell proteins, *see* Hapten-
protein conjugates for T-cell-
dependent immunization

Immunoglobulin class switching,
analysis,
lipopolysaccharide induction,
154–156
proliferation monitoring with
carboxyfluorescein
diacetate, 154

splenic B-cell isolation, 153, 154,
   staining of isotypes,
      fixation, 155, 156
      intracellular staining, 155–157
      surface staining, 156, 157
   mitogen activation, 150
   recombination, 149, 150
Immunoglobulin gene rearrangement,
      see also V-region genes,
   assay, 17
   loss of germline assay, 15–17
   pre-B-1 cells, 272, 273
   somatic hypermutation, 226
   somatic recombination, 225, 226

Life-span, bromodeoxyuridine with
      flow cytometry analysis,
   advantages over tritiated thymidine
      determination, 59, 60
   bromodeoxyuridine labeling, in vivo,
      62, 63–65
   monoclonal antibody staining, 60, 61
   staining, 63–65
Lipid rafts,
   B-cell signaling, 213, 214
   detergent solubility, 214, 215
   elemental rafts, 214
   isolation and characterization,
      cell preparation and activation,
         216, 217, 220
      cholesterol repletion, 219, 222
      discontinuous sucrose gradient
         sedimentation, 217, 218, 220
      lysate preparation, 217, 220
      Western blotting of proteins, 219,
         220, 222
   protein segregation, 214
   structure, 213, 214
Lymphopoiesis, see also Hematopoietic
      stem cell,
   cell cycle analysis,
      bromodeoxyuridine staining, 21
      propidium iodide staining, 20,
         21

cell transfer assay of lineage
      potential, 10, 11
common lymphoid progenitor
      features, 67
fetal thymic organ culture assay,
   culture, 14
   lobe analysis, 14
   plate preparation, 13
   thymic lobe dissection, 13, 14
flow cytometry analysis,
   cell preparation, 5
   8 to 12-color analysis, 9
   four or five-color analysis, 6, 9
   peripheral immune system
         analysis,
      cell suspensions and staining,
         26, 27, 31, 34
      flow cytometry, 27, 28, 30, 31
      four-color staining, 27
   staining, 5, 6, 10, 21, 22
gene expression analysis, see
      Cytoplasmic staining;
      Reverse transcriptase-
      polymerase chain reaction;
      Western blot
gene rearrangement, see
      Immunoglobulin gene
      rearrangement
gene targeting studies, see Cre-loxP
      system; RAG blastocyst
      complementation system
hematopoietic stem cell
      differentiation, 67, 68
high oxygen submersion culture/
      multilineage progenitor
      assay, 14, 15
methylcellulose culture assay, 11, 12
pre-B-1 cell plasticity studies,
   differentiation into immature B-
      cells, 278
   immunoglobulin gene
      rearrangements, 272, 273
   in vivo development studies, 279,
      280

Pax-5 knockout,
  cell line establishment, 278, 279
  effects on plasticity, 273, 279,
    280
  pre-B-1 cell culture,
    initiation, 275, 276
    propagation, 276, 277
  retroviral gene transduction, 277
  stromal cells,
    density and pretreatment for
      pre-B-1 cell growth, 275
    maintenance and amplification,
      274, 275
  stromal cell culture assay, 12, 13, 70,
    72–75
  surface markers on B-cells, 271, 272

Memory B-cells, *see* T-cell-dependent
  affinity maturation
Methylcellulose culture assay, B-cell
  lineage potential, 11, 12
Migration, B-cells,
  assays,
    B-cell isolation,
      fluorescence-activated cell
        sorting, 165
      immunomagnetic separation,
        165
      suspension preparation from
        tissues, 164
    chemokine-induced actin
      polymerization, 166
    filter assay, 162, 165, 166, 169
    in vivo assays,
      carboxyfluorescein diacetate
        labeling, 168
      flow cytometry, 168, 169
      fluorescence microscopy of
        tissue sections, 169
      fluorescent dyes for cell
        labeling, 167, 168
      labeled cell injection in mice,
        168
      lipophilic tracer labeling, 168

Zigmond chamber, 166, 167, 169
  development, 161, 162

Pax-5, knockout effects on pre-B-1 cell
  plasticity, 273, 279, 280
PCR, *see* Polymerase chain reaction
2-Phenyl-oxazolone-coupled proteins for
  T-cell-dependent immuni-
  zation studies, *see* Hapten-
  protein conjugates for T-cell-
  dependent immunization
Plasma cell,
  development and immunoglobulin
    secretion, 111, 112
  4-hydroxy-3-nitro-phenylacetyl-
    coupled proteins for T-cell-
    dependent immunization, *see*
    Hapten-protein conjugates for
    T-cell-dependent
    immunization
Polymerase chain reaction (PCR),
  memory B-cell repertoire studies
    with single-cell resolution,
    184, 186, 187
  single-cell polymerase chain reaction
    for B-cell clonality analysis,
    first-round polymerase chain
      reaction, 231, 234, 235
    histological staining, 229, 230, 233
    microdissection, 230, 231, 233, 234
    product purification and
      sequencing, 232, 233, 235
    proteinase K digestion, 231
    second-round polymerase chain
      reaction,
      V_H, 231, 232
      V_κ, 232, 235
  transgenic mouse screening, 251,
    252, 258
  V-region genes in plasma cells, 118–
    123, 134
Pre-B-1 cells, *see* Lymphopoiesis
Propidium iodide, cell cycle analysis,
  20, 21

RAG blastocyst complementation
system (RBCS),
advantages, 77–80
blastocyst microinjection, 83
chimera screening and analysis,
peripheral blood collection and
preparation, 84
staining and flow cytometry
analysis, 84, 85
complementation efficiency, 85, 86
embryo harvest, 82
embryonic stem cell preparation, 82,
83
mutant embryonic stem cell generation
and culture, 80, 81
RAG 2-deficient mouse colony
maintenance, 81
superovulation, 81, 82
yields, 81, 83, 85
RBCS, *see RAG* blastocyst
complementation system
Rearrangement, *see* Immunoglobulin
gene rearrangement
Retroviral gene transduction,
pre-B-1 cells, 277
primary splenic B-cells,
application to B-cells from other
tissues, 146
B-cell purification, 142, 143
infection, 143, 144, 146
lipopolysaccharide treatment,
140, 143
rationale, 139
retrovirus preparation, 141, 142
Reverse transcriptase-polymerase chain
reaction (RT-PCR), gene
expression analysis in B-cell
development,
complementary DNA synthesis, 19
polymerase chain reaction, 19, 20
RNA isolation, 18, 19
RT-PCR, *see* Reverse transcriptase-
polymerase chain reaction

Signal transduction, *see* B-cell
receptor
Southern blot,
screening of homologous
recombinants, 93, 94
transgenic mouse screening, 248–
251, 257, 258
Sphingolipid rafts, *see* Lipid rafts
Stromal cell culture assay, B-cell
lineage potential, 12, 13, 70,
72–75

T-cell-dependent affinity maturation,
germinal center reaction, 128
4-hydroxy-3-nitro-phenylacetyl-
coupled proteins for T-
cell-dependent
immunization, *see* Hapten-
protein conjugates for T-
cell-dependent
immunization
lymphocyte compartmentalization in
spleen, 127, 128
memory development, 173, 174
2-phenyl-oxazolone-coupled
proteins for T-cell-
dependent immunization
studies, *see* Hapten-protein
conjugates for T-cell-
dependent immunization
plasma cell development and
immunoglobulin secretion,
111, 112, 174
Tolerance,
development, 239, 240
transgenic mouse studies,
breeding, 254, 255
construct design, 240, 241
flow cytometry analysis, 255,
258, 259
founder screening,
polymerase chain reaction,
251, 252, 258

Southern blot, 248–251, 257, 258
hen egg lysozyme transgene generation,
cloning, 243
constructs, 243
modified insert preparation, 243, 256, 257
restriction digestion, 245
transfection, 245, 257
IgM^a assays,
enzyme-linked immunosorbent assay, 256
enzyme-linked immunospot assay, 255, 256, 259
microinjection, 248
transgene expression screening,
immunohistochemistry, 252, 253, 258
flow cytometry, 254
lysozyme bioassay, 254
transgene preparation for microinjection, 246–248, 257
Transgenic mice, *see* Cre-loxP system; *RAG* blastocyst complementation system; Tolerance

V-region genes,
clonality marker, 226
DNA purification and sequencing, 232, 233, 235

plasma cell analysis after immunization,
polymerase chain reaction, 118–123
sequencing and analysis, 121, 123
rearrangement and diversity, 226
single-cell polymerase chain reaction for clonality analysis,
first-round polymerase chain reaction, 231, 234, 235
histological staining, 229, 230, 233
microdissection, 230, 231, 233, 234
proteinase K digestion, 231
second-round polymerase chain reaction,
V_H, 231, 232
V_κ, 232, 235

Western blot,
gene expression analysis in B-cell development, 20
lipid raft-associated proteins, 219, 220, 222
phosphoprotein analysis in B-cell signaling, 200, 201, 204–206, 208, 266, 267

Zigmond chamber, B-cell migration assay, 166, 167, 169

5984
5679 2 yards
3.649 feet